Iterative Algebra and Dynamic Modeling

Springer

New York
Berlin
Heidelberg
Barcelona
Hong Kong
London
Milan
Paris
Singapore
Tokyo

Iterative Algebra and Dynamic Modeling

A Curriculum for the Third Millennium

Kurt Kreith
Don Chakerian

Department of Mathematics
University of California, Davis

 Springer

Library of Congress Cataloging-in-Publication Data
Kreith, Kurt.
    Iterative Algebra and Dynamic Modeling: A Curriculum for the
  Third Millennium / Kurt Kreith, G.D. Chakerian.
        p.   cm. — (Textbooks in mathematical sciences)
    Includes bibliographical references (p.   –   ) and index.
    ISBN 0-387-98758-4 (alk. paper)
      1. Algebra—Data processing.   2. Iterative methods—Data
  processing.   3. Microsoft Excel (Computer file)   I. Chakerian, G.D.
  II. Title.   III. Series.
  QA155.7.E4K74   1999
  512'.0285—dc21   98-55409

Printed on acid-free paper.

Microsoft® Excel 4.0 for the Macintosh® is a registered trademark of Microsoft Corporation in the United States. Copyright©1985–1995 Microsoft Corporation.

Macintosh is a registered trademark of Apple Computer, Inc.

STELLA® II software, ©1985, 1987, 1988, 1990–1996 High Performance Systems, Inc.
All rights reserved

Production managed by Mary Ann Cottone; manufacturing supervised by Nancy Wu.
Typeset in TeXtures from the authors' WordPerfect files by TechBooks, Fairfax, VA.
Printed and bound by R.R. Donnelley and Sons, Harrisonburg, VA.
Printed in the United States of America.

9 8 7 6 5 4 3 2 1

ISBN 0-387-98758-4 Springer-Verlag New York Berlin Heidelberg   SPIN 10711310

To the memory of our sisters,
Gladys Chakerian and Susanne Culler

Contents

Preface

This book is aimed at high-school and college students who have developed facility in precalculus mathematics and are now interested in exploring the interaction of algebra with computer technology. It is also aimed at teachers who seek to enlarge, enrich, and apply the curriculum they teach by reconciling it with the technology that is having such profound impact on our lives and civilization. And finally, it is aimed at "miscellaneous readers" who may have completed their study of algebra some years ago but remain curious about the interaction of computer technology with the world of mathematics.

The book's medium for linking mathematics and technology is a spreadsheet program called Excel. However, the procedures we describe in Excel can be implemented by any medium that facilitates "iteration." Such media include graphing calculators and simple programming languages, as well as other spreadsheet programs. Our use of Excel is based on a desire to provide a transparent programming tool and convenient graphical interface while still challenging the reader to demonstrate a mathematical understanding of the procedures being used. The book's final chapter also makes use of a simulation package called Stella that enables the user to formulate iterative procedures in an icon-based format.

The format of the book is geared toward "active learning." Most of the figures are based on spreadsheet programs that the reader is asked to construct in Excel, or in some other medium that provides graphical representations of data. Many of the spreadsheets in these figures include cells with dark boundaries. These cells contain numerical values (parameters) that can be changed, whereupon Excel recalculates its outputs and redraws the corresponding graphs. Teachers are urged to consider such use of technology as a way of injecting a dynamic quality into their students' mathematical thinking and learning.

Intermingled with the text are "transitional problems" that call on the reader to fill gaps in the book's presentation. Some of these problems address mathematical techniques central to the ideas being developed. Each chapter ends with a collection of "Problems, Exercises, and Projects" that address related mathematical topics and applications at a range of levels. For readers familiar with calculus, there are occasional starred problems that set forth connections between iterative algebra and calculus-based techniques.

The goal of the book is to enrich the standard algebra curriculum by taking account of the computer's capacity for rapid and reliable iteration, thereby bringing school algebra into the computer age. To reinforce the notion that our conception of algebra (and indeed of mathematics) is subject to change, Chapter 1 includes some discussion of algebra's roots in Babylonian reckoning, Greek geometry, and the Arabic formulation of *al-jabr*. Against this background, we go on to develop a broad range of mathematical topics and applications that have traditionally been regarded as "calculus-based."

Among the reasons for making such tools and applications accessible at the high-school level is their connection to issues of "global change." Accordingly, many applications are drawn from ecology, economics, demography, and related subjects. The models so developed are intended to place contemporary issues concerning the global system into an analytical framework, one in which the issues confronting our society and civilization may take on a more coherent form.

In reflecting on how "iterative algebra" came to be, we are aware of many individuals to whom we are deeply indebted. Prominent among these is our Bulgarian colleague Blagovest Sendov whose conferences on "Children in the Information Age" first made us aware of the role that technology can play in fostering mathematical learning and creativity. Through these events we learned of the work of others, including Erich Neuwirth whose contagious enthusiasm for spreadsheets is reflected throughout this book. We also profited from the suggestions of mathematicians affiliated with Russian specialized schools, most notably Nadezhda Alfutova and Vladimir Dubrovsky of Moscow's School No. 18.

Closer to home, it was *Quantum* editors Mark Saul and Tim Weber who provided an outlet for "iterative algebra" in its earliest forms. Jaimie Cloud's interest in "The Mathematics of Global Change" led to a series of summer workshops for New York City teachers, and it was in the course of these that many of the ideas underlying this book emerged. Crucial encouragement was also provided by Matthias Ruth of Boston University who first suggested putting these ideas into book form.

Finally, we want to acknowledge the support of those who made it possible to field test curricula based on parts of this book. Davis High School mathematics chair Steve Legé participated in our summer workshop in NYC, following which principal Howard Cohen provided us with the opportunity to offer an after school course based on an early draft. Thanks to teachers Tonya Richardson, Angela Blaha, and Tom Blozy, a course based on "iterative algebra" is currently under way at Brooklyn Technical High School. Their spirit of mathematical adventure, and that of assistant principal Elizabeth McCollough, are gratefully acknowledged.

Kurt Kreith
Don Chakerian
Davis, California
February 1999

1

POLYNOMIAL EQUATIONS

1.1. Introduction   Welcome to a mathematical excursion dealing with the topics of "iterative algebra" and "dynamic modeling." Defining these terms is what this book is all about, so we will refrain from trying to provide you with a mathematical description of our planned journey at this time. Instead, we simply appeal to the reader to tolerate our interest in bringing the subject of algebra into the computer age, that is, in developing "an algebra for the third millennium."

Our efforts to formulate such a curriculum are guided by the following observations:

1. Like other cultural pursuits, the study of mathematics is subject to change.
2. Algebra, as currently taught in secondary schools, underwent its last major overhaul about 500 years ago, with a substantial tune-up 200 years ago.
3. Computer technology has profoundly affected many aspects of our lives, and it also affects the study of mathematics.
4. Bringing computer technology to bear on algebra can provide new insights into our world (that is where dynamic modeling will enter the picture).

By way of reinforcing the first two observations, Chapter 1 revisits the topic of polynomial equations in both historical and futuristic contexts. It deals with the evolution of some ideas that are central to algebra and the fact that people's conception of this "science of transposition and cancellation" has, over the course of time, undergone profound changes. With such hindsight, we will be in a better position to speculate on changes that the body of mathematics once called *al-jabr* is likely to undergo in the future.

1

1.2. The Quadratic Formula

One of the most prominent mileposts along the algebraic highway is the quadratic formula. It expresses our ability to solve the general quadratic equation

$$(1.2.1) \qquad ax^2 + bx + c = 0,$$

where $a \neq 0$.

As you may recall from your first encounter with algebra, the solution

$$(1.2.2) \qquad x = \frac{-b \pm \sqrt{b^2 - 4ac}}{2a}$$

is based on the algebraic identity

$$(A + B)^2 = A^2 + 2AB + B^2$$

and the process of "completing the square." Dividing through (1.2.1) by a and then transposing the term c/a leads to

$$x^2 + \frac{b}{a}x = -\frac{c}{a}.$$

Now adding $b^2/(4a^2)$ to both sides, we obtain

$$(1.2.3) \qquad x^2 + \frac{b}{a}x + \frac{b^2}{4a^2} = \frac{b^2}{4a^2} - \frac{c}{a}.$$

Problem 1.2.1 Noting that the left side of (1.2.3) is a perfect square, complete the derivation of the litany, "*minus bee plus or minus the square root of bee squared minus four ay cee, all over two ay.*"

Problem 1.2.2 Note that (1.2.3) is equivalent to

$$x^2 + \frac{b}{a}x + \frac{b^2}{4a^2} - \frac{b^2 - 4ac}{4a^2} = 0.$$

(a) Use the identity $A^2 - B^2 = (A - B)(A + B)$ to deduce the quadratic formula.

(b) Assuming $b^2 - 4ac \neq 0$, explain why (1.2.1) has exactly two solutions.

So where did such quadratic equations come from and how were they solved before the quadratic formula was established? Babylonian clay tablets and Egyptian papyri show that problems calling for the solution of quadratic equations go back to ancient times. Many of these problems had a geometric formulation, such as the following:

The result of subtracting the length of the side of a square from the area of the square is 870. Find the length of the side of the square.

Problem 1.2.3 Solve this problem by means of the quadratic formula and reconcile your answer with that given by the Babylonians:

> Take half of 1, which is $\frac{1}{2}$; multiply $\frac{1}{2}$ by $\frac{1}{2}$ which is $\frac{1}{4}$; add $\frac{1}{4}$ to 870, which is $870\frac{1}{4}$. This last is the square of $29\frac{1}{2}$. Now add $\frac{1}{2}$ to $29\frac{1}{2}$. The result is 30, which is the side of the square.

One of the most famous geometric formulations of a quadratic equation involves the "golden ratio" as defined by the Alexandrian Greeks. Here we consider a rectangle with height 1 and base $x > 1$ and having the following property: If a vertical line divides this rectangle into a unit square and a smaller rectangle, then the smaller rectangle is similar to the original rectangle.

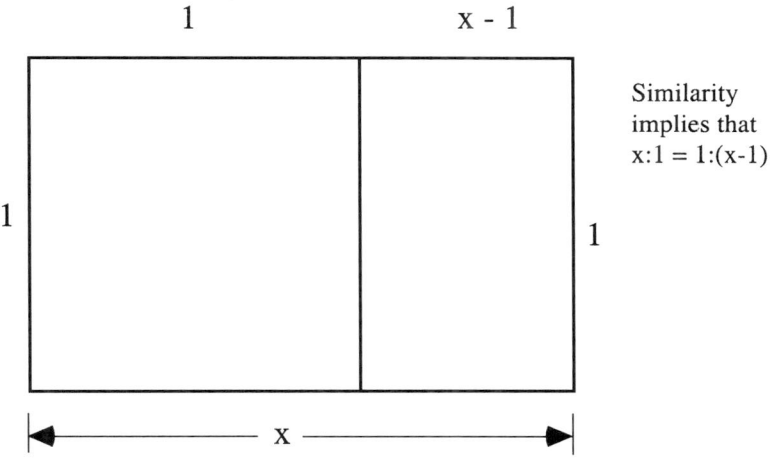

Similarity implies that x:1 = 1:(x-1).

Figure 1.2.1 A "Golden Rectangle"

From Figure 1.2.1 it follows that $x : 1 = 1 : (x - 1)$, which algebra enables us to transform into the quadratic equation $x^2 - x - 1 = 0$. The quadratic formula in turn enables us to write the positive solution as $x = \frac{1+\sqrt{5}}{2}$.

Lacking both algebraic notation and the quadratic formula, the Greeks studied this "golden ratio" by very different means. They were able to show the "irrationality" of $\frac{1+\sqrt{5}}{2}$ without a symbolic representation for its numerical value. As shown by problems at the end of this chapter, Greek geometers were also able to associate the golden ratio with shapes other than rectangles.

An approach based on geometry also provides a basis for solving special quadratic equations of the form

$$x^2 + bx - c^2 = 0,$$

where b and c are given positive numbers. To solve this equation in the Greek spirit, we construct an isosceles triangle with base b and altitude c. Then, drawing the circle centered at the top vertex and passing through the other two vertices, we obtain the picture shown in Figure 1.2.2.

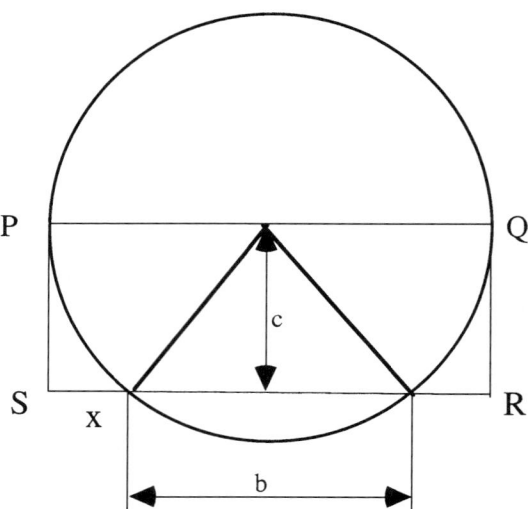

Figure 1.2.2 Geometric Solution of a Quadratic Equation

Here we have also drawn the diameter PQ that is parallel to the base of the triangle and constructed the rectangle $PQRS$ as above.

Problem 1.2.4 Referring to Figure 1.2.2, show that the short segment labeled x corresponds to a root of the quadratic equation $x^2 + bx - c^2 = 0$. [Hint: Use the Pythagorean theorem to express the radius of the circle in terms of b and c.]

Such ad hoc methods for solving quadratic equations persisted throughout the middle ages. One reason is that before modern notation became widely accepted, the identity

$$(A + B)^2 = A^2 + 2AB + B^2$$

that underlies the quadratic formula would have been written

$$A + B \text{ } quadr \text{ } aequalia \text{ } A \text{ } quadr + A \text{ } in \text{ } B.2 + B \text{ } quadr.$$

It was only with the development of less ponderous notation for such algebraic expressions that systematic methods of solution evolved.

François Viète was a master algebraist of the late sixteenth century who pioneered the use of letters to represent both numerical coefficients and unknowns in algebraic expressions. Viète solved the general quadratic equation by:

(a) Showing that if x satisfies $ax^2 + bx + c = 0$ $(a \neq 0)$ and y is defined by

(1.2.4) $$y = x + \frac{b}{2a},$$

then y satisfies the equation

(1.2.5)
$$ay^2 - \frac{b^2}{4a} + c = 0.$$

(b) Solving the last equation for y.

In this way Viète arrived at the solution of (1.2.1), but not at the litany that we have come to associate with the "quadratic formula."

Problem 1.2.5 Use Viète's method to solve $x^2 - x - 1 = 0$.

Problem 1.2.6 Use (1.2.4) and (1.2.5) to arrive at the quadratic formula.

The fact that the quadratic formula served to simplify and unify a disparate collection of techniques for solving $ax^2 + bx + c = 0$ has made it a centerpiece of our contemporary algebra curriculum. Without denying its importance and usefulness, our first task in developing iterative algebra is to supplement the litany *"minus bee plus or minus the square root of bee squared minus four ay cee, all over two ay"* with one that is more appropriate for the computer age. With this goal in mind, we turn to ways in which computer technology can be harnessed in the solution of quadratic equations.

1.3. Solution by Iteration   Iteration is the process of "doing the same thing over and over." Experience suggests that machines tend to outperform humans in carrying out such repetitive processes. Especially when iteration involves repeated numerical calculations, computers tend to be faster, more accurate, and less complaining than people.

Given the relatively recent arrival of computers on the scene, it is interesting to speculate on how their earlier invention might have affected the development of mathematics. By way of example, we have seen how the concept of a "golden rectangle" gave rise to the identity

(1.3.1)
$$x : 1 = 1 : (x - 1).$$

Had the Greeks had access to computers, they might well have approached (1.3.1) by noting that the solution lies between two numbers L and R and then computing both $x : 1$ and $1 : (x - 1)$ for a range of values satisfying $L \leq x \leq R$. Choosing smaller and smaller intervals $[L, R]$, they could have "zeroed in" on the solution to $x^2 - x - 1 = 0$.[1]

The spreadsheet of Figure 1.3.1 enables us to implement such a program by calculating the ratios $x : 1$ and $1 : (x - 1)$ for 21 equally spaced values of x in an interval $L \leq x \leq R$ (initially $1.5 \leq x \leq 2$). It then uses these data to draw graphs of

---

[1] We now tend to call such methods of solution "guess and check." However, they have a long history in mathematics, including the "method of false position" that once occupied a prominent place in the algebra curriculum.

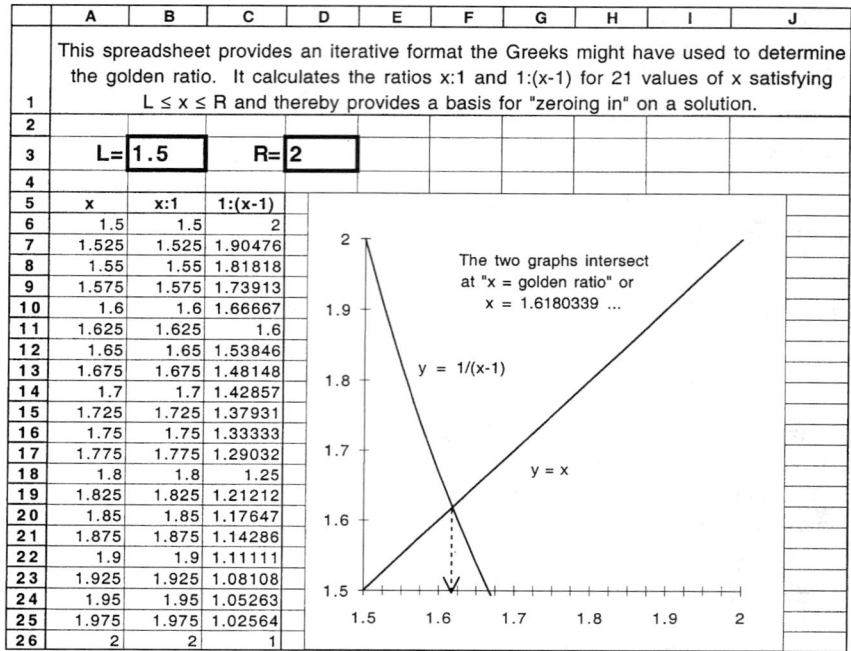

Figure 1.3.1 Using Technology to Estimate the Golden Ratio

$y = x$ and $y = 1/(1-x)$ for $L \leq x \leq R$. The intersection of these two graphs corresponds to the (positive) solution of $x^2 - x - 1 = 0$ and thereby provides a basis for estimating its value.

Given such an estimate, L and R can be chosen closer to the point of intersection of these graphs. With L and R redefined, Excel will repeat its calculations and redraw the graphs.

Problem 1.3.1 Expand on the format of Figure 1.3.1 to construct a spreadsheet that enables us to approximate solutions to $x : a = 1 : (x - b)$ for arbitrary constants a and b.

With such technology at our disposal, other ways of estimating the solutions of quadratic equations come to mind. For example, one could simply ask Excel to make a table of values for $y = ax^2 + bx + c$ for $L \leq x \leq R$ and then draw the corresponding graph (Figure 1.3.2). Noting the point at which this graph crosses the x-axis, we could again "zero in" on the real values of $\frac{-b \pm \sqrt{b^2 - 4ac}}{2a}$ (if any).

Problem 1.3.2 Construct the "quadratic equation solver" of Figure 1.3.2. For the equation $7x^2 - 2x - 3 = 0$, compare its results with those given by application of the quadratic formula.

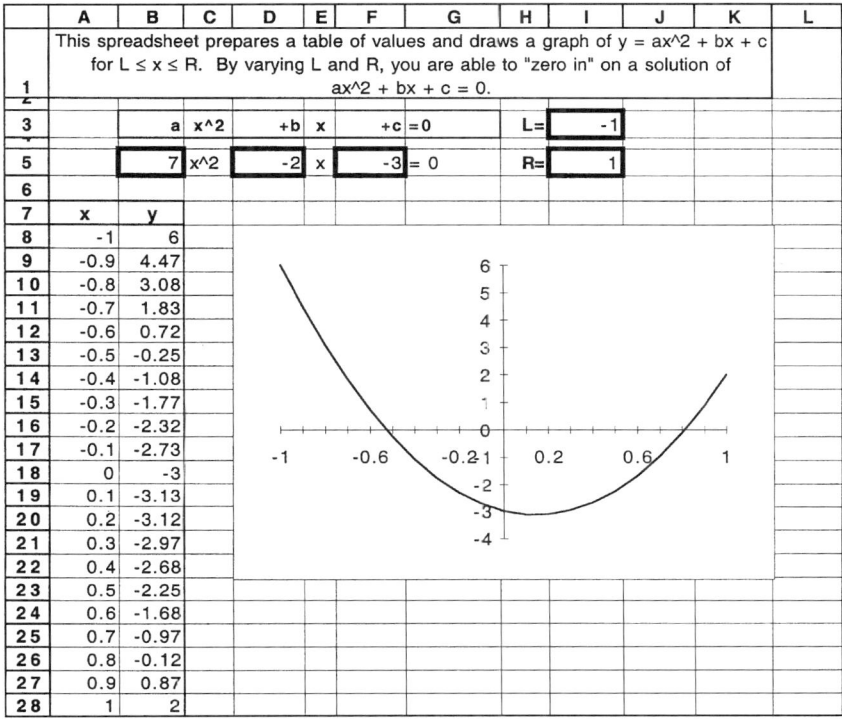

	A	B	C	D	E	F	G	H	I	J	K	L
	\multicolumn{11}{}{This spreadsheet prepares a table of values and draws a graph of y = ax^2 + bx + c}	
	\multicolumn{11}{}{for L ≤ x ≤ R. By varying L and R, you are able to "zero in" on a solution of}	
1	\multicolumn{11}{}{ax^2 + bx + c = 0.}	
3			a	x^2		+b	x	+c	=0		L=	-1	
5		7		x^2		-2	x		-3	= 0	R=	1	
6													
7	x	y											
8	-1	6											
9	-0.9	4.47											
10	-0.8	3.08											
11	-0.7	1.83											
12	-0.6	0.72											
13	-0.5	-0.25											
14	-0.4	-1.08											
15	-0.3	-1.77											
16	-0.2	-2.32											
17	-0.1	-2.73											
18	0	-3											
19	0.1	-3.13											
20	0.2	-3.12											
21	0.3	-2.97											
22	0.4	-2.68											
23	0.5	-2.25											
24	0.6	-1.68											
25	0.7	-0.97											
26	0.8	-0.12											
27	0.9	0.87											
28	1	2											

Figure 1.3.2 A Quadratic Equation Solver

While such methods of solution may constitute impressive applications of technology, they are rather pedestrian in terms of the mathematical ideas that are brought to bear. Lacking the means to implement such "brute force" methods of solution, earlier civilizations were forced to rely on human intellect to develop other techniques for solving polynomial equations. Iterative algebra, as developed in the remainder of this book, represents a marriage between such intellect and technology in a variety of mathematical settings.

1.4. Revisiting the Quadratic Formula

We begin our efforts to harness technology in developing an iterative approach to the solution of polynomial equations with a reexamination of the quadratic formula. Applying

(1.4.1)
$$x = \frac{-b \pm \sqrt{b^2 - 4ac}}{2a}$$

to the equation

$$x^2 - x - 1 = 0$$

yields the "solution"

(1.4.2) $$x = \frac{1 \pm \sqrt{5}}{2}.$$

Similarly, applying (1.4.1) to the equation

$$x^2 - 2x - 21 = 0$$

yields the "solution"

(1.4.3) $$x = 1 \pm \sqrt{22}.$$

Our reason for using quotation marks in writing "solution" is to call attention to a curious fact. While (1.4.2) and (1.4.3) do indeed describe the result of applying the quadratic formula to specific equations, neither "solution" provides us with a numerical value for x. That is, it still remains to determine that

$$\frac{1 + \sqrt{5}}{2} \approx 1.618034$$

and that

$$1 + \sqrt{22} \approx 5.690416.$$

Such examples show that our fascination with the quadratic formula has led us to identify "solution" with "an expression in terms of radicals"—even though such expressions may fail to provide us with an explicit decimal representation for the solution being sought.

In order to locate such "solutions" of quadratic equations on the number line, one also needs to be able to evaluate radicals. In the cases at hand we need to be able to determine that $\sqrt{5} \approx 2.236068$ and $\sqrt{22} \approx 4.690416$. And while such tasks can now be performed by pushing the $\sqrt{}$ key on a hand calculator, there remains the following mathematical fact:

> Rather than "solving" the quadratic equation $ax^2 + bx + c = 0$, the quadratic formula merely reduces the problem to the solution of a simpler quadratic equation of the form $x^2 - k = 0$.

On this basis, we raise the following questions:

1. Can we use iteration to find the square root of a number?
2. If so, might the iterative technique for solving $x^2 - k = 0$ be extended to solving $ax^2 + bx + c = 0$ *without* using the quadratic formula?
3. And if so, might the iterative technique for solving $ax^2 + bx + c = 0$ be extended to more general polynomial equations?

In subsequent sections of this chapter we provide affirmative answers to all three of these questions.

But first, let us make use of our mastery of traditional algebra by creating a "spreadsheet machine" that implements the quadratic formula. Since Excel has a built-in

	A	B	C	D	E	F	G	H	I	J	K
1	For b^2 - 4ac > 0, this spreadsheet uses the quadratic formula and Excel's SQRT function to calculate a numerical value for solutions of the quadratic equation ax^2 + bx + c = 0.										
2											
3		a	x^2	+b	x	+c	= 0				
5		7	x^2	-2	x	-3	= 0				
6											
7	We use SQRT to find √(b^2 - 4ac) = ± 9.38083										
8											
9	Adding	2	dividing by	14	yields	0.81292	or	-0.5272			

Figure 1.4.1 Solving $ax^2 + bx + c = 0$

function SQRT that corresponds to the $\sqrt{\ }$ button on your hand calculator, we can use the quadratic formula to improve substantially on the quadratic equation solver based on "zeroing in on a solution."

Problem 1.4.1 Using or improving upon the format of Figure 1.4.1, create your own "quadratic equation solver" based on the built-in Excel function SQRT.

The fact that we have relied on Excel to find $SQRT(b^2 - 4ac)$ should serve as a reminder that the quadratic formula stops short of giving us a numerical solution of $ax^2 + bx + c = 0$. Creating our own SQRT machine is what the next section is all about.

1.5. Babylonian Iteration
Archaeological records suggest that iterative techniques for finding square roots go back several thousand years. Evidence for this includes a clay tablet from Yale University's Babylonian collection, one that depicts a square and its diagonals (Figure 1.5.1).

Now, it is known that the Babylonians used a positional, base 60 number system. Their "digits" 1, 2, ..., 59 were composed of cuneiform (wedge shaped) markings with different orientations. In particular, ∨ denoted "one" and < denoted "ten," so that the cuneiform markings on the upper left of this tablet indicate that the square has sides of length <<<, or 30. Other cuneiform markings on the diagonal indicate a scribe's belief that a square with side 30 has a diagonal of length

$$
\begin{array}{ccc}
 & & < \vee \vee \\
< < \vee & < \vee & \vee & < \vee \\
< < \vee & < \vee \vee \vee & & < \vee \vee \\
\end{array}
$$

In the context of a base 60 number system, these "chicken scratchings" can be read as

$$42 + \tfrac{25}{60} + \tfrac{35}{60^2} \approx 42.426389.$$

Figure 1.5.1 A Babylonian Clay Tablet

Now, even without the Pythagorean theorem[1] it is easy to see that a square with side $s = 30$ has diagonal of length $d = 30\sqrt{2}$. This follows from a geometric "doubling of a square" with sides of length s and diagonal of length d. From Figure 1.5.2 we see that a second square, built on the diagonal of the square with side s, has twice the area of the first square.

This gives

$$d^2 = 2s^2,$$

or $d = \sqrt{2s^2} = \sqrt{2}s$.

Returning to the Babylonian clay tablet, we now interpret its message as

$$30\sqrt{2} \approx 42.426389.$$

Problem 1.5.1 Use a hand calculator to compare decimal representations of $\sqrt{2}$ and $(1/30) \cdot (42.426389)$. To how many decimal places is the Babylonian value for $\sqrt{2}$ correct?

How did the Babylonians arrive at such a remarkably good approximation to the solution to $x^2 - 2 = 0$? Historians believe that they used an iterative scheme

---

[1] That the Babylonians knew the Pythagorean theorem is strongly suggested by another clay tablet in the Plimpton collection at Columbia University. For a proof of the Pythagorean theorem, see Problem 1.11.5 at the end of this chapter.

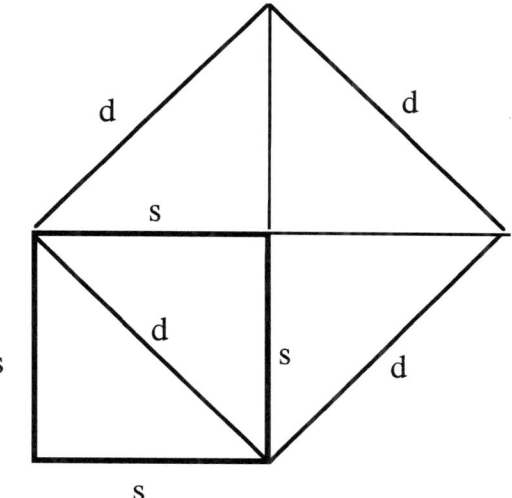

Figure 1.5.2 Doubling the Area of a Square

such as the following:

The "square root of two" is characterized by $\sqrt{2} \cdot \sqrt{2} = 2$. If we make a guess that $\sqrt{2} \approx \frac{3}{2}$ and then compute $(\frac{3}{2}) \cdot (\frac{3}{2}) = \frac{9}{4} \neq 2$, we conclude that $\sqrt{2} \neq \frac{3}{2}$ and, in fact, since $2 < \frac{9}{4}$ that $\sqrt{2} < \frac{3}{2}$. Well, since $\frac{3}{2}$ is too large, and since $(\frac{3}{2}) \cdot (\frac{4}{3})$ *does* equal 2, we also conclude that $\frac{4}{3} < \sqrt{2}$ and that

$$\tfrac{4}{3} < \sqrt{2} < \tfrac{3}{2}.$$

The heart of the process is now to assume that the arithmetic mean of $\frac{4}{3}$ and $\frac{3}{2}$ (which also lies between $\frac{4}{3}$ and $\frac{3}{2}$) will be a better guess than was $\frac{3}{2}$. Accordingly, our second guess is $(\frac{4}{3} + \frac{3}{2})/2 = 17/12$.

Another check now shows that $(17/12) \cdot (17/12) = 189/144 \neq 2$, so that $17/12 \neq \sqrt{2}$. However, $(17/12) \cdot (24/17)$ does equal 2, leading to our third guess of $(17/12 + 24/17)/2 = 577/408 \approx 1.414216$. Thus we have achieved five-decimal-place accuracy, which is almost as good as that of the Babylonians.

Stepping back to analyze this process, we see that the first guess of $x_1 = \frac{3}{2}$ was subject to the operation $(x_1 + 2/x_1)/2$ to get $x_2 = 17/12$. In other words, if

$$F(x) = \frac{\left(x + \dfrac{2}{x}\right)}{2},$$

then $x_2 = F(x_1) = 17/12$, while $x_3 = F(x_2) = 577/408$.

Having reduced the Babylonian calculation of $\sqrt{2}$ to an iterative scheme, one that is based on successive applications of the "iterator" $F(x) = (x + 2/x)/2$, it

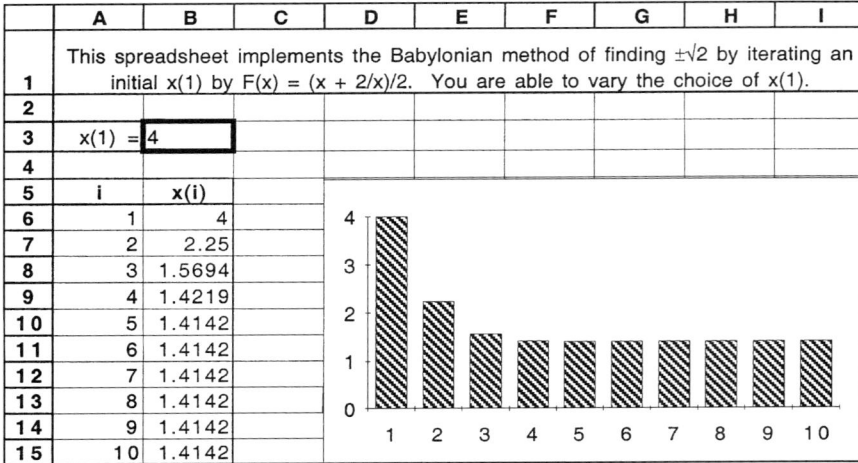

Figure 1.5.3 The Babylonian Method for Finding $\sqrt{2}$

becomes natural to relegate its implementation to a machine. Aside from doing the required calculations very efficiently, the spreadsheet format of Figure 1.5.3 will enable you to explore the use of different initial values $x_1 > 0$ in generating subsequent values $x_2, x_3, \ldots$ that are so remarkably resolute in their approach to $\sqrt{2}$.

Problem 1.5.2 Create a spreadsheet that implements the Babylonian method for calculating $\sqrt{2}$.

An important next step in using iteration to solve polynomial equations is the design of a spreadsheet that provides a "square root machine," one that corresponds to the $\sqrt{}$ button on a hand calculator. This involves generalizing the "Babylonian method" to one that approximates $\sqrt{k}$ along the following lines:

> Given $k > 0$, we have $(\sqrt{k}) \cdot (\sqrt{k}) = k$. For an initial guess $x_1 > 0$, we seek a function $F(x)$ that transforms x_1 to something closer to $\sqrt{k}$. Then, letting $x_2 = F(x_1)$, $x_3 = F(x_2)$, $x_4 = F(x_3)$, etc., we continue until the desired degree of accuracy has been achieved.

What is the required function $F(x)$? We shall soon see that there are many functions one could choose. However, in emulating the "Babylonian method" we would note that both $(\sqrt{k}) \cdot (\sqrt{k})$ and $(x_1) \cdot (k/x_1)$ are equal to k. Therefore, if $x_1 \neq \sqrt{k}$, then $\sqrt{k}$ lies between x_1 and k/x_1.

Problem 1.5.3 Find a function $F(x)$ that transforms an initial guess $x_1 > 0$ into an improved guess $x_2 = F(x_1)$. (See Figure 1.5.4.)

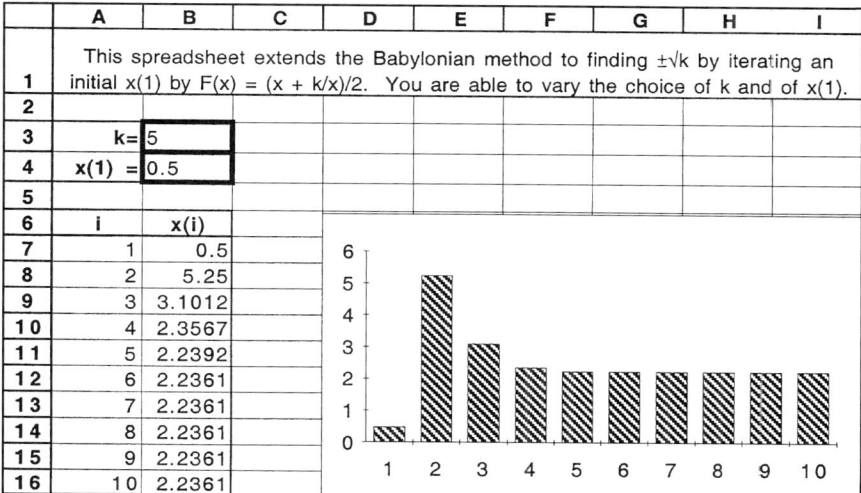

Figure 1.5.4 A Babylonian Square Root Machine

Problem 1.5.4 Using or improving upon the format of Figure 1.5.4, create a spreadsheet "square root machine."

Problem 1.5.5 Recalling that $\sqrt{k}$ denotes the *positive* square root of k, confirm that a starting value $x_1 > 0$ leads to an approximation of $\sqrt{k}$, while $x_1 < 0$ leads to an approximation of $-\sqrt{k}$.

1.6. The Staircase Method

Before trying to extend such techniques to the general quadratic equation, it will be useful to develop some geometric insight into the Babylonian method for finding $\sqrt{2}$. Recalling that this method was based on the iterator $F(x) = (x + 2/x)/2$ and the iterative scheme

$$(1.6.1) \qquad\qquad x_{i+1} = F(x_i),$$

we can use Excel to graph $y = F(x)$ and then trace the above iterative process in relation to this graph (Figure 1.6.1).

Choosing $x_1 = 4$, there are two equivalent techniques for locating the values

$$x_2 = 2.25, \quad x_3 \approx 1.569, \quad x_4 \approx 1.422, \quad \text{etc.}$$

via the graph of $F(x) = (x + 2/x)/2$ in the figure:

First Technique:

Draw the graph of $y = F(x)$ for $1 \le x \le 4$;

Starting with $x_1 = 4$, determine the ordinate $F(x_1) = 2.25$ and call this x_2;

For $x_2 = 2.25$, determine the ordinate $F(x_2) \approx 1.56944$ and call this x_3;

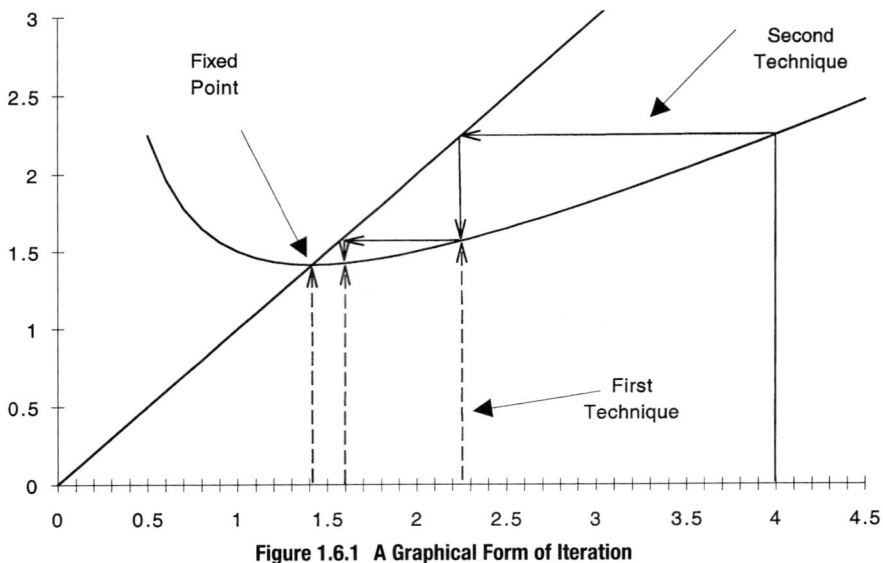

For $x_3 \approx 1.56944$, determine the ordinate $F(x_3) \approx 1.42189$ and call this x_4;
For $x_4 \approx 1.42189$, determine the ordinate $F(x_4) \approx 1.41423$ and call this x_5,
   etc.

Second Technique:

Draw the graphs of $y = F(x)$ *and of* $y = x$ for $1 \leq x \leq 4$;
Starting with $x_1 = 4$, draw a vertical line from $(4, 0)$ to $(4, 2.25)$;
Starting at $(4, 2.25)$, draw a horizontal line to $(2.25, 2.25)$.
Starting at $(2.25, 2.25)$, draw a vertical line to $(2.25, 1.56944)$.
Starting at $(2.25, 1.56944)$, draw a horizontal line to $(1.56944, 1.56944)$;
Starting at $(1.56944, 1.56944)$, draw a vertical line to $(1.56944, 1.42189)$, etc.

The second technique described above will be referred to as the "staircase method"
for representing the iterative scheme (1.6.1). It provides some important insights
into what it is about $F(x) = (x + 2/x)/2$ that makes it into such an effective
iterator for finding $\sqrt{2}$.

First, we note that the "staircase" generated by any $x_1 > 0$ approaches the intersection of the graphs of $y = F(x)$ and $y = x$. In the case at hand, this point of
intersection can be found by solving

$$\frac{1}{2}\left(x + \frac{2}{x}\right) = x,$$

which is, for $x \neq 0$, equivalent to $x^2 - 2 = 0$. Then, from the fact that it lies
on the graph of $y = x$, we conclude that the point of intersection of these graphs
has coordinates $(\sqrt{2}, \sqrt{2})$. This leads us to a formulation of the first essential

property for a function $F(x)$ to serve as an iterator for finding $\sqrt{2}$:

1. $\sqrt{2}$ must be a "fixed point" for F in the sense that $\sqrt{2}$ is a solution of $F(x) = x$.

This concept of fixed point will be central to much of what follows. Formally, x_0 is called a *fixed point* of a function F if it satisfies $F(x_0) = x_0$, i.e., if the value x_0 is unchanged under the application of the function F.

However, having $\sqrt{2}$ as a fixed point is not sufficient for a function to serve as an iterator for finding $\sqrt{2}$. This is illustrated by the function $G(x) = x^2/2 + x - 1$.

Problem 1.6.1 Show that $\sqrt{2}$ is a fixed point of $G(x) = x^2/2 + x - 1$.

While $\sqrt{2}$ is indeed a fixed point of $G(x)$, an initial $x_1 = \frac{3}{2}$ now leads to

$$G(x_1) = x_2 = \tfrac{13}{8}, \quad G(x_2) = x_3 \approx 1.945, \quad G(x_3) = x_4 \approx 2.837, \quad \text{etc.}$$

To see why these iterates are moving *away* from the fixed point $x = \sqrt{2} \approx 1.414$ we draw a staircase diagram based on the graphs of $y = x^2/2 + x - 1$ and $y = x$.

What we see from Figure 1.6.2 is that the *slope* of an iterator at a fixed point x_0 is crucial to determining its effects. The fact that the slope m of $G(x)$ satisfies $m > 1$

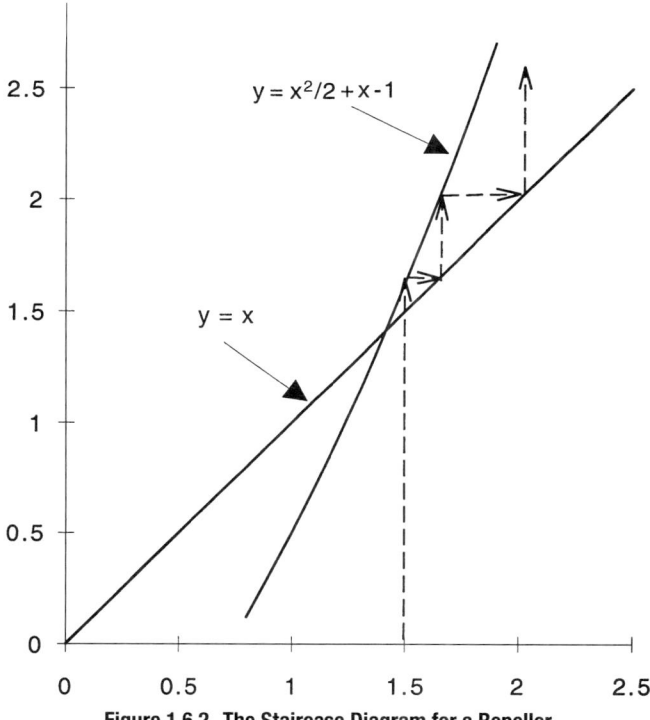

Figure 1.6.2 The Staircase Diagram for a Repeller

at $x = \sqrt{2}$ causes successive iterates to move *away* from the fixed point $x = \sqrt{2}$, and we refer to $\sqrt{2}$ as a *repeller* for G.

Problem 1.6.2 Show that $H(x) = -2x^2 + x + 4$ also has $\sqrt{2}$ as a fixed point. Use a staircase diagram to represent iterations of $x_1 = 1$ by H. Explain why these iterates fail to approach $\sqrt{2}$.

As indicated by Problem 1.6.2, if an iterator has slope $m < -1$ at a fixed point x_0, then that fixed point will also be a repeller. However, if $|m| < 1$ in some interval $L \leq x \leq R$ about the fixed point x_0, and if x_1 is chosen in this interval, then a staircase diagram starting at $x = x_1$ will approach the point (x_0, x_0). In this case x_0 is called an *attractor* for the function.

These geometrically intuitive observations lead us to formulate a second essential condition for a function F to serve as an effective iterator for finding $\sqrt{2}$.

2. The slope of the graph of $F(x)$ must satisfy $|m| < 1$ in an interval about $x = \sqrt{2}$.

If $F(x)$ satisfies both conditions 1 and 2 then for an initial x_1 in the interval specified above, the iterations $x_2 = F(x_1)$, $x_3 = F(x_2)$, etc. will approach $\sqrt{2}$.

These observations provide a basis for some further insights into the Babylonian method for finding $\sqrt{2}$ and the process of iteration. First, it is possible to generalize from $\sqrt{2}$ to any fixed point x_0. As indicated in Figure 1.6.3, for any smooth function F whose slope satisfies $|m| < 1$ in an interval containing a fixed point x_0, the graph $y = F(x)$ lies inside the region determined by $|y - x_0| < |x - x_0|$.

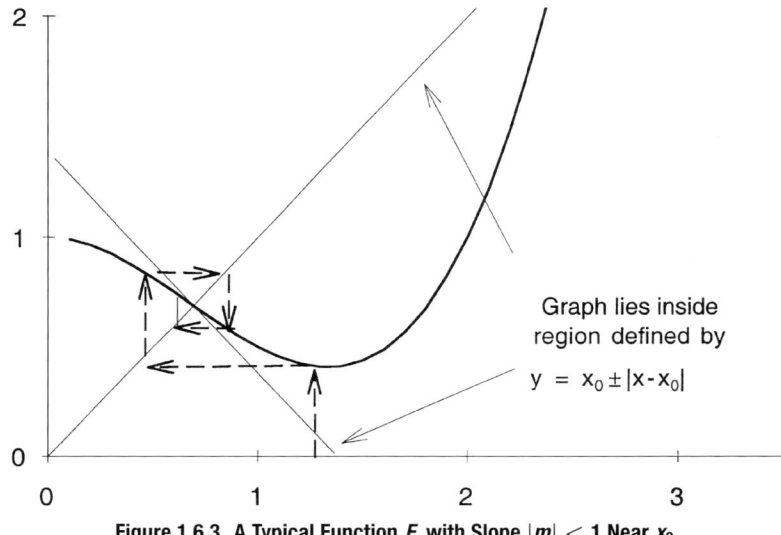

Figure 1.6.3 **A Typical Function** *F* **with Slope** $|m| < 1$ **Near** x_0

Problem 1.6.3 Sketch the graph of a typical function $F(x)$ whose slope satisfies $0 < m < 1$ in an interval near a fixed point x_0. Given an x_1 that lies in that interval, illustrate the approach of the iterates of x_1 toward x_0. Repeat for a function whose slope satisfies $-1 < m < 0$.

Second, it is the "flatness" of the iterator $F(x) = (x + 2/x)/2$ near $x_0 = \sqrt{2}$ that makes it such an effective iterator. Not only does the slope of $F(x)$ satisfy $|m| < 1$ for $x > \sqrt{\frac{2}{3}}$, but, as shown in Problem 1.6.7 below, at $x = \sqrt{2}$ we have $m = 0$. (A calculus-based proof of this fact is contained in the starred Problem 1.11.13 at the end of this chapter.)

But before turning to such proofs, let us harness technology to make a compelling argument in support of this fact. In Figure 1.6.4 we construct a table of values for $y = (x + 2/x)/2$ in an interval $L \leq x \leq R$. Recalling the Babylonian approximation $\sqrt{2} \approx 1.414213$, and shrinking the interval from $1 \leq x \leq 2$ to $1.4 \leq x \leq 1.5$ to $1.41 \leq x \leq 1.42$ to ..., we can provide a graphic illustration of the flatness of $F(x) = (x + 2/x)/2 = (x^2 + 2)/(2x)$ at $x = \sqrt{2}$.

Recalling the Babylonian iterator $G(x) = (x + k/x)/2$ for approximating $\sqrt{k}$, there arises the question of whether the graph of G also satisfies $m = 0$ at $x = \sqrt{k}$.

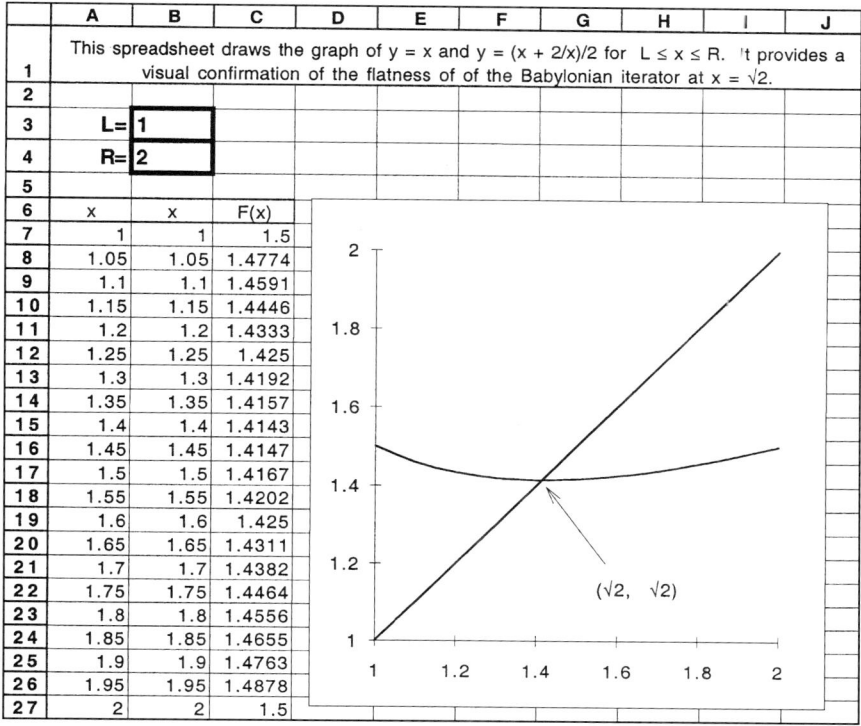

Figure 1.6.4 Testing for Flatness of $F(x) = (x + 2/x)/2$ at $\sqrt{2}$

	A	B	C	D	E	F	G	H	I	J
1	This spreadsheet draws the graph of y = x and y = (x + k/x)/2 for L ≤ x ≤ R. It provides a visual confirmation of the flatness of the Babylonian iterator at x = √k.									
2										
3	L= 1				k= 5					
4	R= 5									
5										
6	x	x	F(x)							
7	1	1	3							
8	1.2	1.2	2.6833							
9	1.4	1.4	2.4857							
10	1.6	1.6	2.3625							
11	1.8	1.8	2.2889							
12	2	2	2.25							
13	2.2	2.2	2.2364							
14	2.4	2.4	2.2417							
15	2.6	2.6	2.2615							
16	2.8	2.8	2.2929							
17	3	3	2.3333							
18	3.2	3.2	2.3813							
19	3.4	3.4	2.4353							
20	3.6	3.6	2.4944							
21	3.8	3.8	2.5579							
22	4	4	2.625							
23	4.2	4.2	2.6952							
24	4.4	4.4	2.7682							
25	4.6	4.6	2.8435							
26	4.8	4.8	2.9208							
27	5	5	3							

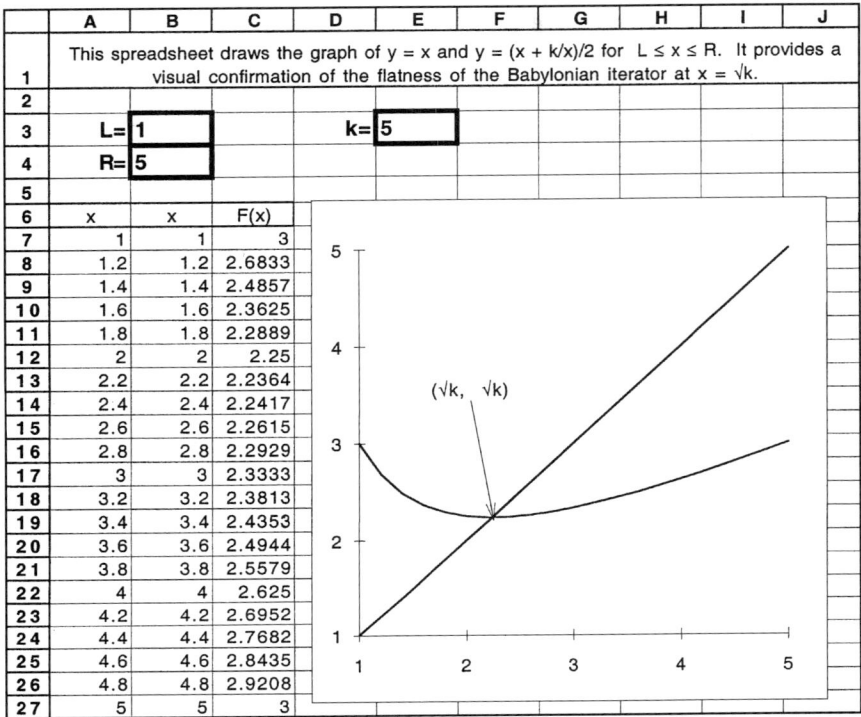

Figure 1.6.5 Testing for Flatness of $F(x) = (x + k/x)/2$ at $\sqrt{k}$

Problem 1.6.4 Verify that $\sqrt{k}$ is a fixed point for $G(x) = (x + k/x)/2$.

Problem 1.6.5 Construct a spreadsheet that displays a table of values for $(x + k/x)/2$ (Figure 1.6.5). Using ten equally spaced values in a variable interval $L \leq x \leq R$, and relying on the spreadsheet to calculate $\sqrt{k}$ by means of the built-in function SQRT(k), test the iterator $G(x) = (x + k/x)$ for flatness at $\sqrt{3}$, $\sqrt{5}$, and $\sqrt{163}$.

Consider now any iterator $F(x)$ whose graph is a smooth curve. Showing that its graph is flat (i.e., has $m = 0$) at a fixed point x_0 is equivalent to showing that its graph does not dip below the horizontal line $y = F(x_0)$ near x_0 and, as a consequence, that this horizontal line is *tangent* to the graph of $y = F(x)$ at the point $(x_0, F(x_0))$. Problem 1.6.7 uses this observation[1] in an elementary proof that the graph of $F(x) = (x + k/x)/2$ is flat at $(\sqrt{k}, \sqrt{k})$. This proof is based on the following special case of the *arithmetic and geometric means inequality* (the A–G inequality): If a and b are positive numbers, then

$$(1.6.2) \qquad \frac{a+b}{2} \geq \sqrt{ab}.$$

[1] See also p. 23 of *Surely You're Joking Mr. Feynman* (Bantam Books, 1986).

The left side of (1.6.2) is called the *arithmetic mean* of a and b, and the right side is called the *geometric mean*.

Problem 1.6.6 By expanding $(\sqrt{a} - \sqrt{b})^2$, establish (1.6.2).

Problem 1.6.7 Suppose $k > 0$ and that $F(x) = (x + k/x)/2$. Use (1.6.2) to show that

$$F(x) \geq \sqrt{k}$$

for all $x > 0$. On this basis, explain why $F(x) \geq F(\sqrt{k})$ for all $x > 0$ and, as a consequence, why $F(x)$ has zero slope at $x = \sqrt{k}$.

Problem 1.6.8 Show that if c and d are negative numbers, then

(1.6.3) $$\frac{c + d}{2} \leq -\sqrt{cd}.$$

On this basis, explain why for $k > 0$, the graph of $F(x) = (x + k/x)/2$ has zero slope at $x = -\sqrt{k}$.

What does all this have to with solving the quadratic equation $ax^2 + bx + c = 0$? Well, if we could find an iterator $F(x)$ whose fixed points are the same as the solutions of this equation and whose slope satisfies $|m| < 1$ near those fixed points, then we could approximate those fixed points by iteration. Our ideal would be an iterator satisfying $m = 0$ at these fixed points. Finding such an iterator is the goal of Section 1.7 to follow.

1.7. Quadratic Equations
Having extended the Babylonian technique for solving $x^2 - 2 = 0$ to $x^2 - k = 0$ and having attained some geometric insight into what makes $(x + 2/x)/2$ and $(x + k/x)/2$ such effective iterators in this regard, we now seek further extensions to the general quadratic equation

(1.7.1) $$ax^2 + bx + c = 0.$$

As a starting point, let us consider the special case where $a \neq 0, b = 0$, and $c/a < 0$. Here

(1.7.2) $$ax^2 + c = 0$$

can be put in the form $x^2 - k = 0$ with $k = -c/a > 0$. Recalling the iterator

$$F(x) = (x + k/x)/2$$

for approximating solutions of $x^2 - k = 0$, we are led to try

(1.7.3) $$F(x) = \frac{1}{2}\left(x - \frac{c}{ax}\right) = \frac{ax^2 - c}{2ax}$$

as an iterator for approximating solutions of (1.7.2).

Problem 1.7.1 Confirm that the solutions of $ax^2 + c = 0$ are fixed points of $F(x) = \frac{ax^2 - c}{2ax}$.

Going on to confirm that the slope of the graph of F satisfies $|m| < 1$ near $x = \pm\sqrt{\frac{-c}{a}}$ will require additional effort (e.g., see Problem 1.11.13 at the end of this chapter). However, we can also use a spreadsheet to study this question.

Problem 1.7.2 For variable values of a and c, generalize Problem 1.6.3 by building a spreadsheet that graphs both $F(x)$ and $ax^2 + c$ in an interval containing $\sqrt{(-c/a)}$. For several values of $a > 0$ and $c < 0$, confirm that the graph of $F(x)$ is flat at the zeros of $ax^2 + c$ (see Figure 1.7.1).

Having developed insight into the solution of $ax^2 + c = 0$ by iteration, we turn to the general quadratic equation

$$(1.7.4) \qquad\qquad ax^2 + bx + c = 0,$$

with $a \neq 0$. Our objective is to modify the iterator $F(x) = \frac{ax^2 - c}{2ax}$ so as to obtain a $G(x)$ whose fixed points are the same as solutions of $ax^2 + bx + c = 0$.

Beginning with the fixed point condition

$$(1.7.5) \qquad\qquad \frac{ax^2 - c}{2ax} = x,$$

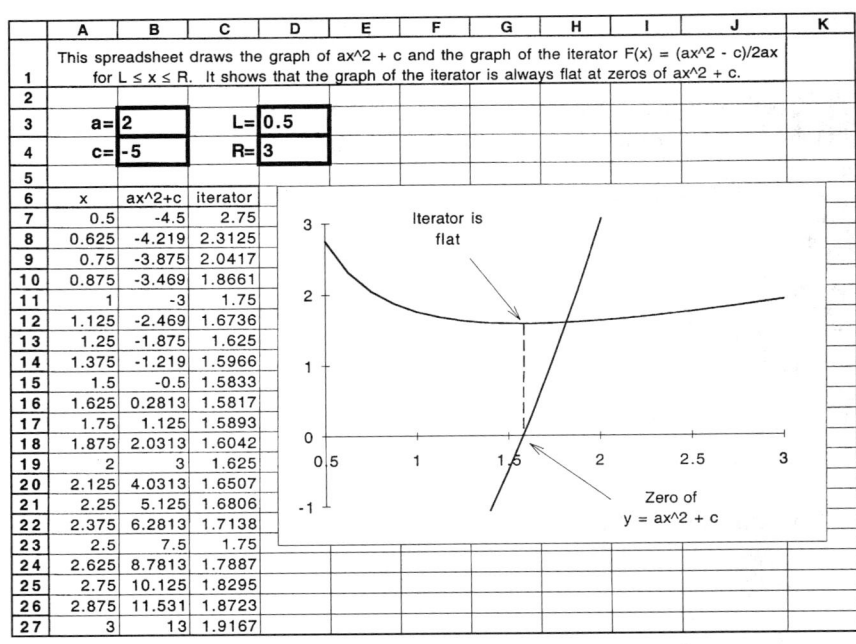

Figure 1.7.1 Confirming the Flatness of $F(x)$ at Zeros of $ax^2 + c$

we can modify the *numerator* of $F(x)$ to obtain the function

(1.7.6) $$\frac{ax^2 - bx - c}{2ax}.$$

Problem 1.7.3 Confirm that the fixed points of (1.7.6) are the same as solutions of $ax^2 + bx + c = 0$.

We could also modify the *denominator* of $F(x)$ to obtain

(1.7.7) $$\frac{ax^2 - c}{2ax + b}.$$

Problem 1.7.4 Confirm that the fixed points of (1.7.7) are the same as solutions of $ax^2 + bx + c = 0$.

In fact, it is possible to modify both numerator and denominator of $F(x)$ to obtain functions with the desired fixed points. However, since it is the flatness of the iterator that is at issue, let us pause to test the two candidates for $G(x)$ that have already been noted. This can be done with a spreadsheet that graphs $G(x)$ and $ax^2 + bx + c$, while allowing one to alter the values of a, b and c. (See Figure 1.7.2.)

Problem 1.7.5 Construct a spreadsheet in the format of Figure 1.7.2 to test (1.7.6) and (1.7.7) for flatness at points where $ax^2 + bx + c = 0$. By examining several cases, confirm that whenever $b^2 - 4ac > 0$, the graph of $G(x)$ appears to be flat at zeros of $ax^2 + bx + c$.

What happens if $b^2 - 4ac = 0$? Setting $c = b^2/(4a)$, the iterator becomes

$$G(x) = \frac{4a^2 x - b^2}{4a(2ax + b)}.$$

Problem 1.7.6 Show that if $b^2 - 4ac = 0$, the iterator $G(x)$ has slope $\frac{1}{2}$ for all $x \neq -b/(2a)$. Explain why even in this case the iterative scheme $x_{i+1} = \frac{2ax_i - b}{4a}$ leads to a solution of $ax^2 + bx + c = 0$.

Well, we hope that Problems 1.7.4–1.7.6 provide compelling reasons for agreeing that it is

(1.7.8) $$G(x) = \frac{ax^2 - c}{2ax + b}$$

that has all the attributes of a highly effective iterator for solving $ax^2 + bx + c = 0$. Not only does it have the right fixed points, but its slope seems to satisfy $m = 0$

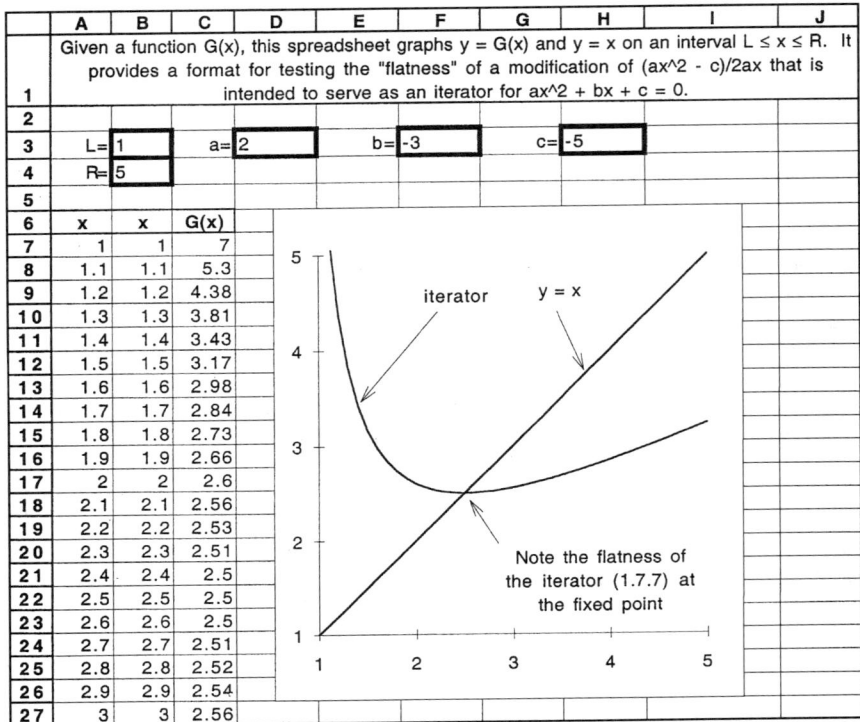

Figure 1.7.2 Testing a Possible Iterator for Flatness

wherever $ax^2 + bx + c = 0$ and $b^2 - 4ac > 0$. The latter observation will be confirmed in Problems 1.11.12 and 1.11.13 at the end of this chapter.

With our observations so confirmed, we assert that the litany *ay ex squared minus cee, all over two ay ex plus bee* deserves a place alongside the quadratic formula in our "algebra for the third millennium." For those who still have doubt, we hope to win you over in Section 1.8, where we will extend these ideas to cubic equations.

But first, by way of celebrating our discovery of the iterator (1.7.8), let us embody it in a spreadsheet that implements its iterative approach for solving the general quadratic $ax^2 + bx + c = 0$, at least whenever $b^2 - 4ac \geq 0$. (Whether iterative techniques apply in case $b^2 - 4ac < 0$ is a question we return to in Chapter 4.)

Problem 1.7.7 Construct a "quadratic equation solver" in the format of Figure 1.7.3.

Problem 1.7.8 For the quadratic equation $x^2 - x - 1 = 0$, generate an accurate graph of the iterator $F(x) = \frac{x^2+1}{2x-1}$ for $0 \leq x \leq 4$. Generate a staircase diagram

	A	B	C	D	E	F	G	H	I	J
1	This spreadsheet approximates a solutions of ax^2 + bx + c = 0 by repeated application of the iterator G(x) = (ax^2 - c)/(2ax + b) to an initial value x(1) ≠ -b/2a. The solution approximated depends on whether x(1) > -b/2a or x(1) < -b/2a.									
2										
3	a=	3		b=	1		(-b+√(b^2-4ac))/2a=		1.135042	
4	c=	-5		x(1)=	2		(-b-√(b^2-4ac))/2a=		-1.468375	
5								x(10)=	1.13504161	
6	i	x(i)								
7	1	2								
8	2	1.30769								
9	3	1.14515								
10	4	1.13508								
11	5	1.13504								
12	6	1.13504								
13	7	1.13504								
14	8	1.13504								
15	9	1.13504								
16	10	1.13504								

Figure 1.7.3 A Quadratic Equation Solver

starting at $(x_1, y_1) = (4, 0)$ and confirm that it agrees with the output of Figure 1.7.3 when $a = 1, b = -1, c = -1$, and $x_1 = 4$.

Problem 1.7.9 Explain why $x_1 = -b/(2a)$ is not an allowable choice for x_1.

Problem 1.7.10 Sketch a "generic graph" of the iterator $F(x) = \frac{ax^2-c}{2cx+b}$. Use this sketch to predict whether iterates generated by Figure 1.7.3 will approximate $x_1 = \frac{-b+\sqrt{b^2-4ac}}{2a}$ or $x_2 = \frac{-b-\sqrt{b^2-4ac}}{2a}$.

1.8. Cubic Equations
Given the presence of quadratic equations throughout the development of mathematics, it is not surprising that cubic equations also played an important role. Babylonian mathematics dating back to 1600 B.C. contains evidence of efforts to find solutions of some special cubic equations. Around A.D. 1100 the Persian poet, astronomer, and mathematician Omar Khayyam gave a geometric method for solving

$$(1.8.1) \qquad\qquad x^3 + b^2 x - b^2 c = 0,$$

where b and c are given numbers. The following problem makes use of analytic geometry to convey the underlying idea.

Problem 1.8.1 In the (x, y)-plane, consider the parabola with equation $y = x^2/b$ and the circle with equation $x^2 + y^2 - cx = 0$. If these curves intersect in a point $(x_1, y_1) \neq (0, 0)$, then show that x_1 is a root of the cubic equation above.

Problem 1.8.2 Use Omar Khayyam's method to give a geometric representation of the real solution of $x^3 + x - 1 = 0$.

However, the general cubic

$$(1.8.2) \qquad\qquad ax^3 + bx^2 + cx + d = 0$$

with $a \neq 0$ turned out to be much more challenging. The method used to solve quadratic equations suggests that one try[1]

$$(1.8.3) \qquad\qquad (A + B)^3 = A^3 + 3A^2B + 3AB^2 + B^3$$

as a basis for "completing the cube."

Problem 1.8.3 Use the identity (1.8.3) to solve $x^3 + 3bx^2 + 3b^2x + d = 0$.

Unfortunately, the technique of Problem 1.8.3 does not extend to the general case of $ax^3 + bx^2 + cx + d = 0$.

Making use of his improved notation, Viète found substitutions that could be used to transform many algebraic equations to simpler forms. For example,

Problem 1.8.4 Show that if x satisfies the general cubic (1.8.2) and $y = x + b/(3a)$, then y satisfies a cubic equation in which the quadratic term is missing, that is, an equation of the form $py^3 + qy + r = 0$.

It was a sequence of such transformations that finally led to a solution of the general cubic in the sixteenth century in terms of what is now called Cardano's method, as described in problems at the end of this chapter. The discovery that the general cubic also lends itself to "solution in terms of radicals" has had a profound effect on algebra for almost 500 years.

While Cardano's method enables us to reduce the general cubic to an expression involving radicals, it does not provide the kind of "formula" or "litany" that we associate with the quadratic case. That is, even after this discovery, there remained distinct differences between the solution of quadratic and cubic equations. By contrast, the iterative algebra litany for solving quadratic equations

   ay ex squared minus cee, all over two ay ex plus bee

will allow for a very direct generalization—not only to cubics but to general polynomial equations as well.

To extend our iterative techniques to solving the general cubic, we begin by asking whether the Babylonian method for finding $\sqrt{k}$ might be extended to finding $\sqrt[3]{k}$. Once we have established an algorithm for solving $x^3 - k = 0$ by iteration, it will be easy to find an iterator for solving $ax^3 + d = 0$. And furthermore, based on

---

[1] "A + B cubo aequalia A cubus + B in A quadr.3 + A in B quadr.3 + B cubo" in sixteenth-century notation.

our experience in Section 1.7, this may be just one propitious guess away from devising an iterator for approximating solutions of the general cubic (1.8.2).

To be specific, let us set $k = 2$ and recall that $\sqrt[3]{2}$ is defined by the property that

$$(\sqrt[3]{2})(\sqrt[3]{2})(\sqrt[3]{2}) = 2.$$

Now, a guess that $\sqrt[3]{k} = \frac{3}{2}$ is clearly too large, i.e., $(\frac{3}{2})(\frac{3}{2})(\frac{3}{2}) > 2$. However, $(\frac{3}{2})(\frac{3}{2})(\frac{8}{9}) = 2$, showing that $\frac{8}{9} \le \sqrt[3]{2} < \frac{3}{2}$. In other words, choosing $x_1 = \frac{3}{2}$ leads us to look for an x_2 satisfying

$$\frac{8}{9} < x_2 < \frac{3}{2}.$$

At this point we could choose x_2 as the arithmetic mean of the *three* numbers $\frac{3}{2}$, $\frac{3}{2}$, and $\frac{8}{9}$, whose product is 2, arriving at

$$x_2 = (\tfrac{3}{2} + \tfrac{3}{2} + \tfrac{8}{9})/3 = \frac{35}{27}.$$

Or else we could also choose x_2 as the arithmetic mean of $\frac{3}{2}$ and $\frac{8}{9}$, arriving at

$$x_2 = (\tfrac{3}{2} + \tfrac{8}{9})/2 = \frac{43}{36}.$$

To help us choose between these options we construct a spreadsheet that implements *both* of these procedures and compares their effectiveness. The spreadsheet in Figure 1.8.1 compares the iterative schemes $x_{i+1} = F(x_i)$ vs. $x_{i+1} = G(x_i)$ when

$$F(x) = (2x + 2/x^2)/3 = \frac{2x^3 + 2}{3x^2} \quad \text{and} \quad G(x) = (x + 2/x^2)/2 = \frac{x^3 + 2}{2x^2}.$$

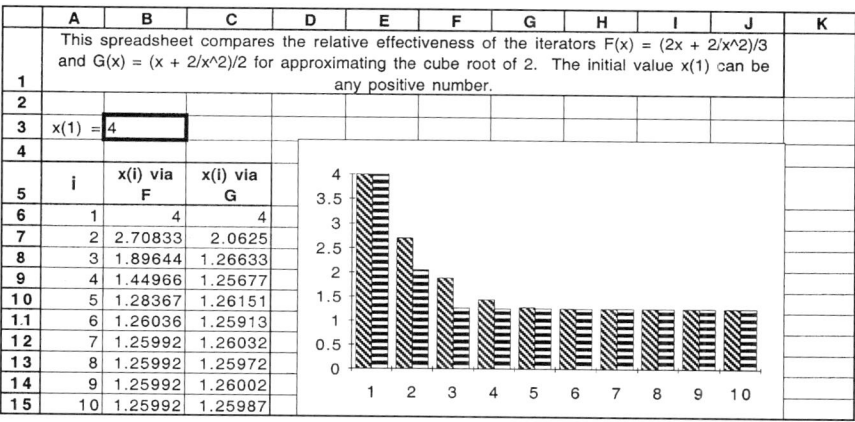

Figure 1.8.1 **Comparing Two Iterators for Solving** $x^3 - 2 = 0$

In the specific example considered in Figure 1.8.1, the iterates generated by F "settle down" at $\sqrt[3]{2} \approx 1.25992$ more quickly than those generated by G. This observation is pursued in the problems below.

Problem 1.8.5 Confirm that $x = \sqrt[3]{2}$ is a fixed point for both $F(x)$ and $G(x)$.

Problem 1.8.6 Recalling the format of Problem 1.6.5, create a spreadsheet that compares the "flatness" of the iterators F and G near $\sqrt[3]{2} \approx 1.259921$.

On the basis of such comparisons, one is likely to develop a marked preference for

$$(1.8.4) \qquad\qquad F(x) = (2x + 2/x^2)/3$$

over

$$G(x) = (x + 2/x^2)/2$$

as an iterator for approximating $\sqrt[3]{2}$. Replacing 2 by k, we will also find that the iterator $F(x) = (2x + k/x^2)/3$ provides the basis for a highly effective "Babylonian cube root machine."

Problem 1.8.7 Recalling that $\sqrt[3]{k}$ is characterized by $(\sqrt[3]{k})(\sqrt[3]{k})(\sqrt[3]{k}) = k$, explain why $F(x) = (2x + k/x^2)/3$ is a likely choice as an iterator for approximating $\sqrt[3]{k}$.

Problem 1.8.8 Build a cube root machine in the format of Figure 1.8.2.

	A	B	C	D	E	F	G	H	I	J
1	This spreadsheet calculates the cube root of k by means of the iterator F(x) = (2x + k/x^2)/3 applied to an initial value x(1) ≠ 0. The entries for both k and x(1) can be varied.									
2										
3	k=	2		x(10)=		1.259921				
4	x(1)=	1.5								
5										
6	i	x(i)								
7	1	1.5		1.5						
8	2	1.2963		1.45						
9	3	1.2609		1.4						
10	4	1.2599		1.35						
11	5	1.2599		1.3						
12	6	1.2599		1.25						
13	7	1.2599		1.2						
14	8	1.2599		1.15						
15	9	1.2599		1.1						
16	10	1.2599		1 2 3 4 5 6 7 8 9 10						
17										

Figure 1.8.2 A Babylonian Cube Root Machine

In problems at the end of this chapter, it is shown that the A–G inequality $\frac{a+b}{2} \geq \sqrt{ab}$ extends to three (and more!) positive numbers in the form

(1.8.5)
$$\frac{a+b+c}{3} \geq \sqrt[3]{abc}.$$

Problem 1.8.9 Assuming that the A–G inequality (1.8.5) is correct, show that $F(x) \geq \sqrt[3]{k}$ for all $x > 0$. On this basis, explain why we might expect F to be a more effective iterator than G.

Turning now to the general cubic (1.8.2), let us agree on $F(x) = (2x + k/x^2)/3$ as our choice of iterator for approximating solutions of $x^3 - k = 0$. This leads us to

(1.8.6)
$$F(x) = \frac{2ax^3 - d}{3ax^2}$$

as the corresponding iterator for solving

(1.8.7)
$$ax^3 + d = 0.$$

Problem 1.8.10 Derive (1.8.6) from $F(x) = (2x + k/x^2)/3$.

As before, it is useful to check that the fixed points of F, characterized by

(1.8.8)
$$\frac{2ax^3 - d}{3ax^2} = x,$$

do indeed correspond to solutions of (1.8.7).

The challenge now is to devise a modification of (1.8.6), one that transforms F into an iterator for the general cubic (1.8.2). Recalling that in the case of quadratic equations it was modification of the *denominator* that led to *ay ex squared minus cee, all over two ay ex plus bee*, this is a natural approach to try.

Problem 1.8.11 Show that a modification of the denominator of (1.8.6) leads to an iterator *two ay ex cubed minus dee, all over three ay ex squared plus bee ex plus cee* whose fixed points coincide with the solutions of $ax^3 + bx^2 + cx + d = 0$.

Problem 1.8.12 Create a spreadsheet that graphs both $ax^3 + bx^2 + cx + d = 0$ and

(1.8.9)
$$F(x) = \frac{2ax^3 - d}{3ax^2 + bx + c}$$

for $L \leq x \leq R$. Is it again the case that the graph of $y = F(x)$ is flat whenever $ax^3 + bx^2 + cx + d = 0$? [Hint: You will have to avoid asking Excel to draw the

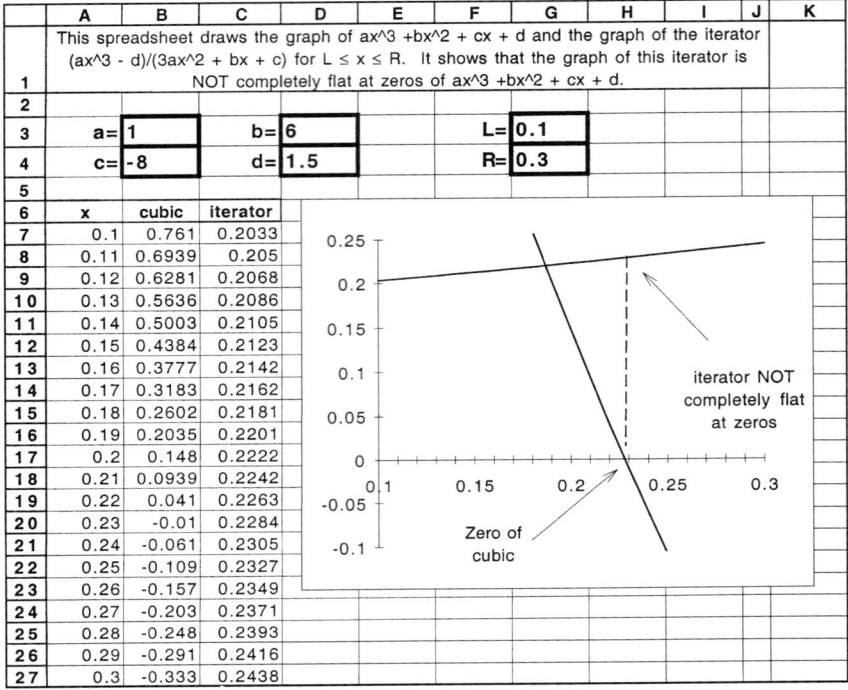

Figure 1.8.3 Testing an Iterator for Flatness at Zeros

graph of $F(x)$ near zeros of $3ax^2 + bx + c$. Careful choice of L and R provides one way of doing this.]

While modifying just the denominator of (1.8.6) provides a reasonably effective tool for solving cubics, Figure 1.8.3 suggests that the resulting iterator is far from optimal. In particular, it leaves open the question of whether we can find a modification of (1.8.6) that, like *ay ex squared minus cee, all over two ay ex plus bee*, in the quadratic case, *does* always have slope $m = 0$ at the zeros of (1.8.2). By modifying both denominator *and* numerator of (1.8.6), we can obtain the following iterator, whose fixed points coincide with the zeros of the general cubic (1.8.2).

$$(1.8.10) \qquad\qquad \frac{2ax^3 + bx^2 - d}{3ax^2 + 2bx + c}.$$

Problem 1.8.13 Confirm that the fixed points of (1.8.10) coincide with the zeros of $ax^3 + bx^2 + cx + d$.

Problem 1.8.14 Use the format of Figure 1.8.3 to test the flatness of the graph of (1.8.10) at zeros of $ax^3 + bx^2 + cx + d$ (Figure 1.8.4). [Hint: You will have to avoid asking Excel to draw the graph of $F(x)$ near zeros of $3ax^2 + 2bx + c$.]

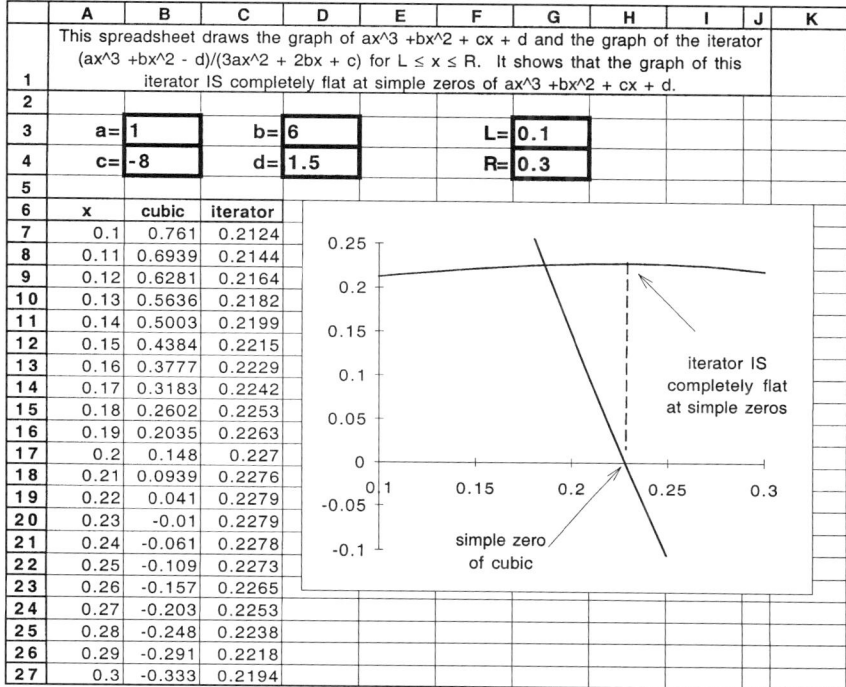

	A	B	C	D	E	F	G	H	I	J	K
1	This spreadsheet draws the graph of ax^3 +bx^2 + cx + d and the graph of the iterator (ax^3 +bx^2 - d)/(3ax^2 + 2bx + c) for L ≤ x ≤ R. It shows that the graph of this iterator IS completely flat at simple zeros of ax^3 +bx^2 + cx + d.										
2											
3	a=	1		b=	6			L=	0.1		
4	c=	-8		d=	1.5			R=	0.3		
5											
6	x	cubic	iterator								
7	0.1	0.761	0.2124								
8	0.11	0.6939	0.2144								
9	0.12	0.6281	0.2164								
10	0.13	0.5636	0.2182								
11	0.14	0.5003	0.2199								
12	0.15	0.4384	0.2215								
13	0.16	0.3777	0.2229								
14	0.17	0.3183	0.2242								
15	0.18	0.2602	0.2253								
16	0.19	0.2035	0.2263								
17	0.2	0.148	0.227								
18	0.21	0.0939	0.2276								
19	0.22	0.041	0.2279								
20	0.23	-0.01	0.2279								
21	0.24	-0.061	0.2278								
22	0.25	-0.109	0.2273								
23	0.26	-0.157	0.2265								
24	0.27	-0.203	0.2253								
25	0.28	-0.248	0.2238								
26	0.29	-0.291	0.2218								
27	0.3	-0.333	0.2194								

Figure 1.8.4 Confirming the Flatness of an Iterator for Cubics

Further discussion of why (1.8.10) is the more effective iterator in the cubic case is contained in problems at the end of this chapter. There we also show how calculus-based techniques can be used to determine this iterator.

On the basis of experimentation with Figure 1.8.4, we now turn to actually solving cubic equations.

Problem 1.8.15 In the format of Figure 1.8.5, use (1.8.10) to construct a "solver" for the general cubic equation $ax^3 + bx^2 + cx + d = 0$.

Problem 1.8.16 Use the spreadsheet from Problem 1.8.15 to find the real solution(s) of $x^3 - x^2 - 1 = 0$ to eight decimal places.

1.9. Higher-Order Equations

Having succeeded in solving quadratic and cubic equations in terms of radicals, it was natural for mathematicians to attempt to conquer quartic (i.e., fourth-degree, or biquadratic) polynomials in a similar fashion. During the sixteenth century Ludovico Ferrari, a student of Cardano, managed to find a general method for finding the roots of a quartic

	A	B	C	D	E	F	G	H	I	J
1	This spreadsheet approximates a solution of ax^3 + bx^2 + cx + d = 0 by repeated application of the iterator (2ax^3 +bx^2 - d)/(3ax^2 + 2bx +c) to an initial x(1) for which 3ax^2 + 2bx +c ≠ 0.									
2										
3	a=	3		b=	1		x(1)=	2		
4	c=	-2		d=	1		x(20)=	-0.824123		
5										
6	i	x(i)								
7	1	2								
8	2	1.2439								
9	3	0.7369								
10	4	0.3056								
11	5	-0.506								
12	6	-1.176								
13	7	-0.928								
14	8	-0.837								
15	9	-0.824								
16	10	-0.824								
17	11	-0.824								
18	12	-0.824								
19	13	-0.824								
20	14	-0.824								
21	15	-0.824								
22	16	-0.824								
23	17	-0.824								
24	18	-0.824								
25	19	-0.824								
26	20	-0.824								

Figure 1.8.5 A General Cubic Equation Solver

equation

(1.9.1) $$ax^4 + bx^3 + cx^2 + dx + e = 0$$

in terms of radicals involving the coefficients. Ferrari's method begins as follows.

Problem 1.9.1 Verify that the substitution $y = x + \frac{b}{4a}$ transforms (1.9.1) into a fourth-order equation for y in which there is no cubic term.

On the basis of Problem 1.9.1, it suffices to consider quartic equations of the form

(1.9.2) $$x^4 + px^2 + qx + r = 0.$$

Expanding

$$(x^2 + p + t)^2 = x^4 + 2px^2 + 2tx^2 + p^2 + 2pt + t^2$$

and substituting for x^4 from (1.9.2), one finds a number t that makes the resulting expression a perfect square. It turns out that this is always possible by solving a cubic equation in t, something that can be done using Cardano's method. Further insight into Ferrari's method is provided by problems at the end of this chapter.

Problem 1.9.2 Confirm that $F(x) = \frac{3x + k/x^3}{4}$ is an effective iterator for solving $x^4 - k = 0$. Use this fact to find an iterator for solving $ax^4 + e = 0$.

Problem 1.9.3 Starting with the iterator from Problem 1.9.2, experiment with modifications of the numerator and denominator of this $F(x)$ to obtain an effective iterator for the general quartic equation $ax^4 + bx^3 + cx^2 + dx + e = 0$. Create a "quartic equation solver" in spreadsheet format.

The next historical step, of course, was to consider polynomials of degree higher than four. Especially in light of the successes of sixteenth-century algebraists, the result here is somewhat surprising. In the early nineteenth century, as a culmination of the work of many mathematicians, Niels Henrik Abel and Évariste Galois proved that for equations of degree 5 and higher, *no* general formulas exist that express their solutions in algebraic form, i.e., in terms of radicals based on their coefficients. The genius and tragedy of Abel and Galois is underscored by the fact that both died before age 30.

One might still think that while there is no "general formula" for such polynomial roots, there could still exist algebraic formulas that are tailored to specific forms of quintic (i.e., fifth-degree) equations. That this is not the case is illustrated by

(1.9.3) $$2x^5 - 5x^4 + 5 = 0,$$

none of whose roots is expressible in terms of standard algebraic operations applied to the coefficients of this equation. However, by generalizing the ideas underlying Problems 1.9.2 and 1.9.3, you would have no difficulty in using iterative techniques to find a solution of this equation to any desired degree of accuracy.

Problem 1.9.4 Use a spreadsheet to find a solution of (1.9.3) to eight decimal places.

For the past two hundred years, the development of algebra has been profoundly influenced by this distinction between polynomial equations of degree less than or equal to four and those of degree greater than four. However, iterative methods, propelled by the power of computers, enable us to find solutions of polynomial equations with arbitrary accuracy by techniques that are independent of the degree of the polynomial. Will such a unifying capability influence our future conception of what constitutes *al-jabr*? And if so, what other mathematical ideas will become part of "the science of transposition and cancellation"?

1.10. Analytic Geometry   In developing iterative techniques for solving polynomial equations, we have also paid homage to more classical methods and some of their historical roots. In so doing, however, we have failed to acknowledge a historical development of central importance. Here we refer to the concepts from *analytic geometry* that have been used repeatedly in this chapter and will continue to appear in succeeding ones.

During the seventeenth century, Reńe Descartes revolutionized mathematics with the introduction of a "method of coordinates" that provided a unification of algebra and geometry. When applied to a polynomial such as $x^3 + 3x^2 + 5$, this method associates with x *cubo* $+ x$ *quadr.*$3 + 5$ a geometric shape called its graph. In Figure 1.3.2 this tool made it possible to give the solution of the general quadratic equation a vivid geometric representation in terms of the points at which the graph of $p(x) = ax^2 + bx + c$ crosses the x-axis. In Figure 1.6.3 analytic geometry made it possible to study the effectiveness of an iterator in terms of the steepness of its graph.

Using this tool, many problems in Euclidean geometry can also be reduced to the solution of algebraic equations. By way of example, the following problem shows how analytic geometry enables us to derive a classic result by algebraic means. The theorem of Thales (Thay-leez) asserts that "any angle inscribed in a semicircle is a right angle."

To establish this result using analytic geometry, consider a circle of radius r centered at the origin in the (x, y)-plane. This circle has equation $x^2 + y^2 = r^2$. With $A = (-r, 0)$ and $B = (r, 0)$, let $P = (a, b)$ be an arbitrary point on the circle different from A and B.

Problem 1.10.1 Referring to Figure 1.10.1, show that the angle APB (an arbitrary angle inscribed in a semicircle) is a right angle. [Hint: Use the fact that

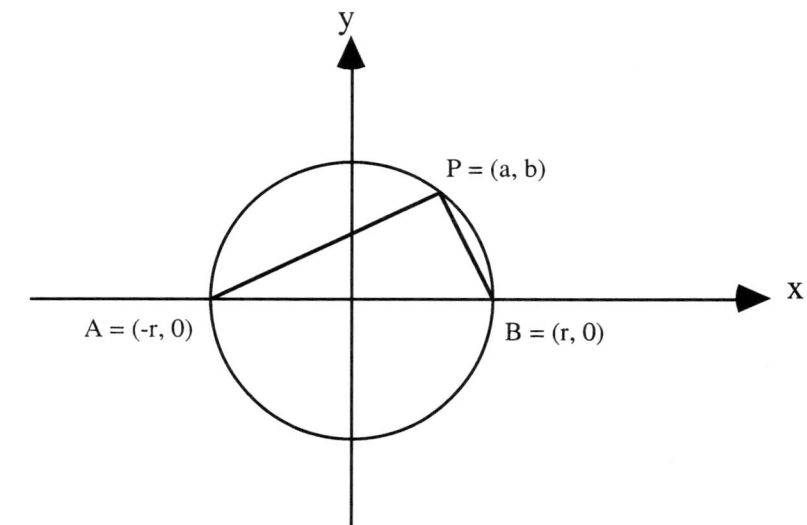

Figure 1.10.1 Proving the Theorem of Thales

two lines in the (x, y)-plane are perpendicular if the slope of one is the negative reciprocal of the other.]

Because analytic geometry will play an important role throughout our development of iterative algebra, it will be useful to develop a number of its relevant ideas at this time. In Chapters 2 and 3 the expression $ax^2 + bx + c$ will reappear not in terms of a polynomial equation to be solved, but rather as an iterator in its own right. To analyze the effect of $ax^2 + bx + c$ as an iterator, it will be important to determine the steepness of its parabolic graph at various points. This, it turns out, can done by using analytic geometry to express some classical Greek concepts.

The Greeks defined a parabola as the locus of points equidistant from a fixed line (called the directrix) and a fixed point (called the focus), where the point is not on the line. Assuming, for example, that the directrix is the line $y = -1$ and the focus has coordinates $(1, 2)$, Figure 1.10.2 illustrates the use of Cartesian coordinates to arrive at an algebraic definition of the resulting parabola.

The fact that the distance of an arbitrary point (x, y) on the parabola is equidistant from the point $(1, 2)$ and from the line $y = -1$ now takes the form

$$\sqrt{(x - 1)^2 + (y - 2)^2} = |y + 1|.$$

Squaring and simplifying, this leads to an equation of the form

$$(1.10.1) \qquad\qquad y = ax^2 + bx + c$$

with $a = \frac{1}{6}$, $b = -\frac{1}{3}$, and $c = \frac{2}{3}$. Indeed, it can be shown that *any* parabola with a vertical axis of symmetry can be written in the form (1.10.1).

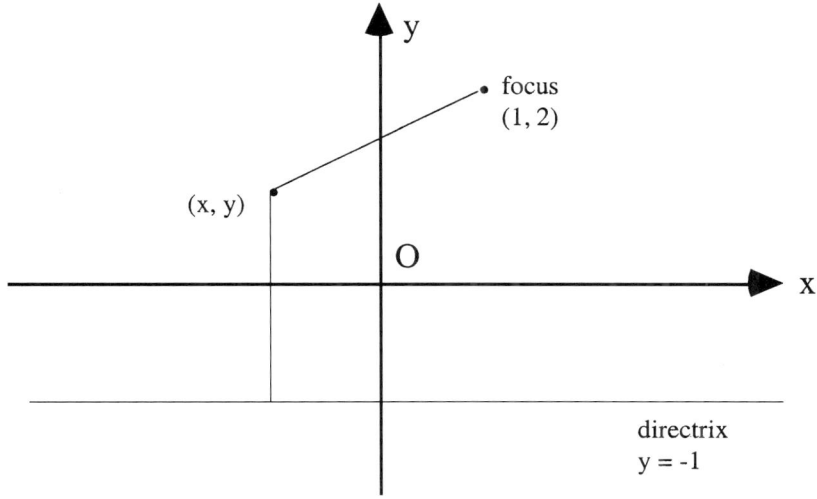

Figure 1.10.2 The Cartesian Coordinate Definition of a Parabola

Problem 1.10.2

(a) Suppose a parabola has its focus at (h, k) and its directrix is given by $y = d$. Show that the equation of the parabola is $y = ax^2 + bx + c$, where

$$a = \frac{1}{2(k - d)}, \quad b = -\frac{h}{k - d}, \quad c = \frac{h^2 + k^2 - d^2}{2(k - d)}.$$

(b) Conversely, suppose we are given a, b, c, with $a \neq 0$. Show that the curve $y = ax^2 + bx + c$ is a parabola with focus (h, k) and directrix $y = d$, where

$$h = -\frac{b}{2a}, \quad k = \frac{1}{4a} + \frac{4ac - b^2}{4a}, \quad d = -\frac{1}{4a} + \frac{4ac - b^2}{4a}.$$

(c) Use part (b) to find the focus and directrix of

(i) $y = x^2$.   (ii) $y = x^2 - 1$.   (iii) $y = x^2 - x - 1$.

In light of Problem 1.10.2, it will be interesting to construct a spreadsheet that graphs a parabola with given focus and directrix (Figure 1.10.3).

The *latus rectum* of a parabola is the chord obtained by intersecting the parabola with a line through its focus and perpendicular to the axis of symmetry (Figure 1.10.4).

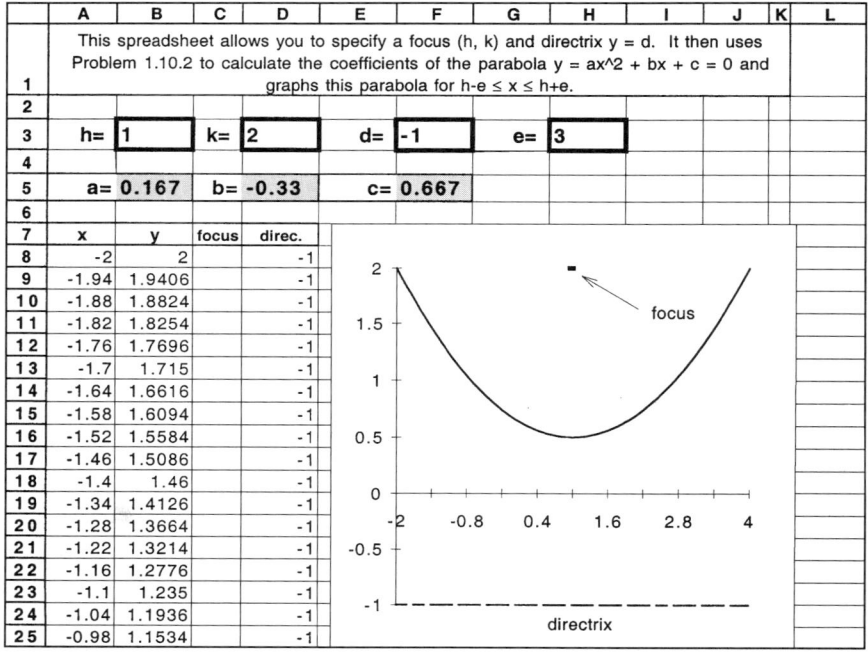

Figure 1.10.3 **Graphing a Parabola**

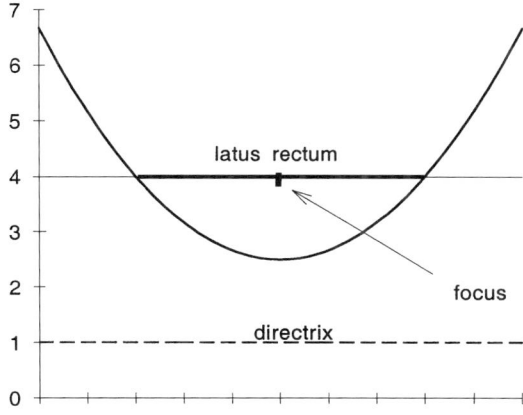

Figure 1.10.4 The Latus Rectum of a Parabola

The chord so obtained will play an important role in our later study of quadratic iterators. As shown in problems at the end of this chapter, the slope m of the parabola $y = ax^2 + bx + c$ satisfies $|m| \leq 1$ on the arc joining the ends of the latus rectum, and $|m| > 1$ elsewhere.

Problem 1.10.3 Assuming the preceding statement, characterize the points on the graph of the parabola $y = x^2$ where the slope m satisfies $|m| < 1$.

Another important application of analytic geometry arises in the Cartesian interpretation of a system of linear equations. For example, the system of two linear equations in two unknowns

(1.10.2)
$$2x - y = 3,$$
$$3x + 4y = -5$$

can be viewed as an algebraic representation of two straight lines (Figure 1.10.5). In this context, the solution of (1.10.2) corresponds to the coordinates of the point (x_0, y_0) at which these lines intersect.

Problem 1.10.4 Find the solution of (1.10.2) and show that it corresponds to the intersection of lines in Figure 1.10.5.

Through analytic geometry we can obtain geometric interpretations of a variety of algebraic phenomena. For example, the fact that the system

$$2x - y = 3,$$
$$-4x + 2y = 7$$

has no solution is demonstrated algebraically by multiplying the first equation by -2. This gives $-4x + 2y = -6$, contradicting the second equation. But the striking geometric reason for this failure is that the corresponding lines in the

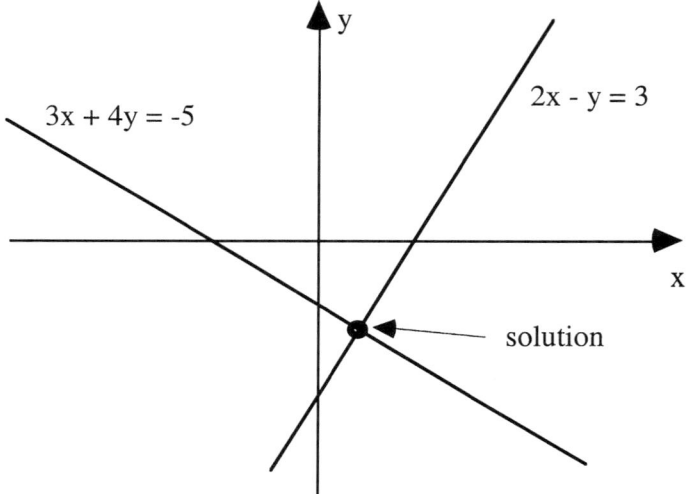

Figure 1.10.5 A System of Linear Equations in the Cartesian Plane

(x, y)-plane are parallel (Figure 1.10.6). As a result, there is no point (x_0, y_0) that belongs to both lines.

Some of the problems to follow elaborate on the role of analytic geometry in establishing geometric properties by means of algebraic tools, and conversely the use of geometry to represent algebraic relationships.

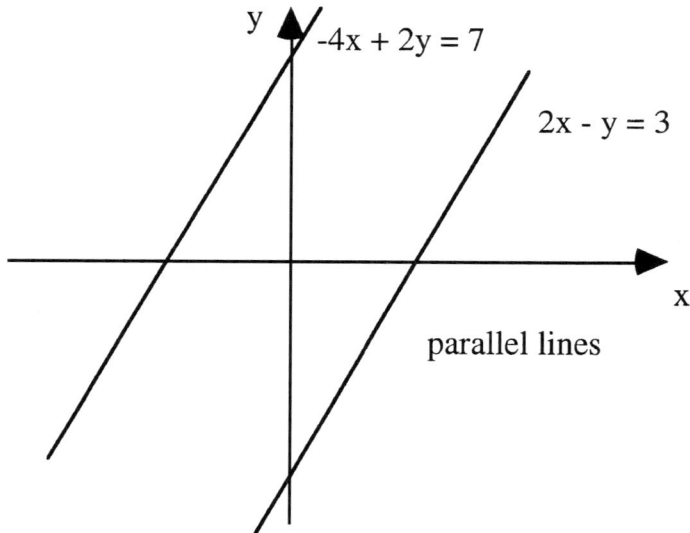

Figure 1.10.6 A Linear System With No Solutions

1.11. Problems, Exercises, and Projects This chapter has provided a rather different view of algebra from that which is traditionally encountered. In bringing computer technology to bear on the solution of polynomial equations, we have also indicated some of the historical roots of the techniques developed. This led us to a number of mathematical topics (golden ratio, arithmetic–geometric mean inequality, properties of parabolas, analytic geometry) that arise in connecting the old with the new.

The problems that follow elaborate on these topics. They also provide further insight into the ways in which algebra was practiced in earlier times, thereby setting the stage for a dramatically different view of algebra, as contained in succeeding chapters.

Another Geometric Solution of a Quadratic Equation

The following geometric solution of a special quadratic equation was given by René Descartes in his *La Geometrie* (1637).

Problem 1.11.1 Given the equation $x^2 + bx - c^2 = 0$, where $b > 0$ and $c > 0$, construct a circle with center O and radius $b/2$ (see Figure 1.11.1). Let QR be tangent to the circle at Q, with $QR = c$, and let S and T be the points where the line through O and R intersects the circle.

(a) Use the quadratic formula to show that $x^2 + bx - c^2 = 0$ always has a positive root and a negative root.

(b) Show that RS is the positive root of the equation and that RT is the absolute value of the negative root.

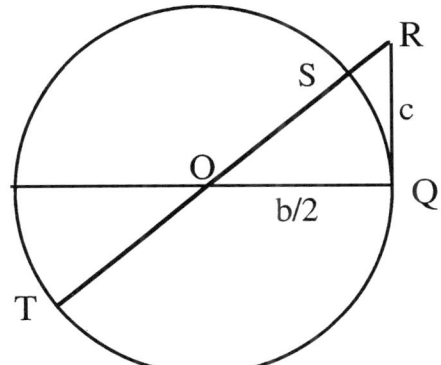

Figure 1.11.1 Descartes's Solution of $x^2 + bx - c^2 = 0$

More on the Golden Ratio

In Section 1.2 we asserted that Greek geometers were able to associate the golden ratio (and hence the polynomial equation $x^2 - x - 1 = 0$) with shapes other than rectangles. The following problem shows how the golden ratio arises in certain isosceles triangles and in regular pentagons.

Problem 1.11.2 The rectangle in Figure 1.2.1 has the property that the small rectangle, left after cutting off a square, is similar to the large (original) rectangle. The discussion accompanying this figure shows that in this case the ratio of long side to the short side is the golden ratio, viz., $(1 + \sqrt{5})/2$.

(a) Show *conversely* that any rectangle for which the ratio of long to short side is the golden ratio has the property that removal of a square leaves a rectangle similar to the original.

(b) Suppose a point P is chosen on a line segment AB in such a way that

$$\frac{AB}{AP} = \frac{AP}{PB}$$

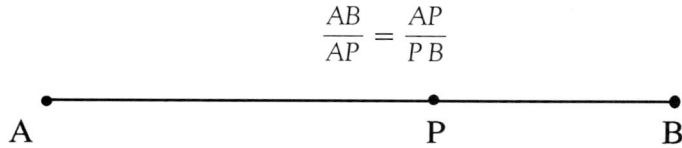

If $AB/AP = x$, show that $x^2 - x - 1 = 0$, and so deduce that

$$\frac{AB}{AP} = \frac{AP}{PB} = \frac{1 + \sqrt{5}}{2}.$$

(c) Consider the isosceles triangle ABC with base angles $72°$ and apex angle $36°$.

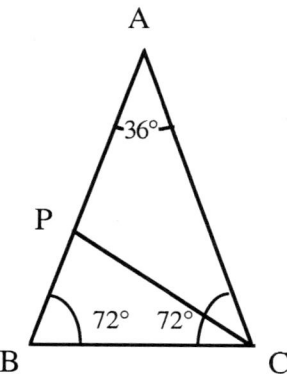

Draw the angle bisector at C, intersecting side AB at point P, and explain why $\triangle CPB$ is similar to $\triangle ABC$. Then use similar triangles to show that

$$\frac{AC}{BC} = \frac{CP}{PB}.$$

Also show that $BC = CP = AP$, and apply this to the preceding equation to get

$$\frac{AB}{AP} = \frac{AP}{PB}.$$

Use part (b) to explain why this shows that $\frac{AB}{BC}$ is the golden ratio.

(d) If the polygon $ABCDE$ in the figure below is a *regular pentagon*, show that the ratio BD/AB of the diagonal of the pentagon to its side is the golden ratio.

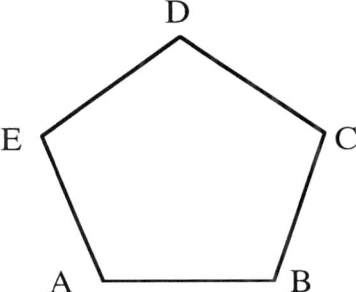

[Suggestion: Analyze the triangle $\triangle ABD$.]

The Golden Ratio is Irrational

A number is rational if it can be expressed in the form m/n, where m and n are integers and $n \neq 0$. From an algebraic point of view, rational numbers are those that arise as solutions of linear equations of the form $ax + b = 0$, where a and b are integers and $a \neq 0$. For us, the fact that the golden ratio $\tau = (1 + \sqrt{5})/2$ is "irrational" is closely related to the fact that $\sqrt{5}$ is also not rational.

Exercise Show that if the golden ratio τ were rational, then $\sqrt{5}$ would also be rational.

As shown in the following problem, Greek geometers established the irrationality of the golden ratio by very different means.

Problem 1.11.3 Proving that a number is not rational often relies on the method of contradiction: *Assuming the result is valid, we arrive at an absurdity.* In this way we conclude that the result cannot be true.

(a) Show that if $\tau = m/n$, where m and n are positive integers, then it would be possible to build a golden rectangle using a total of mn unit squares, i.e., it would be possible dissect such a golden rectangle into mn squares of size 1×1.

(b) Assuming $\tau = m/n$ and that a golden rectangle has been built with a total of mn unit squares, show that we can remove n^2 unit squares to obtain a smaller golden rectangle containing $mn - n^2$ unit squares.

(c) If it were possible to build a golden rectangle from mn unit squares, the method of part (b) shows that it is also possible to build a golden rectangle *using a smaller number of unit squares*. But then we could repeat part (b) and build another golden rectangle using even fewer unit squares. Explain how these observations lead to a contradiction.

(d) Explain why the preceding shows that it is not possible to build a golden rectangle using unit squares. How does this show that the golden ratio τ is not expressible in the form m/n, where m and n are integers?

The Babylonian Estimate of $\sqrt{2}$

Like $\sqrt{5}$ and the golden ratio, $\sqrt{2}$ is not expressible in the form m/n, where m and n are integers. However, the Babylonian approximation for $\sqrt{2}$ that is embodied in Figure 1.5.1, provides a remarkable rational approximation to $\sqrt{2}$.

Problem 1.11.4 The Babylonian approximation for $\sqrt{2}$ was

$$B = \left(\frac{1}{30}\right)\left(42 + \frac{25}{60} + \frac{35}{60^2}\right).$$

(a) Show that $B = \frac{30547}{21600} = \frac{141279875}{99900000} = 1 + \frac{41421}{100000} + \frac{296}{99900000}$.

(b) Show that $\frac{296}{999} = 0.296296296\cdots = 0.\overline{296}$.

(c) Show that the exact decimal representation of B is in fact

$$B = 1.41421\overline{296},$$

and compare this with the correct value of $\sqrt{2}$ (whose decimal expansion begins as $1.41421356\cdots$).

The Pythagorean Theorem

To relate the Babylonian clay tablet of Figure 1.5.1 to $\sqrt{2}$, we employed a special case of the Pythagorean theorem. Using Figure 1.5.2, areas were used to prove that in an isosceles right triangle with legs of length s and hypotenuse d, we have a special case of the Pythagorean theorem, namely

$$s^2 + s^2 = d^2.$$

The following problem shows how these ideas can be extended to establishing other special cases of the Pythagorean theorem and to establishing the great theorem itself.

Problem 1.11.5 The Pythagorean Theorem asserts that for any right triangle with legs of length a and b and hypotenuse c,

$$a^2 + b^2 = c^2.$$

(a) Use areas in the figure below to establish a special case of the Pythagorean theorem, namely that a right triangle with legs of length 1 and 2 has hypotenuse $\sqrt{5}$.

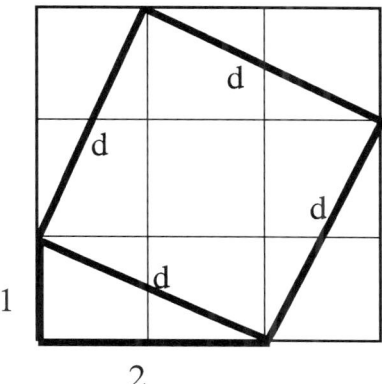

(b) The figure below shows a square with sides of length c. Each of its sides is the hypotenuse of a right triangle with legs of length a and b. Together, they comprise a square whose sides have length $a + b$. Show that the area of the inside square is $a^2 + b^2$, and use this fact to establish $a^2 + b^2 = c^2$.

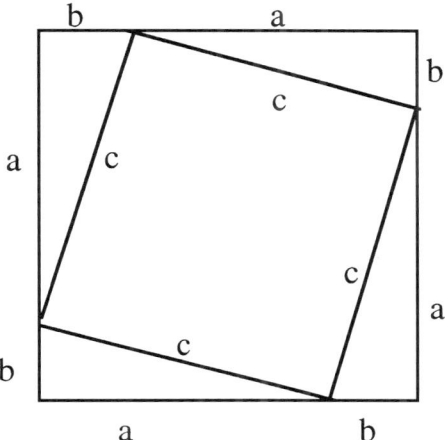

[Hint: c^2 is the area left over after we subtract the areas of the four triangles from the area of the outer square.]

The Arithmetic and Geometric Means and Babylonian Iteration

The Babylonian technique for finding square roots and roots of higher order is closely related to the concepts of *arithmetic mean* and *geometric mean*. The following four problems describe this connection and the proof of the arithmetic–geometric mean inequality. As noted in Problems 1.6.8 and 1.8.9, this connection can be used to establish the "flatness" of certain iterators at their fixed points.

Problem 1.11.6 Recall from Section 1.6 that if $a > 0$ and $b > 0$, the arithmetic mean of a and b is defined by $\frac{a+b}{2}$ and the geometric mean by $\sqrt{ab}$.

(a) Show that both the arithmetic and geometric means lie between a and b. [Hint: Consider the cases $a < b$, $a = b$, and $a > b$ separately. For example, if $a < b$, then you must show that $a < \frac{a+b}{2} < b$ and $a < \sqrt{ab} < b$.]

(b) If $k > 0$ and $x_1 > 0$, use (a) to conclude that both $(x_1 + k/x_1)/2$ and $\sqrt{k}$ are between x_1 and k/x_1.

(c) Explain why, in view of part (b), we would expect $(x_1 + k/x_1)/2$ to be a better approximation for $\sqrt{k}$ than x_1.

Problem 1.11.7 The arithmetic and geometric means of two numbers can be generalized as follows. If $a_1, a_2, \ldots a_n$ are n given positive numbers, the *arithmetic mean* of $a_1, a_2, \ldots a_n$ is defined by

$$A = (a_1 + a_2 + \cdots + a_n)/n$$

and their *geometric mean* by

$$G = (a_1 \cdot a_2 \cdots a_n)^{1/n}.$$

The arithmetic–geometric mean inequality now asserts that $A \geq G$, or

$$(a_1 + a_2 + \cdots + a_n)/n \geq (a_1 \cdot a_2 \cdots a_n)^{1/n},$$

with equality holding if and only if $a_1 = a_2 = \cdots = a_n$.

(a) Recalling Problem 1.6.6, prove the A–G mean inequality in the case $n = 2$.

(b) Show that for any positive numbers x, y, z we have

$$x^3 + y^3 + z^3 - 3xyz = \tfrac{1}{2}(x + y + z)\left[(x - y)^2 + (y - z)^2 + (z - x)^2\right].$$

(c) Using the fact that right side of the above identity is nonnegative, establish the A–G mean inequality for $n = 3$. [This form of $A \geq G$ was the basis for Problem 1.8.9.]

Problem 1.11.8 In Problem 1.11.7 we stated the A–G mean inequality for general n, but gave proofs only for the cases $n = 2$ and $n = 3$.

(a) By applying the case $n = 2$ to $A_1 = \frac{a_1 + a_2}{2}$ and $A_2 = \frac{a_3 + a_4}{2}$, establish the A–G mean inequality for $n = 4$.

(b) Using the preceding idea, establish the A–G mean inequality for $n = 8$.

(c) Explain how the A–G mean inequality can be established for any $n = 2^k$, where k is a positive integer.

Using Backward Induction to Prove the A–G Mean Inequality for Arbitrary Values of n

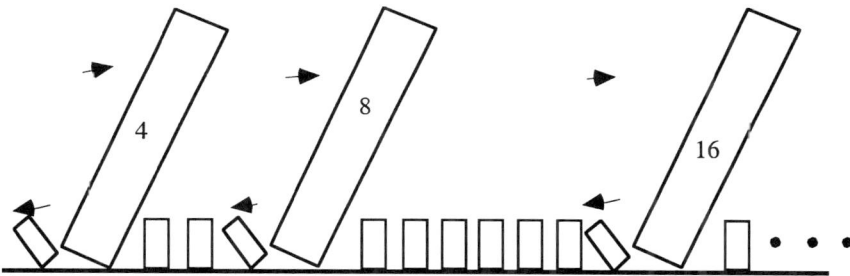

Problem 1.11.9 In order to establish the A–G mean inequality for arbitrary $n \geq 2$, we employ a principle of "backward induction" as suggested by the figure above.

(a) Letting $A = \frac{a_1 + a_2 + \cdots + a_{n-1}}{n-1}$, show that if the A–G mean inequality holds for $a_1, a_2, \ldots, a_{n-1}, A$, then it also holds for $a_1, a_2, \ldots, a_{n-1}$.

(b) Combine the above observation with Problem 1.11.8(c) to establish the A–G mean inequality for arbitrary n.

[This ingenious proof was given by A. L. Cauchy in the early nineteenth century.]

The Effect of Slope Near a Fixed Point

In Section 1.6 we noted the crucial role that the slope of an iterator plays in determining whether a fixed point x_0 is an attractor or a repeller. Further insights into the relationship between slope and the corresponding staircase diagram can be obtained by considering a *linear* iterator of the form $F(x) = ax + b$. Here we can establish a formula for x_n in terms of an initial value for x_1 and establish a direct connection between the slope of the iterator $F(x)$ and the pattern created by the corresponding staircase diagram.

Problem 1.11.10 Consider the iterator $F(x) = ax$ with slope $m = a$ for all values of x.

(a) Show that $x_0 = 0$ is a fixed point of F and that given a value for x_1, we can describe the iterative process by the formula

$$x_{n+1} = a^n x_1, \qquad n = 0, 1, 2, \ldots.$$

(b) Explain why for $a > 1$ and $x_1 > 0$ the values of x_n increase without bound. For $x_1 = 1$ and $a = 1.1$, calculate x_2, x_3, and x_4 and reconcile these values with the corresponding staircase diagram.

(c) Explain why for $0 < a < 1$ and $x_1 > 0$ the values of x_n decrease to zero. For $x_1 = 1$ and $a = 0.9$, calculate x_2, x_3, and x_4 and reconcile these values with the corresponding staircase diagram.

(d) Explain why for $-1 < a < 0$ the x_n alternate in sign while the values of $|x_n|$ decrease to zero. For $x_1 = 1$ and $a = -0.1$, calculate x_2, x_3, and x_4 and reconcile these values with the corresponding staircase (or cobweb) diagram.

(e) Explain why for $a < -1$ the x_n alternate in sign while the values of $|x_n|$ grow without bound. For $x_1 = 1$ and $a = -1.1$, calculate x_2, x_3, and x_4 and reconcile these values with the corresponding staircase (or cobweb) diagram.

Problem 1.11.11 Consider the iterator $G(x) = ax + b$ with slope $m = a$ for all values of x.

(a) Assuming $a \neq 1$, show that $x_0 = b/(1 - a)$ is a fixed point of F and that given a value for x_1, we can describe the iterative process by the formula

$$x_{n+1} = a^n x_1 + \left(\frac{1 - a^n}{1 - a} \right) b, \qquad n = 0, 1, 2, \ldots.$$

(b) Explain why for $a > 1$ and $x_1 > x_0 = b/(1 - a)$, the values of x_n increase without bound. For $x_1 = -5, a = 1.1$, and $b = 1$ calculate x_2, x_3, and x_4 and reconcile their values with the corresponding staircase diagram. What happens if $x_1 < b/(1 - a)$?

(c) Explain why for $0 < a < 1$ and $x_1 > b/(1 - a)$ the values of x_n decrease to $b/(1 - a)$. For $x_1 = 1, a = 0.9$, and $b = -1$, calculate x_2, x_3, and x_4 and reconcile their values with the corresponding staircase diagram.

(d) Explain why for $-1 < a < 0$ the values of $x_n - \frac{b}{1-a}$ alternate in sign while the values of x_n approach $b/(1 - a)$. For $x_1 = 1$ and $a = -0.1$ and $b = 1$, calculate x_2, x_3, and x_4 and reconcile their values with the corresponding staircase (or cobweb) diagram.

(e) Explain why for $a < -1$ the values of $x_n - \frac{b}{1-a}$ alternate in sign while the values of $|x_n - \frac{b}{1-a}|$ grow without bound. For $x_1 = 1$, $a = -1.1$, and $b = 1$, calculate x_2, x_3, and x_4 and reconcile their values with the corresponding staircase (or cobweb) diagram.

An Effective Iterator for the General Quadratic

In Section 1.7 we found several iterators whose fixed points correspond to the zeros of the general quadratic $ax^2 + bx + c$. Our choice of $G(x) = \frac{ax^2-c}{2ax+b}$ was based on spreadsheet experiments that indicated the "flatness" of its graph at those fixed points. The following two problems establish this property mathematically, first by using the A–G mean inequality and then by bringing calculus to bear.

Problem 1.11.12 Assuming $a > 0$ and $b^2 - 4ac > 0$, let $x_0 = \frac{-b+\sqrt{b^2-4ac}}{2a}$ denote the larger solution of $ax^2 + bx + c = 0$. Our goal is to show that the graph of $G(x) = \frac{ax^2-c}{2ax+b}$ is horizontal at x_0 (similar considerations apply at the smaller root $\frac{-b-\sqrt{b^2-4ac}}{2a}$).

(a) Use the A–G mean inequality to prove that for all x,

$$x^2 + x_0^2 \geq 2|xx_0|,$$

and deduce from this that

$$ax^2 + ax_0^2 \geq 2axx_0.$$

(b) Use the last inequality and the fact that $ax_0^2 + bx_0 + c = 0$ to show that

$$ax^2 - c \geq 2axx_0 + bx_0,$$

so that

$$ax^2 - c \geq x_0(2ax + b).$$

(c) If x is near x_0, then $2ax + b$ is near $2ax_0 + b = \sqrt{b^2 - 4ac} > 0$. Thus $2ax + b > 0$ for x sufficiently close to x_0. Use this fact, together with (b), to deduce that $G(x) \geq x_0 = G(x_0)$ for all x close to x_0.

(d) Use (c) to explain why the graph of $G(x)$ must be horizontal at x_0.

*Problem 1.11.13 The power of the calculus makes it easy to establish the "flatness" of the iterator $G(x) = \frac{ax^2-c}{2ax+b}$ at zeros of $ax^2 + bx + c$.

(a) Confirm that for $G(x)$ as above, $G'(x) = \frac{2a^2x^2+2abx+2ac}{(2ax+b)^2}$.
(b) Use (a) to establish the fact that the graph of $G(x)$ is horizontal at simple zeros of $ax^2 + bx + c$.

Faster Iterators

Throughout Chapter 1 we have come to value iterators whose graphs are horizontal at their fixed points. It was on this basis that we rejected $G(x) = \frac{x^3+2}{2x^2}$ as an iterator for finding $\sqrt[3]{2}$, choosing instead $F(x) = \frac{2x^3+2}{3x^2}$, whose graph is flat at $x = \sqrt[3]{2}$. However, there are also "degrees of flatness" among iterators whose graphs have this desired property. The next two problems illustrate this by improving on the Babylonian iterator $F(x) = \frac{x^2+2}{2x}$, whose graph has slope zero at $x = \sqrt{2}$ (see Problem 1.6.7).

Problem 1.11.14 Given an iterator $F(x)$ for which the iterative scheme $x_{i+1} = F(x_i)$ approaches a fixed point x_0, one can always find a "faster iterator" of the form $F(F(x))$ for which $x_{i+1} = F(F(x_i))$ approaches x_0 "twice as fast."

   (a) Calculate $F(F(x))$ when $F(x) = \frac{x^2+2}{2x}$.
   (b) Construct a spreadsheet that approximates $\sqrt{2}$ by repeated applications of $F(x)$ and $F(F(x))$ to the same initial value x_1. By comparing the lists of iterates produced by F and $F(F)$, confirm the fact that $F(F)$ is a more effective iterator.
   (c) Use a spreadsheet to graph both $F(x)$ and $F(F(x))$ for $1 \le x \le 2$. Confirm that while the graphs of both functions are horizontal at $x = \sqrt{2}$, the graph of $F(F(x))$ is "flatter" than that of $F(x)$ at $x = \sqrt{2}$.

Problem 1.11.15 Construct a spreadsheet that implements the iterative scheme $x_{i+1} = G(x_i)$, where $G(x) = \frac{x^3+6x}{3x^2+2}$.

   (a) Show that $x = \sqrt{2}$ is a fixed point of G.
   (b) Show that the iterative scheme $x_{i+1} = G(x_i)$ approaches $\sqrt{2}$ more rapidly than the scheme $x_{i+1} = F(F(x_i))$ in the preceding problem.
   (c) Compare the graphs of $F(F(x))$ and $G(x)$ near $x = \sqrt{2}$.

More on Cubic Equations

While the solution of quadratic equations has provided an interesting context in which to develop iterative techniques, cubic equations provide a more compelling reason for seeking such alternatives. The following problems develop some of the classical solution techniques against which the "cubic equation solver" of Figure 1.8.5 should be compared.

A method for solving the general cubic "in terms of radicals" was first published by Girolamo Cardano in his treatise *Ars Magna* ("The Great Art"), which appeared in 1545. The technique is usually referred to as Cardano's method, even though (as noted in Cardano's book) its main part originated with Tartaglia.[1]

---

[1] An enthralling account of the life of Cardano and the controversy surrounding his publication of the solution of the cubic can be found in William Dunham's *Journey Through Genius*, Wiley, 1990, Chapter 6.

Problem 1.11.16 In Problem 1.8.4 it is shown that the problem of solving the general cubic can be reduced to solving an equation of the form

$$(*) \qquad\qquad x^3 + px + q = 0.$$

(a) Set $x = u + v$ in $(*)$ and show that the resulting equation for u and v can be put in the form

$$u^3 + v^3 + (u + v)(3uv + p) = -q.$$

(b) Use (a) to show that if we can find u and v such that

$$u^3 + v^3 = -q \quad \text{and} \quad uv = -\frac{p}{3},$$

then $x = u + v$ satisfies $(*)$.

(c) The two equations in (b) can be put in the form

$$(**) \qquad\qquad u^3 + v^3 = -q \quad \text{and} \quad u^3 v^3 = -\frac{p^3}{27}.$$

Show that if t_1 and t_2 are the solutions of the quadratic equation

$$t^2 + qt - \frac{p^3}{27} = 0,$$

then $t_1 + t_2 = -q$ and $t_1 t_2 = p^3/27$, and deduce from this that $u = t_1^{1/3}$ and $v = t_2^{1/3}$ are solutions of the above pair of equations $(**)$.

(d) Use the preceding to show that

$$x = \left(-\frac{q}{2} + \sqrt{\left(\frac{q}{2}\right)^2 + \frac{p^3}{27}} \right)^{1/3} + \left(-\frac{q}{2} - \sqrt{\left(\frac{q}{2}\right)^2 + \frac{p^3}{27}} \right)^{1/3}$$

is a solution of $x^3 + px + q = 0$.

Problem 1.11.17 Use the "cubic equation solver" of Figure 1.8.5 to obtain an approximate solution of $x^3 + x - 1 = 0$ (the equation solved by Omar Khayyam). Note that this equation has a real root slightly larger than 0.68.

(a) Use Cardano's method from Problem 1.11.16 to show that

$$x = \left(\frac{1 + \sqrt{31/27}}{2} \right)^{1/3} + \left(\frac{1 - \sqrt{31/27}}{2} \right)^{1/3}$$

is a solution of $x^3 + x - 1 = 0$.

(b) Check that the root found in (a) is approximately $x_0 = 0.6823278$.

(c) Create a spreadsheet graph of $y = x^3 + x - 1$ and confirm that $x^3 + x - 1 = 0$ has only one *real* solution (the other two roots are complex numbers).

(d) What is the iterator that underlies the approximations of x_0 produced by the cubic equation solver in Figure 1.8.5?

Complex Solutions

If $b^2 - 4ac < 0$, then the solutions of the quadratic equation $ax^2 + bx + c = 0$ are complex numbers. In Chapter 4 we will learn how to determine complex roots by iteration, but for the moment such tools are not at hand—neither for quadratic equations nor those of higher order. We can, however, determine complex solutions for $ax^2 + bx + c = 0$ by the quadratic formula, and as shown in the next two problems, this approach can help us find complex solutions of cubic equations as well.

Problem 1.11.18 Recall that in the preceding problem one root of $x^3 + x - 1 = 0$ was found to be $x_0 = 0.6823278\ldots$.

(a) Show that if $x_0 = 0.6823278\ldots$ is the real solution of $x^3 + x - 1 = 0$, then

$$(x - x_0)(x^2 + x_0 x + x_0^2 + 1) = x^3 + x - 1.$$

[Hint: Consider the "long division" of $x^3 + x - 1$ by $x - x_0$.]

(b) Show that the two complex roots of $x^3 + x - 1 = 0$ are given by

$$\frac{-x_0 \pm i\sqrt{3x_0^2 + 4}}{2},$$

where $i = \sqrt{-1}$.

(c) Show that if x_0 is a real solution of $x^3 + px + q = 0$, then the other two roots are

$$\frac{-x_0 \pm \sqrt{-3x_0^2 - 4p}}{2}.$$

Newton's Method

The staircase diagram in Figure 1.6.1 provides a geometric representation of the Babylonian iterative scheme for approximating $\sqrt{2}$. It describes a geometric construction that enables one to locate successive values of x_i obtained from $x_{i+1} = F(x_i)$, in this case when $F(x) = \frac{x^2 + 2}{2x}$.

Newton's method provides another geometric representation of this same process. While the staircase method requires only the construction of horizontal and vertical lines relative to the graph of an iterator, Newton's method requires the construction of tangent lines. For this reason, the numerical implementation of Newton's method requires tools from the calculus, namely computing derivatives.

*Problem 1.11.19 The figure below contains the graph of $f(x) = x^2 - 2$, which crosses the x-axis at $x = \pm\sqrt{2}$. The steps below describe Newton's method as it corresponds to $x_{i+1} = F(x_i)$ with $F(x) = \frac{x^2+2}{2x}$ and $x_1 = 2$.

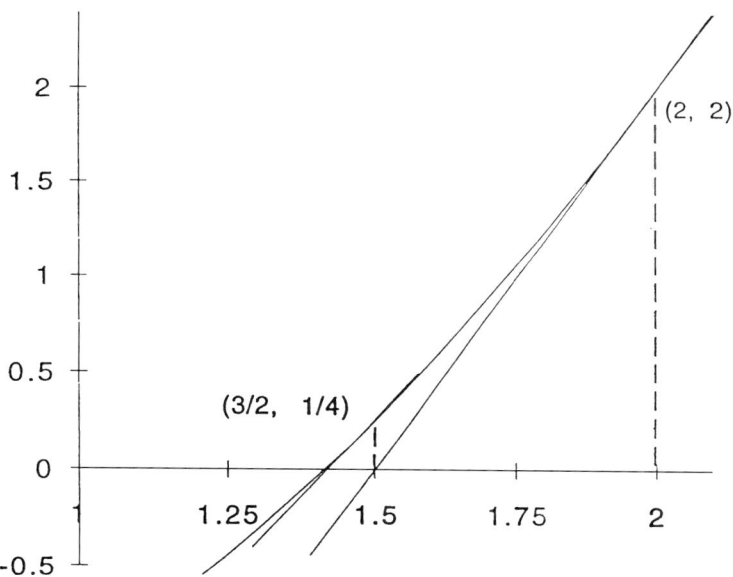

(a) Starting at $x_1 = 2$ on the x-axis, draw a vertical line from x_1 to the graph of $f(x)$, intersecting the graph at $(2, 2)$.

(b) Noting that $f'(x) = 2x$, confirm that the graph has slope $m = 4$ at $(2, 2)$. Recalling the "point–slope" form $y - y_0 = m(x - x_0)$ for a line, draw the line $y - 2 = 4(x - 2)$ that is tangent to the graph at $(2, 2)$.

(c) By setting $y = 0$ in the equation $y - 2 = 4(x - 2)$, confirm that this tangent line crosses the x-axis at $x = \frac{3}{2}$.

(d) On the basis of (c), set $x_2 = \frac{3}{2}$ and repeat the above process. That is, draw a vertical line through $x = x_2$ that intersects the graph of $f(x)$ at $(\frac{3}{2}, \frac{1}{4})$. From $f'(x) = 2x$ deduce that the graph has slope $m = 2(\frac{3}{2}) = 3$ at this point of intersection, and use this fact to determine the tangent line $y - \frac{1}{4} = 3(x - \frac{3}{2})$. Find the point of intersection of this line with the x-axis and denote it as x_3.

(e) Confirm that the values of x_2 and x_3 so obtained are the same as those generated by the iterative scheme $x_{i+1} = F(x_i)$ when $F(x) = \frac{x^2+2}{2x}$ and $x_1 = 2$.

Problem 1.11.20 The method used in Problem 1.11.19 to approximate a zero of $f(x) = x^2 - 2$ can be applied to more general functions.

(a) Let $f(x)$ be a given function. According to the calculus, the graph of $f(x)$ has slope $f'(x)$ at $(x, f(x))$. Show that for a given x_1, the line tangent to the graph of $f(x)$ at $(x_1, f(x_1))$ is given by
$y - f(x_1) = f'(x_1)(x - x_1)$.

(b) Show that the tangent line constructed in (a) crosses the x-axis at
$x_1 - \frac{f(x_1)}{f'(x_1)}$.

(c) Show that Newton's method, as illustrated in Problem 1.11.19, is of the form $x_{i+1} = F(x_i)$, where $F(x) = x - \frac{f(x)}{f'(x)}$.

(d) Show that for $f(x) = x^2 - 2$, Newton's method yields the Babylonian iterator $F(x) = \frac{x^2+2}{2x}$.

(e) Show that for $f(x) = ax^2 + bx + c$, (c) yields $F(x) = \frac{ax^2-c}{2ax+b}$.

Problem 1.11.21 It can be shown that none of the roots of the fifth-degree polynomial

$$f(x) = 2x^5 - 5x^4 + 5$$

is expressible in terms of radicals.

(a) Given that the derivative of $f(x)$ is $f'(x) = 10x^4 - 20x^3$, create a spreadsheet that enables you to approximate solutions of $f(x) = 0$ by Newton's method.

(b) Confirm that an initial value $x_1 = 2.01$ leads to an approximation of the root $x_0 = 2.42807279\ldots$.

(c) Confirm that an initial value $x_1 = 3$ also leads to an approximation of the root $x_0 = 2.42807279\ldots$.

(d) Draw a graph of $f(x)$ that explains why one of these initial values leads to a gradual approximation, while the other leads to a very rapid approximation of this root.

Tangent Line to a Parabola

In Chapter 3 we will need the following fact about parabolas of the form

$$y = ax^2 + bx + c, \quad \text{where} \quad a > 0.$$

Such parabolas have a vertical axis of symmetry and open upward. Their graphs have slope $m < 1$ below the latus rectum and slope $m > 1$ above the latus rectum. While this result can also be established using the calculus, the following problems provide a proof based on synthetic geometry, as practiced by the ancient Greeks.

Problem 1.11.22 Consider the parabola with equation $y = \frac{x^2}{4k}$, where k is a positive constant. As shown in Problem 1.10.2, this parabola has focus $F = (0, k)$ and directrix given by $y = -k$.

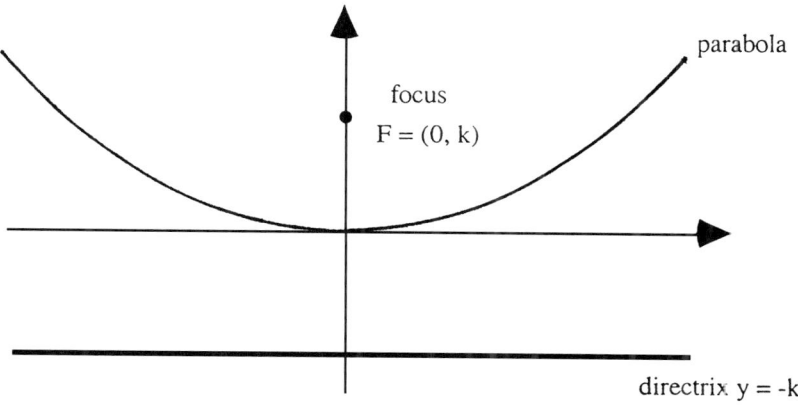

Let $P = (a, b)$ be any point on the parabola, and let $Q = (a, -k)$ be the foot of the perpendicular dropped from P to the directrix:

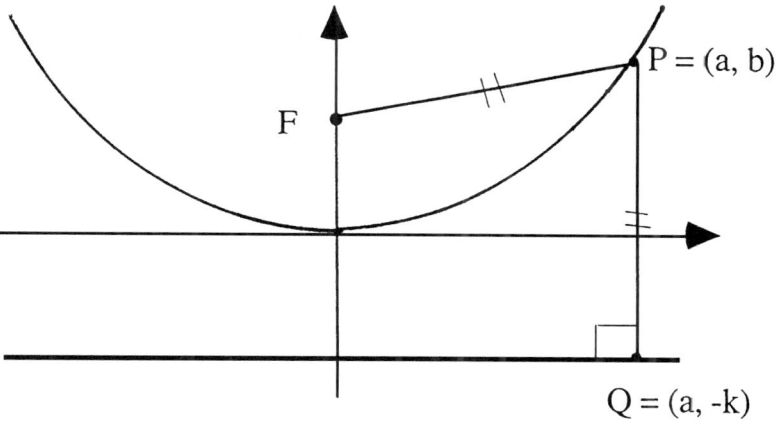

Since $FP = PQ$, the perpendicular bisector t of FQ passes through $P = (a, b)$ (note that $\triangle FPQ$ is an *isosceles* triangle):

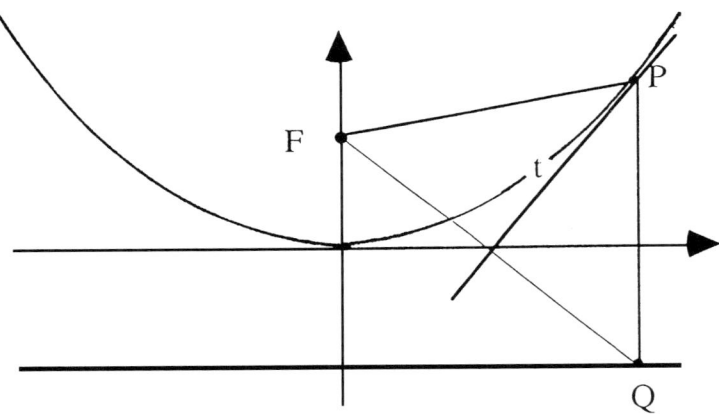

(a) Show that the midpoint of FQ is $(\frac{a}{2}, 0)$.

(b) Letting t denote the perpendicular bisector of FQ, use the fact that t passes through both $(\frac{a}{2}, 0)$ and (a, b) to show that its equation is
$$y = \frac{2b}{a}x - b.$$

(c) Assuming still that (a, b) is on the parabola, show that if (x_1, y_1) is any other point on the parabola, then (x_1, y_1) is *above* the line $y = \frac{2b}{a}x - b$. [Hint: We seek to show that $y_1 > \frac{2b}{a}x_1 - b$ for any $x_1 \neq a$. Since $y_1 = \frac{b}{a^2}x_1^2$ (why?) and $x_1^2 > 2ax_1 - a^2$ (why?), it follows that ....]

(d) Explain why the preceding shows that the line t with equation $y = \frac{2b}{a}x - b$ is *tangent* to the parabola at (a, b).

(e) Suppose (a, b) is a point on the parabola with equation $y = \frac{x^2}{4k}$. Deduce from part (d) that the slope of the parabola at (a, b) is $\frac{2b}{a}$. Recalling that (a, b) lies on the parabola $y = \frac{x^2}{4k}$, show that this is also equal to $\frac{a}{2k}$.

(f) Use (e) to show that the slope m of the parabola $y = \frac{x^2}{4k}$ satisfies
$$|m| < 1 \quad \text{for} \quad -2k < x < 2k,$$
i.e., $|m| < 1$ along the arc joining the endpoints of the latus rectum.

Problem 1.11.23 In the preceding problem we used analytic geometry to prove that given a point $P = (a, b)$ on the parabola $y = \frac{x^2}{4k}$, all other points of the parabola lie above a particular line t containing P. This line was the perpendicular bisector of the line segment determined by the focus F of the parabola and the point Q obtained by dropping a perpendicular from P to the parabola's directrix.

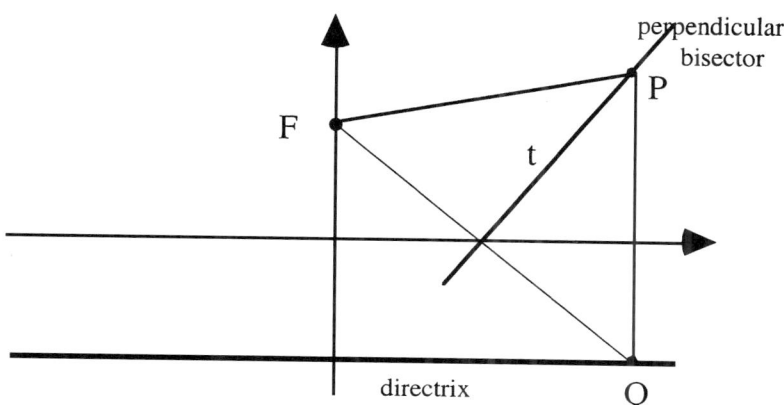

To give a purely *synthetic* proof that all other points of the parabola lie above t, we can show that it is impossible for such a point to lie *below* t.

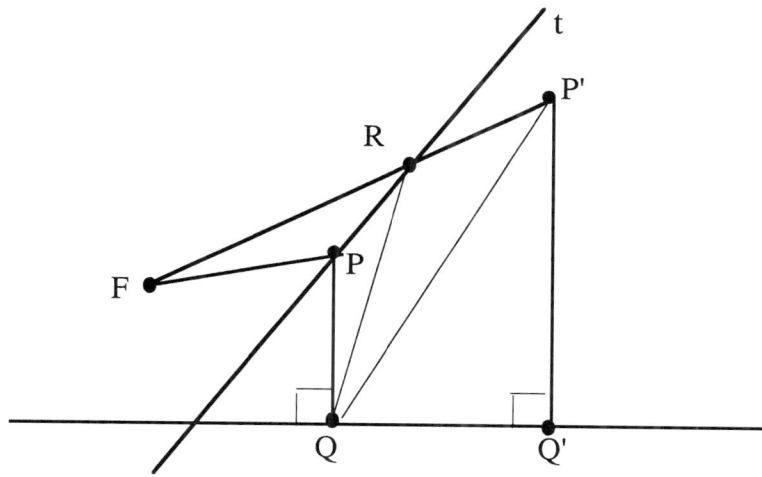

Let P' be any point below t.

   (a) Why is $P'Q \geq P'Q'$?

   (b) Why is $QR + RP' > P'Q$?

   (c) Why is $FR + RP' > P'Q$?

   (d) Explain why (a), (b), (c) show that $FP' > P'Q'$.

(e) Why does (d) show that P' cannot be a point on the parabola and therefore that the parabola lies *above* t, as we wanted to prove?

The Source of Solutions of a Cubic Equation

Problem 1.11.24 The cubic equation $(x+1)(x-1)(x-2) = x^3 - 2x^2 - x + 2 = 0$ has three real roots, -1, 1, and 2. Each of these solutions can be approximated by iterations of the form $x_{i+1} = F(x_i)$, where $F(x) = \frac{2x^3 - 2x^2 - 2}{3x^2 - 4x - 1}$. This problem associates with an initial value x_1 the particular root approached by its iterates. For convenience, we will denote by A, B, and C the sets of initial values whose iterates approach -1, 1, and 2, respectively.

(a) Use a spreadsheet to graph the iterator $F(x) = \frac{2x^3 - 2x^2 - 2}{3x^2 - 4x - 1}$ for $-4 \le x \le 4$ and $-3 \le y \le 5$. Explain the reason for the vertical asymptotes at $x \approx -0.21525$ and at $x \approx 1.5486$.

(b) Based on the graph in part (a), explain why you would expect values of $x_1 < -0.215$ to belong to A and values of $x_1 > 1.549$ to belong to C.

(c) For an arbitrary interval $L \le x \le R$, let $\Delta = (R - L)/100$. Create a spreadsheet with the value L in cell A3, $L + \Delta$ in B3, $L + 2\Delta$ in C3, ..., $L + 100\Delta = R$ in CW3. Regarding these entries as 100 possible choices for x_1, implement 30 iterations of each them, recording the corresponding value for x_{34} in cells A33, B33, ... , CW33. [The numbers appearing in row 33 should be very close to -1, to 1, or to 2.]

(d) For $L = -4$ and $R = 4$, draw a bar graph of the data in cells A33, B33, ... , CW33. Confirm that the nature of this bar graph appears to be consistent with part (b).

(e) Explore the structure of the interval $-0.3 \le x_1 \le 1.6$, classifying subintervals as belonging to A, B, or C.

(f) Can you classify $1.45884112 \le x_1 \le 1.45884115$?

Ferrari's Method for the Quartic

Problem 1.11.25 This problem illustrates Ferrari's method by applying it to a special fourth degree equation, namely

$$(*) \qquad x^4 + 4x - 1 = 0.$$

This is an equation of the form (1.9.2) with $p = 0, q = 4$, and $r = -1$.

(a) By expanding $(x^2 + t)^2$ and substituting from $(*)$, show that

$$(**) \qquad (x^2 + t)^2 = 2tx^2 - 4x + t^2 + 1.$$

(b) Show that if $16 - (4)(2t)(t^2 + 1) = 0$, then the righthand side of $(**)$ is a square of the form $(ax + b)^2$, where a and b may depend on t. [Hint: What conditions on A, B, C assure that $Ax^2 + Bx + C = (Dx + E)^2$?]

(c) Show that the above equation for t is a cubic, one that is equivalent to

$$t^3 + t - 2 = 0$$

and having $t = 1$ as one of its roots.

(d) Setting $t = 1$ in (**), find the four roots of (*).

2 DIFFERENCE EQUATIONS

2.1. Introduction   Having invested considerable time and energy developing new ways of solving polynomial equations, the question, "Why is this important?" may be lurking in the back of the reader's mind. After all, many mathematicians find the questions associated with "solvability in terms of radicals" to be of great interest.[1] Furthermore, as indicated by the starred problems at the end of Chapter 1, a knowledge of calculus provides an alternative approach to using iteration as a tool for solving polynomial equations. Is an ability to solve polynomial equations in the Babylonian spirit really worth all this effort?

Perhaps not in and of itself. However, as we shall see in this and succeeding chapters, concepts such as *iteration* and *fixed point*, combined with the computer skills developed in representing these ideas on spreadsheets, *are* profoundly important. In combination, they enable us to engage mathematical ideas that tend to be regarded as accessible only to specialists armed with advanced computational techniques. However, in the context of computer-assisted forms of algebra, these topics can now be studied, understood, and applied independent of the calculus. In this way iterative algebra provides a window on both mathematics and the world of dynamic modeling.

The present chapter begins such a development with a "rule for change" called a *difference equation*. Rather than approaching difference equations through calculus-based rules for change (i.e., by first studying differential equations), we shall relate difference equations to the iterative processes that we have used to solve

---

[1] See Section 1.9.

polynomial equations. This connection will provide us with an ability to use familiar iteration-based tools to think creatively about topics of central importance to our world and civilization—and perhaps to our future well-being.

The connection between iteration and difference equations is both simple and direct. Until now, our study of iteration has been based on repeated applications of the rule

$$(2.1.1) \qquad\qquad x_{i+1} = F(x_i)$$

for a given iterator $F(x)$. By subtracting x_i from both sides of (2.1.1), we obtain the equivalent *difference equation*

$$(2.1.2) \qquad\qquad x_{i+1} - x_i = F(x_i) - x_i.$$

Defining $f(x) = F(x) - x$, (2.1.2) can now be written in the form

$$(2.1.3) \qquad\qquad x_{i+1} - x_i = f(x_i).$$

By way of example, the Babylonian method for finding $\sqrt{5}$ called for repeated applications of the iterator

$$F(x) = \frac{x + \dfrac{5}{x}}{2}$$

to an initial $x_1 > 0$. In the context of difference equations, this becomes

$$x_{i+1} - x_i = \frac{x_i + \dfrac{5}{x_i}}{2} - x_i$$
$$= \frac{5 - x_i^2}{2x_i}.$$

While difference equations provide an alternative way of approximating solutions of $x^2 - 5 = 0$, such a recasting of Babylonian techniques is not our goal. Rather, we will regard (2.1.3) as a "rule for change," one that is central to many forms of dynamic modeling. Usually, this calls for associating with the index i a unit of time, such as a second, a month, or a year. Iterations then correspond to a prescribed rate, and as a result, difference equations provide a description of how quickly a variable x is changing with time. In this way, an ability to deal with equations of the form (2.1.3) will be an important step toward modeling "change" in a variety of contexts.

2.2. A Banking Analogy
By way of a familiar example of a rule for change, we begin by modeling the "dynamics" of $100 deposited in a bank at 10% annual interest. If x_i denotes the amount of money on deposit at the beginning of the ith year, then the bank's promise to pay 10% interest at the end of each year can be represented by the difference equation

$$(2.2.1) \qquad\qquad x_{i+1} - x_i = f(x_i)$$

i	x_i	$f(x_i) = .1x_i$
1	100	10
2	110	11
3	121	12.1
etc.		

Figure 2.2.1 A Banking Spreadsheet

with $f(x) = 0.1x$ and $x_1 = 100$. Here it will be convenient to introduce the notation

$$(2.2.2) \qquad \Delta x_i = x_{i+1} - x_i,$$

in which the Greek letter Δ (read "delta") represents "change." With this notation, the difference equation (2.2.1) can be represented on a spreadsheet in the format of Figure 2.2.1.

This figure provides a format that we shall use in modeling the general difference equation $\Delta x_i = f(x_i)$. In the case of our banking problem, Δx_i represents the amount of interest to be credited to the balance at the end of the ith year.

Problem 2.2.1 Use the format of Figures 2.2.1 and 2.2.2 to create a spreadsheet that calculates the annual growth of an initial deposit x_1 receiving interest at $100r\%$/year.

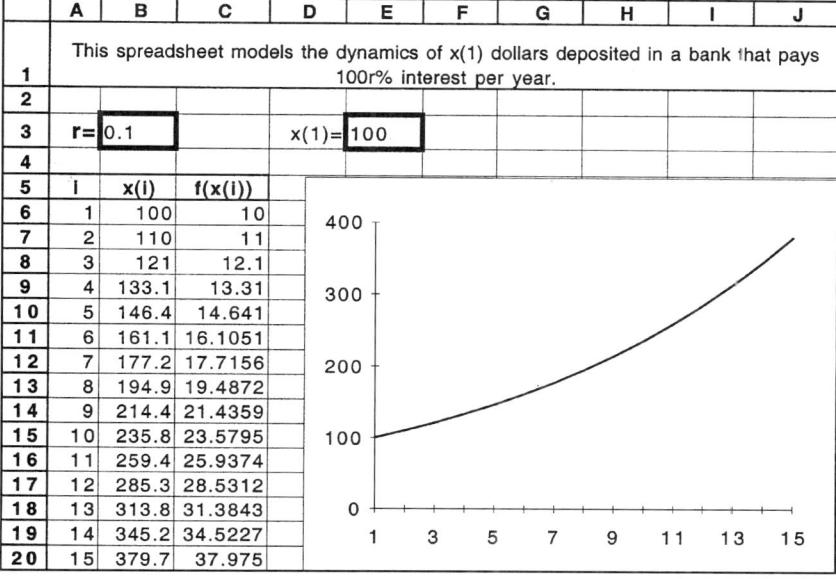

Figure 2.2.2 Calculating Interest by Iteration

As with every such difference equation, (2.2.1) can be transformed into an iterative process of the kind considered in Chapter 1. From $x_{i+1} - x_i = 0.1x_i$ we readily obtain

(2.2.3) $x_{i+1} = 1.1x_i$,

which is of the form $x_{i+1} = F(x_i)$ with $F(x) = 1.1x$. However, for reasons that will soon become clearer, it is often desirable to use the "three column format" of Figure 2.2.1 that displays $f(x_i)$ as well as x_i .

The difference equation $x_{i+1} - x_i = r x_i$, of which (2.2.1) is a special case, is exceptional in that it is relatively easy to find a "closed-form" solution based on a formula for x_n in terms of x_1, r, and n.

Problem 2.2.2 Derive a general formula for the value of x_1 dollars deposited at $100r\%$ for n years. Use this formula to find the value of an initial deposit of \$100 after it has accrued 10% annual interest for 10 years. [Hint: Generalize (2.2.3).]

Because such closed-form solutions are rather unusual, we will often rely on spreadsheets to provide *numerical solutions*, ones that are based on repetitive computations. However, the fact that (2.2.1) is amenable to both numerical- *and* closed-form solution makes it an instructive prototype for more complicated equations. For this reason we will continue to use spreadsheets to analyze (2.2.1), even though for this special equation more direct mathematical techniques are also available.

An important modification of our basic rule for calculating interest arises with more frequent compounding. Rather than crediting interest to your account once a year (annual compounding), many banks compound interest more frequently. Thus a bank that pays 10% interest "compounded quarterly" is in reality paying 2.5% interest every 3 months. Here we shall use the symbol Δt to denote the amount of time between successive compoundings and "$\Delta t \mapsto \Delta t/4$" to denote a fourfold increase in the rate of compounding. However, the index associated with the quarterly balances $y_1, y_2, y_3, \ldots$ will continue to increase by 1 at each compounding.

Because quarterly compounding generates an entirely new sequence of balances, we now use a different symbol, y_j, in place of x_i. The symbol y_j denotes the balance at the beginning of the jth quarter, whereas x_i was used to denote the balance at the beginning of the ith year.

To implement such quarterly compounding we can think of changing the unit of time between iterations from i (each Δi corresponds to 12 months) to j (each Δj corresponds to 3 months). This leads to the difference equation

(2.2.4) $\Delta y_j = y_{j+1} - y_j = 0.025y_j$

j	Time	x_j	$f(x_j)$
1	0	100	10
2	.25	102.50	10.25
3	.5	105.0625	10.50625
etc.			

Figure 2.2.3 A Format for Quarterly Compounding

for an initial deposit y_1. Note that the right-hand side of (2.2.4) is just $f(y_j)/4$, where f corresponds to the original "10% rule," i.e., $f(y) = 0.1y$.

In making such changes, it remains important to keep track of "years" as the underlying unit of time, and this will be done by inserting an additional column for "Time" into our spreadsheet. Under annual compounding "Time $= (i - 1)$," whereas under quarterly compounding "Time $= (j - 1)/4$."

We now create a spreadsheet to implement such quarterly compounding. However, to prevent a proliferation of symbols, we revert to using x rather than y for the sequence of balances obtained. Note that this sequence of x_j's is *not* the same as that obtained from annual compounding. With these conventions understood, quarterly compounding can be represented in the spreadsheet format of Figure 2.2.3.

Problem 2.2.3 Use the format of Figure 2.2.3 to create a spreadsheet that calculates the dynamics of an initial deposit x_1 accruing $100r\%$ annual interest compounded quarterly. In so doing:

   a. Let the index j count the number of times the balance has been updated.
   b. Introduce a new column that keeps track of the Time elapsed (measured in years).
   c. Note that it is $f(x_j)/4$ that is added to the balance each quarter, not $f(x_j)$.

Problem 2.2.4 Use the format of Figure 2.2.4 below to create a spreadsheet that compounds $100r\%$ interest N times a year, where N is an integer ≥ 1.

There is one more remark to be made concerning notation. In changing from annual to quarterly compounding, we also changed the index of iteration from i to j. However, Figure 2.2.3 works just as well if the original index i is used throughout.[1] Accordingly, in Figure 2.2.4 we will continue to denote the index by i, even as the process undergoes more frequent compounding.

---

[1] Because of this property, the index of iteration is sometimes referred to as a "dummy index."

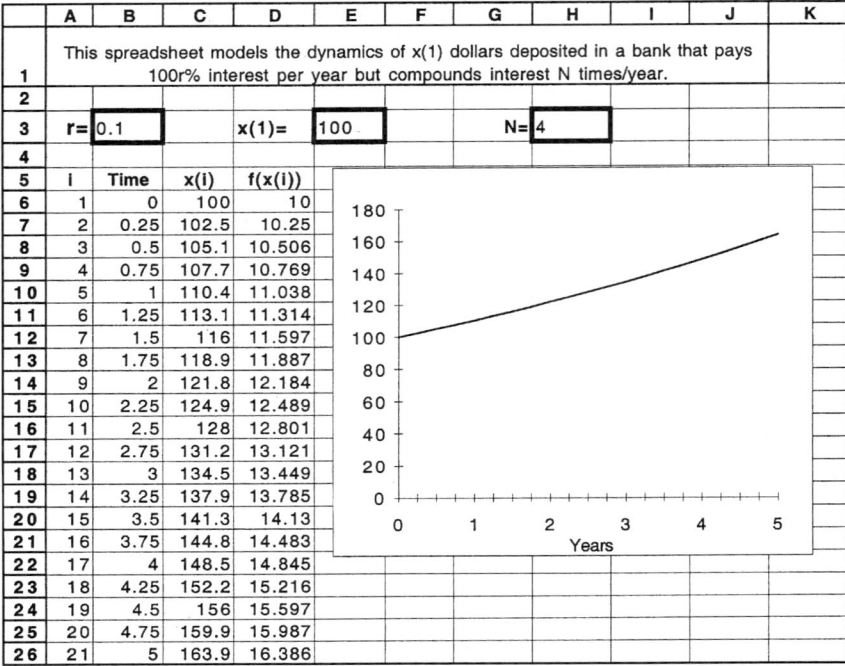

Figure 2.2.4 Compounding Interest _N_ Times a Year

Figure 2.2.4 embodies the important concept of *refi nement* as applied to the difference equation

(2.2.5) $$x_{i+1} - x_i = r\,x_i .$$

This process of "compounding N times as often" will be denoted by

(2.2.6) $$\Delta t \mapsto \frac{\Delta t}{N}$$

and applied to more general difference equations as well. But before leaving our familiar banking analogy, let us use Figure 2.2.4 to explore the phenomenon of refinement as it applies to (2.2.5).

Problem 2.2.5 Compare the value of $100 invested at 10% annual interest after 10 years when $N = 1$, $N = 4$, and $N = 10$.

Problem 2.2.6 Recalling the closed-form solution for annual compounding derived in Problem 2.2.2, find a closed-form solution for calculating the value of x_1 dollars deposited at $100r\%$ for n years when interest is compounded N times a year.

Problem 2.2.7 Given $1 invested at 100% for a year, annual compounding will lead to a balance of $2 at the end of the year. Compare this with:

(a) Compounding 4 times a year.

(b) Compounding 10 times a year.

(c) Compounding 100 times a year.

Will more frequent compounding ever yield $3 at the end of the year?

As indicated by the problems above, more frequent compounding will lead to faster annual growth in the balance, but such increases in growth rate are subject to diminishing returns. This reflects the fact that the difference equation

$$x_{i+1} - x_i = r x_i$$

is closely related to the "differential equation"

$$\frac{dx}{dt} = r x,$$

and that it is the solution of the latter that is being approximated by $x_{i+1} - x_i = r x_i / N$ as N increases. More generally, it is the process of refinement that connects the difference equation

$$x_{i+1} - x_i = f(x_i)$$

to its calculus-based cousin

$$\frac{dx}{dt} = f(x).$$

But before leaving our monetary analogy for other examples, let us use some of Excel's features to simulate some rather unusual banking practices, ones that are not readily represented by "$dx/dt = f(x)$." In our spreadsheet format, these correspond to programming a rule other than "$f(x_i) = r x_i$" into the last column of the spreadsheet (here you may want to consult the User's Guide for a description of some of Excel's "built-in" functions.)

For example, consider a bank that offers an initial interest rate of 3% compounded annually. However, for its loyal customers it increases this rate by 1% annually until the rate reaches 10%. This would correspond to the rule

$$\Delta x_i = [MIN(2 + i, 10)/100] \cdot x_i.$$

A bank that randomizes your interest between 7% and 8% would correspond to

$$\Delta x_i = [(7 + RAND())/100] \cdot x_i.$$

Here the command "RAND()" causes Excel to generate a random number between 0 and 1.

Finally, consider a bank that pays its new customers 10% interest for the first three years, and thereafter pays its loyal customers 12% interest *on their balance of three years ago*. This would correspond to

$$\Delta x_i = 0.1 x_i \quad \text{for} \quad i = 1, 2, 3.$$
$$\Delta x_i = 0.12 x_{i-3} \quad \text{for} \quad i = 4, 5, \ldots.$$

Does loyalty pay at such a bank?

Problem 2.2.8 Create spreadsheets that simulate the above banking practices with annual compounding.

These examples typify difference equations more general than (2.1.3), including equations of the form

$$(2.2.7) \qquad\qquad x_{i+1} - x_i = f(i, x_i)$$

and

$$(2.2.8) \qquad\qquad x_{i+1} - x_i = f(i, x_{i-d}),$$

where d represents a "delay." We shall encounter further examples of such equations in later applications.

2.3. Radioactive Decay and Economic Equilibrium   Throughout Section 2.2 we studied

$$(2.3.1) \qquad\qquad x_{i+1} - x_i = r x_i$$

while assuming the constant r to be positive. This assumption was motivated by the example of money in the bank, one in which the balance is always expected to grow. However, from a mathematical point of view we could also study (2.3.1) for $r < 0$.

The case $r < 0$ yields an approximation to the decay of a radioactive substance, a constant percentage of which undergoes a transmutation[1] during any given interval of time. Given the percentage of the substance that undergoes such change between i and $i + 1$, an initial amount x_1 would be subject to the rule for change (2.3.1) with $r < 0$.

Spreadsheets such as those constructed to study compound interest allow us to study this phenomenon as well. That is, by simply inserting a negative value for r in the cell B3 in Figure 2.3.1, we obtain a solution to (2.3.1) with $r < 0$.

Problem 2.3.1 For a given initial value x_1, find a closed-form solution of (2.3.1). [Hint: Write (2.3.1) in the form $x_{i+1} = F(x_i)$.]

Problem 2.3.2 Given 100 grams of a radioactive substance that decays at 10% a year, use (2.3.1) and a hand calculator to approximate *in a single step* the amount remaining after 10 years.

As with our banking problems, it is possible to subject (2.3.1) to the process of refinement. Both Figure 2.3.1 and the last two problems readily lend themselves to such modification.

[1] It is transformed into another element.

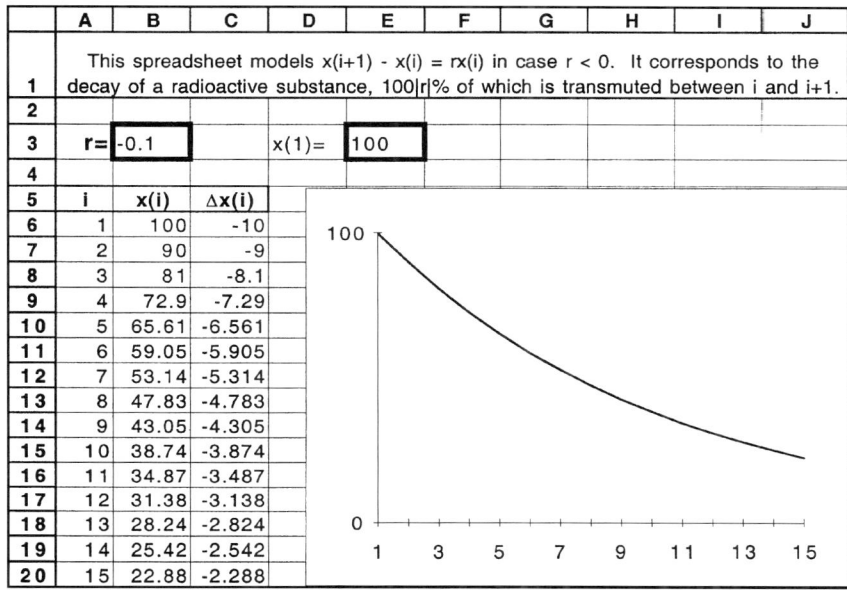

	A	B	C	D	E	F	G	H	I	J
1	This spreadsheet models x(i+1) - x(i) = rx(i) in case r < 0. It corresponds to the decay of a radioactive substance, 100|r|% of which is transmuted between i and i+1.									
2										
3	r=	-0.1			x(1)=	100				
4										
5	i	x(i)	Δx(i)							
6	1	100	-10							
7	2	90	-9							
8	3	81	-8.1							
9	4	72.9	-7.29							
10	5	65.61	-6.561							
11	6	59.05	-5.905							
12	7	53.14	-5.314							
13	8	47.83	-4.783							
14	9	43.05	-4.305							
15	10	38.74	-3.874							
16	11	34.87	-3.487							
17	12	31.38	-3.138							
18	13	28.24	-2.824							
19	14	25.42	-2.542							
20	15	22.88	-2.288							

Figure 2.3.1 Modeling Radioactive Decay

Problem 2.3.3 Repeat Problem 2.3.1 in the case where Δt is replaced by $\Delta t/N$.

Problem 2.3.4 For an element with $r = -0.1$, estimate the number of years required for *half* of the initial amount to undergo transmutation. This is called the "half-life" of this element.

Problem 2.3.5 In the case $r > 0$, refinements of the form $\Delta t \mapsto \Delta t/N$ increase the rate at which the balance grows. Determine the effect of refinement on the solution of (2.3.1) when $-1 < r < 0$.

Before leaving this interpretation of $r < 0$, there is a feature that deserves attention. In the context of radioactive decay, the amount of material transmuted during the ith interval certainly cannot exceed the amount present at the beginning of that interval. In other words, values of r satisfying $r < -1$ do not make physical sense.

There is, however, nothing to prevent us from choosing $r < -1$ in (2.3.1), in Figure 2.3.1, or in the closed-form solution of Problem 2.3.1. For $r = -1.1$, an initial value $x_1 > 0$ would "be reduced by 110%," making x_2 negative, x_3 positive, etc. The possibility of such successive "overshoots" of the equilibrium value $x = 0$ is something we shall return to in Section 2.4 and again in Chapter 3. However, it is of interest even now to extend (2.3.1) beyond the physical interpretation afforded by radioactive decay.

Problem 2.3.6 For $x_1 = 100$, find a closed-form solution of $x_{i+1} - x_i = r x_i$ in the case $r = -1.1$. Confirm your solution with a spreadsheet.

Problem 2.3.7 Repeat Problem 2.3.6 with $r = -2.1$.

An important generalization of (2.3.1) is to consider

$$(2.3.2) \qquad\qquad x_{i+1} - x_i = a x_i + b,$$

where a and b are positive constants. In a strictly formal sense, we can reduce the solution of (2.3.2) to that of (2.3.1).

Problem 2.3.8 Show that if $y_i = x_i + b/a$, then y_i satisfies (2.3.1) with $r = a$.

It will, however, be preferable to approach (2.3.2) directly, thereby setting the stage for more general problems in Section 2.4. Noting that Δx_i is zero whenever $x_i = -b/a$, we shall refer to $x_0 = -b/a$ as an *equilibrium value* for (2.3.2). More generally, x_0 will be called an equilibrium value for

$$(2.3.3) \qquad\qquad x_{i+1} - x_i = f(x_i)$$

whenever $f(x_0) = 0$.

Problem 2.3.9 Letting $F(x) = f(x) + x$, show that x_0 is a fixed point for F if and only if it is an equilibrium value for (2.3.3).

Problem 2.3.10 Given $f(x) = ax + b$, show that $x_0 = -b/a$ is an *attracting* fixed point for $F(x) = f(x) + x$ whenever $-2 < a < 0$.

The spreadsheet of Figure 2.3.2 calculates the solution of $x_{i+1} - x_i = a x_i + b$ for arbitrary values of a, b, and x_1.

By way of application of such a spreadsheet, let us consider a problem central to the subject of microeconomics. Here we deal with the market for a particular commodity (say broccoli) whose price at time i is denoted by x_i. The price determines both how much producers bring to the market between i and $i + 1$, and how much consumers buy at the market between i and $i + 1$.

The economic assumption that "an increase in price will increase production" is reflected by the (unrealistically convenient) formula

$$(2.3.4) \qquad\qquad P_i = c + d x_i,$$

where P_i denotes the amount of broccoli producers bring to market during the ith period in response to the price x_i. Here $c \geq 0$ and $d > 0$ are constants whose size reflects "the behavior of producers."

Similarly, "the behavior of consumers" can be described by the (unrealistically convenient) formula

$$(2.3.5) \qquad\qquad C_i = e - f x_i,$$

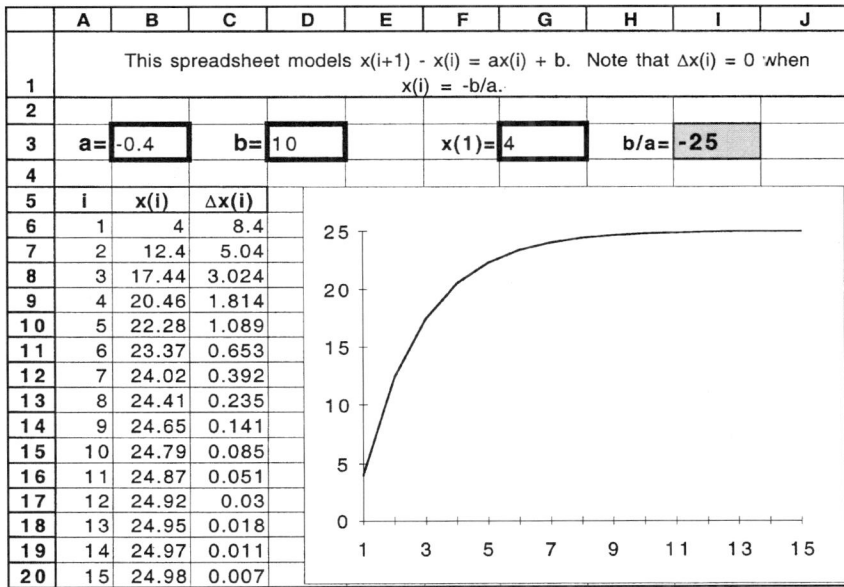

Figure 2.3.2 Solving $x_{i+1} - x_i = ax_i + b$

with $e > 0$ and $f > 0$. Here C_i denotes the amount of broccoli consumers will buy during the ith period, given the price x_i. According to (2.3.5), if broccoli is free, then e units will be consumed. If, however, the price reaches e/f, broccoli consumption will be reduced to zero. In between, there is a price at which production equals consumption (i.e., $P_i = C_i$) and where the market is said to be in equilibrium.

If such a market is to avoid both shortage and spoilage of broccoli, it will need a mechanism that moves x_i toward the *equilibrium price*

$$(2.3.6) \qquad x_0 = \frac{e - c}{d + f}.$$

This value is determined by solving $P_i = C_i$, or equivalently,

$$(2.3.7) \qquad c + d x_i = e - f x_i.$$

If the price is too high ($x_i > x_0$), there will be too much broccoli; if the price is too low ($x_i < x_0$), there will be shortage. For this reason, economists place great emphasis on modeling mechanisms that move such a market toward equilibrium.

To arrive at such a mechanism, economists need an additional assumption. For example, defining "excess demand" during the ith period by

$$(2.3.8) \qquad E_i = C_i - P_i = e - c - (d + f)x_i,$$

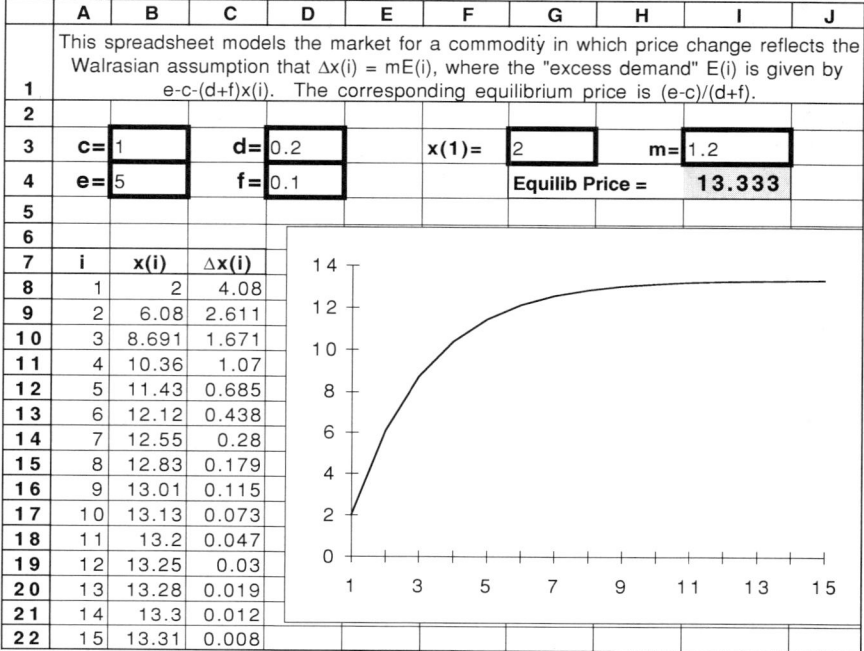

	A	B	C	D	E	F	G	H	I	J
1	This spreadsheet models the market for a commodity in which price change reflects the Walrasian assumption that Δx(i) = mE(i), where the "excess demand" E(i) is given by e-c-(d+f)x(i). The corresponding equilibrium price is (e-c)/(d+f).									
2										
3	c=	1		d=	0.2		x(1)=	2		m= 1.2
4	e=	5		f=	0.1			Equilib Price =		13.333
5										
6										
7	i	x(i)	Δx(i)							
8	1	2	4.08							
9	2	6.08	2.611							
10	3	8.691	1.671							
11	4	10.36	1.07							
12	5	11.43	0.685							
13	6	12.12	0.438							
14	7	12.55	0.28							
15	8	12.83	0.179							
16	9	13.01	0.115							
17	10	13.13	0.073							
18	11	13.2	0.047							
19	12	13.25	0.03							
20	13	13.28	0.019							
21	14	13.3	0.012							
22	15	13.31	0.008							

Figure 2.3.3 Determining Equilibrium Price

the *Walrasian* assumption asserts that prices will change in proportion to this excess. This gives rise to the "rule for change"

$$(2.3.9) \qquad x_{i+1} - x_i = mE_i,$$

where m is a positive constant reflecting the responsiveness of the marketplace to excess demand. In this way, a shortage of broccoli during the ith period will lead to an increase in price during the $(i + 1)$st period, while a surplus (i.e., a "negative excess") will lead to a decrease in price. Furthermore, the size of this price adjustment is assumed proportional to the size of E_i.

This situation can be represented by a spreadsheet such as that of Figure 2.3.3.

Problem 2.3.11 Create a spreadsheet for (2.3.8) and (2.3.9) in the format of Figure 2.3.3.

Such models raise the question of whether for a given starting price x_1 the market will in fact move toward equilibrium. In other words, will the solution of the difference equation

$$(2.3.10) \qquad x_{i+1} - x_i = m(e - c) - m(d + f)x_i$$

approach its equilibrium value $x_0 = (e - c)/(d + f)$? Since (2.3.10) is of the form

$$x_{i+1} - x_i = a x_i + b,$$

the answer can be obtained from Problem 2.3.12.

Problem 2.3.12 Under what conditions will the market described by (2.3.10) move toward equilibrium? If these conditions are not met, what kind of adjustment in the market mechanism will enable the market to stabilize in this way?

Engaging as it is to start out with a problem in physics (radioactive decay) and end up making economic pronouncements (about equilibrium prices), it seems important to point out limitations inherent in the model we have created. For a start, the linear relationships (2.3.4) and (2.3.5) are unrealistically simple. It also seems likely that the behavior of consumers and producers will be influenced by considerations other than the price x_i. For example, the amount of a product that consumers buy is likely to be influenced by their total disposable incomes and the prices of competing products. As for producers, rather than just responding to existing prices, they are likely to try to *anticipate* prices or even influence them, e.g., using advertising to create demand. Also, patterns of consumption and production are not constant, but vary with time (in our case, with changes in the index i).

Some of these limitations can be addressed by refining the model considered above. Thus in Section 2.4 we allow for nonlinear generalizations of (2.3.4) and (2.3.5), and in Section 2.5 we consider ways in which producers can anticipate price change. However, any such mathematical models of human behavior must be taken with a grain of salt (as should broccoli).

To illustrate efforts to inject reality into such models, let us return to the broccoli market. While consumers may well base their decisions on the current price of broccoli, the producers' decision on how much broccoli to plant has to be made months in advance. In the context of our original model, this observation may lead us to modify (2.3.4) to read

(2.3.4)′ $$P_i = c + d x_{i-1}$$

while retaining

(2.3.5) $$C_i = e - f x_i.$$

Now, instead of deriving price from the Walrasian assumption

$$\Delta x_i = m E_i,$$

we will derive our rule for change from the assumption that $E_i = 0$. That is, while production in the ith period is based on x_{i-1}, the actual price during the ith period will be determined by a decision to sell all the broccoli produced on the basis of (2.3.4)'. This value of x_i is determined from $C_i = P_i$, or by solving

$$e - f x_i = c + d x_{i-1}.$$

The x_i so determined will now determine P_{i+1}, etc.

Problem 2.3.13 Show that if $E_{i+1} = 0$, then (2.3.4)' and (2.3.5) imply that

(2.3.11) $$x_{i+1} - x_i = \frac{e - c}{f} - \left(1 + \frac{d}{f}\right) x_i.$$

It is interesting to note that (2.3.11), derived on the basis of assumptions quite different from those for (2.3.10), nonetheless has the same equilibrium value. However, these two difference equations lend themselves to very different interpretations.

In the case of (2.3.10), we can always assure an "approach to equilibrium" by choosing m sufficiently small, i.e., by adjusting the responsiveness of the market to price change. However, in the case of (2.3.11), we have no comparably convenient "handle" with which to adjust market mechanisms. Rather, it is the four variables c, d, e, and f, rooted in the behavior of producers and consumers, that determine the behavior of solutions of (2.3.11).

Problem 2.3.14 Determine conditions under which solutions of (2.3.11) approach the equilibrium price (2.3.6). Interpret these conditions in terms of the behavior of consumers and producers.

2.4. Logistic Equations   In Section 2.3 we considered difference equations of the form

(2.4.1) $$x_{i+1} - x_i = f(x_i)$$

where $f(x)$ is a linear function, i.e., $f(x) = ax + b$. In this section we consider examples in which $f(x)$ is a *quadratic* function and give several applications.

The so-called *logistic* difference equation

(2.4.2) $$x_{i+1} - x_i = a x_i - b x_i^2$$

is best known in the context of population dynamics, where it is also called the Verhulst equation.[1] Here (2.4.2) is regarded as a generalization of

(2.4.3) $$x_{i+1} - x_i = a x_i,$$

which we have until now used to model money growing at a constant annual

---

[1] In honor of the Belgian biologist who first related it to the growth of certain populations.

rate of $100a\%$. However, (2.4.3) also models the dynamics of a population with a constant growth rate.

As we have seen in Problem 2.2.2, (2.4.3) has the closed-form solution

(2.4.4) $$x_n = (1+a)^{n-1} x_1.$$

Here the index n appears as an exponent, and this relationship between x_n and n leads us to refer to (2.4.4) as representing *exponential growth*.

An important characteristic of exponential growth is that it leads to a "fixed doubling time," one that can be determined from (2.4.4) by solving

$$2x_1 = (1+a)^{n-1} x_1$$

for n. In his famous *Essay on the Principle of Population*, Thomas Malthus used the concept of fixed doubling time to argue that unchecked exponential growth among the human species is destined to lead "to the checks of vice and misery."

Problem 2.4.1 Solve the equation $2x_1 = (1+a)^n x_1$ to obtain the doubling time for a population with annual growth rate a.

Problem 2.4.2 Verify Malthus's assertion that "calculated on a mortality of one to thirty-six, if the births be to deaths in the proportion of three to one, the period of doubling will be only twelve years and four fifths."

Problem 2.4.3 Since the mid-1800s, the earth's human population has been doubling approximately every 40 years. Determine the corresponding annual growth rate a.

Questions about the sustainability of exponential growth have created interest in models that give rise to other forms of growth. For example, a population of yeast cells confined to a closed jar may grow exponentially while its size is small in comparison to the jar. However, at some point the population encounters constraints and, in the case at hand, tends to level off in an S-shaped, or "sigmoid," growth curve. (Figure 2.4.1.)

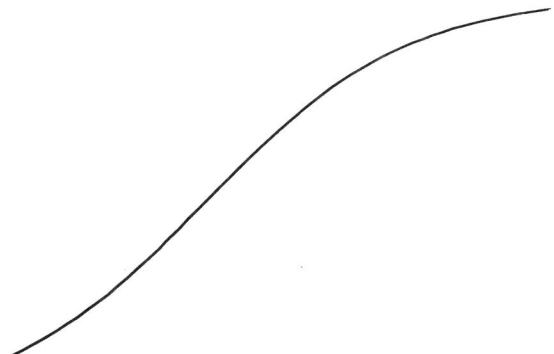

Figure 2.4.1 Sigmoid Growth of Yeast Cells in a Closed Jar

It was in the context of such observations that Verhulst sought to modify (2.4.3), building in a new term that represents the limits imposed by the jar. This led him to replace "$f(x_i) = ax_i$" by "$f(x_i) = ax_i - bx_i^2$." Here the additional term $-bx_i^2$ has the effect of "damping" the exponential growth that would otherwise be generated by the first term ax_i. Since the "growth term" ax_i is linear in x_i and the "damping term" $-bx_i^2$ is *super*linear, it follows that for large values of x_i, the damping term will dominate.

Before trying to determine the effect of Verhulst's modification, let us build a spreadsheet (Figure 2.4.2) that models (2.4.2), one in which varying values can be assigned to the constants a and b and to the initial population x_1.

Problem 2.4.4 Construct a spreadsheet for $x_{i+1} - x_i = ax_i - bx_i^2$ in the format of Figure 2.4.2. Use this spreadsheet to confirm that for $a = 0.1$ and $b < 0.05$:

(a) Solutions of (2.4.2) "level off" as in Figure 2.4.1.
(b) The value of x_0 toward which solutions tend does not depend on the size of the initial population x_1.

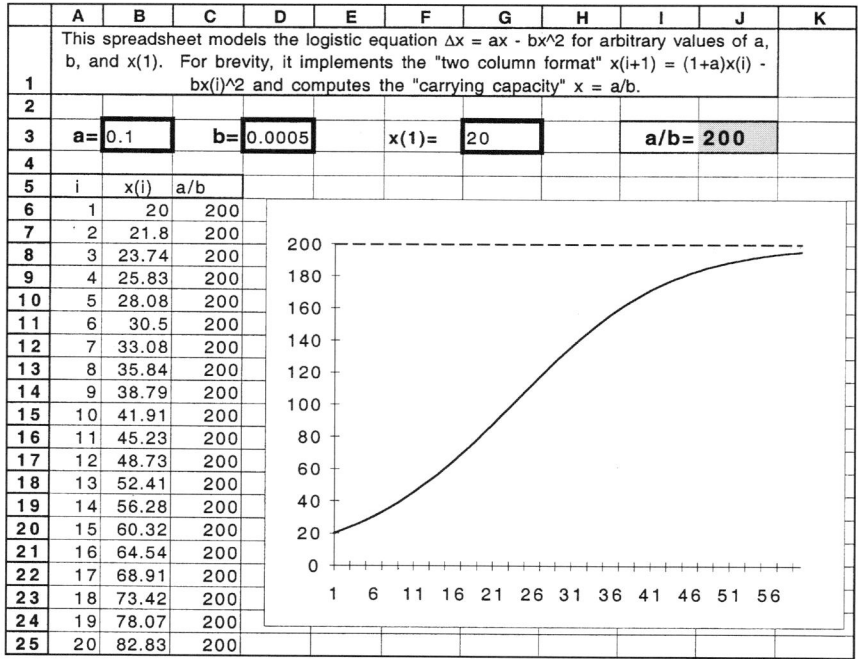

Figure 2.4.2 Modeling Logistic Growth

To determine the value at which solutions of (2.4.2) level off, we need only rewrite the difference equation as

$$(2.4.5) \qquad \Delta x_i = x_{i+1} - x_i = a x_i \left(1 - \frac{b}{a} x_i \right).$$

In this form we see that the population grows whenever $0 < x_i < a/b$ and declines whenever $x_i > a/b$. Accordingly, $x_0 = a/b$ turns out to be an equilibrium value for such a population. The analogy with yeast growing in a closed jar now suggests that this equilibrium value corresponds to the "carrying capacity" of the environment in which a particular population exists.

Problem 2.4.5 Writing the logistic equation (2.4.2) in the form $x_{i+1} = F(x_i)$, confirm that $x_0 = a/b$ is a fixed point for F. Then:

    (a) Use a spreadsheet to graph both $y = x$ and $y = F(x)$.
    (b) Confirm that for $a = 0.1$ and $b = 0.0005$, $x_0 = 200$ is an attractor for F.
    (c) Find a pair of values for a and b such that a/b is a repeller for F.

While the logistic equation is often identified with populations that exhibit sigmoid growth in approaching their natural limits, there are many other difference equations whose solutions exhibit similar behavior. For example, for $k > 1$, the modified logistic equation

$$(2.4.6) \qquad x_{i+1} - x_i = a x_i - b x_i^k$$

may also have solutions with an S-shaped graph such as the one depicted in Figure 2.4.1. However, the equilibrium value toward which solutions of (2.4.6) tend now varies with k as well as with the coefficient of the damping term b.

Problem 2.4.6 Generalize equation (2.4.5) to determine the "carrying capacity" associated with (2.4.6) when $a = 0.1, b = 0.0005$, and $k = 3$. Confirm that your answer corresponds to that given by the spreadsheet in Figure 2.4.3.

The spreadsheet in Figure 2.4.3 enables us to confirm the qualitative nature of solutions of (2.4.6) and illustrates the dependence of the equilibrium value on a, b, and k.

While such models may not provide a literal representation of carrying capacity, they can be instructive in thinking about this concept. And while the growth of yeast in a jar may correspond rather well to the sigmoid solutions of (2.4.2) and (2.4.6), situations involving other species give rise to rather different phenomena. For example, following a year of abundant rainfall a population of deer may overshoot the steady-state food supply of its habitat. Such circumstances can give rise to the phenomena of "overshoot and oscillation to equilibrium" and "overshoot and collapse," as indicated in Figure 2.4.4. Or else the growth of populations may

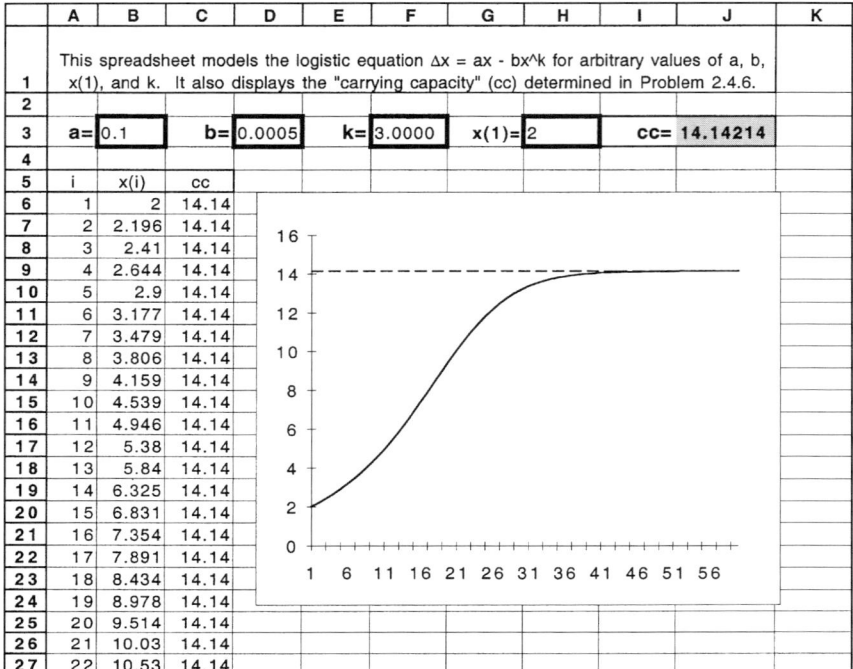

	A	B	C	D	E	F	G	H	I	J	K
1	This spreadsheet models the logistic equation Δx = ax - bx^k for arbitrary values of a, b, x(1), and k. It also displays the "carrying capacity" (cc) determined in Problem 2.4.6.										
2											
3	a=	0.1	b=	0.0005		k=	3.0000		x(1)=	2	cc= 14.14214
4											
5	i	x(i)	cc								
6	1	2	14.14								
7	2	2.196	14.14								
8	3	2.41	14.14								
9	4	2.644	14.14								
10	5	2.9	14.14								
11	6	3.177	14.14								
12	7	3.479	14.14								
13	8	3.806	14.14								
14	9	4.159	14.14								
15	10	4.539	14.14								
16	11	4.946	14.14								
17	12	5.38	14.14								
18	13	5.84	14.14								
19	14	6.325	14.14								
20	15	6.831	14.14								
21	16	7.354	14.14								
22	17	7.891	14.14								
23	18	8.434	14.14								
24	19	8.978	14.14								
25	20	9.514	14.14								
26	21	10.03	14.14								
27	22	10.53	14.14								

Figure 2.4.3 Sigmoid Growth with Superlinear Damping

correspond to something more complicated that reflects combinations of such phenomena.

By way of gaining further insight into Figure 2.4.4, let us consider a population that has access to a store of *non*renewable resources that contributes to its growth (e.g., a population of deer that has been introduced into a new environment with an initial stockpile of hay). These resources can enable a population to overshoot the carrying capacity of its environment by injecting *delays* into the impact of forces that would otherwise dampen the growth of this population.

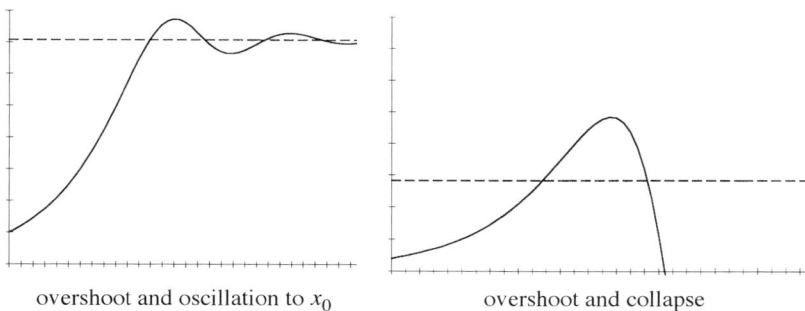

overshoot and oscillation to x_0 overshoot and collapse

Figure 2.4.4 Two Kinds of Overshoot Behaviors

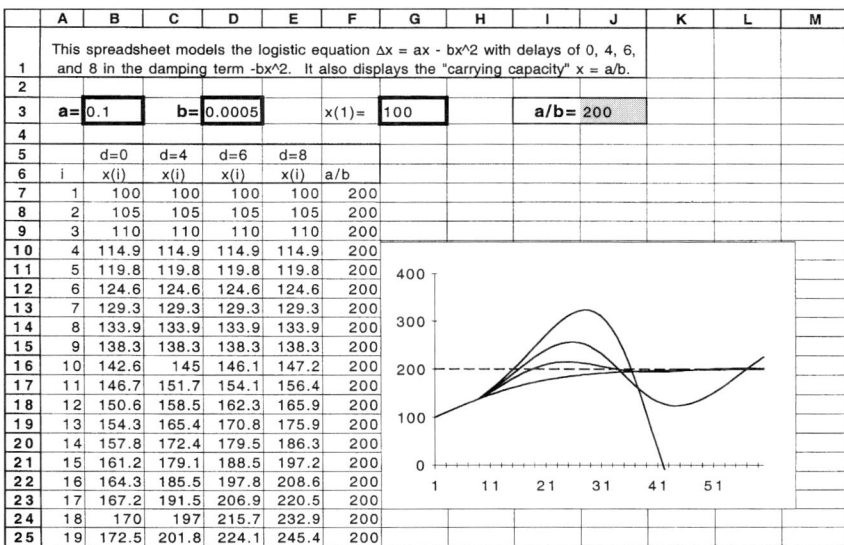

	A	B	C	D	E	F	G	H	I	J	K	L	M
1	This spreadsheet models the logistic equation Δx = ax - bx^2 with delays of 0, 4, 6, and 8 in the damping term -bx^2. It also displays the "carrying capacity" x = a/b.												
2													
3	a= 0.1		b= 0.0005			x(1)=	100			a/b= 200			
4													
5		d=0	d=4	d=6	d=8								
6	i	x(i)	x(i)	x(i)	x(i)	a/b							
7	1	100	100	100	100	200							
8	2	105	105	105	105	200							
9	3	110	110	110	110	200							
10	4	114.9	114.9	114.9	114.9	200							
11	5	119.8	119.8	119.8	119.8	200							
12	6	124.6	124.6	124.6	124.6	200							
13	7	129.3	129.3	129.3	129.3	200							
14	8	133.9	133.9	133.9	133.9	200							
15	9	138.3	138.3	138.3	138.3	200							
16	10	142.6	145	146.1	147.2	200							
17	11	146.7	151.7	154.1	156.4	200							
18	12	150.6	158.5	162.3	165.9	200							
19	13	154.3	165.4	170.8	175.9	200							
20	14	157.8	172.4	179.5	186.3	200							
21	15	161.2	179.1	188.5	197.2	200							
22	16	164.3	185.5	197.8	208.6	200							
23	17	167.2	191.5	206.9	220.5	200							
24	18	170	197	215.7	232.9	200							
25	19	172.5	201.8	224.1	245.4	200							

Figure 2.4.5 Logistic Growth with Delayed Damping

The logistic equation can be modified to accommodate such delays by setting

$$(2.4.7) \qquad
\begin{aligned}
x_{i+1} - x_i &= a x_i - b x_i^2 & \text{for} \quad i &= 1, 2, \dots, 9, \\
x_{i+1} - x_i &= a x_i - b x_{i-d}^2 & \text{for} \quad i &= 10, 11, \dots,
\end{aligned}$$

where d is a positive integer whose value reflects the size of the delay. The spreadsheet in Figure 2.4.5 models the growth of a population satisfying (2.4.7) for delays of $d = 0, 4, 6$, and 8.

The case $d = 0$ corresponds to the classical logistic model (2.4.2) in which the x_i approach the equilibrium value a/b. For $a = 0.1$ and $b = 0.0005$ this can be interpreted as representing a population that is subject to a carrying capacity of $x_0 = a/b = 200$. However, delays corresponding to $d > 0$ lead to an overshoot of this carrying capacity, one that can be followed by a variety of behaviors.

Problem 2.4.7 Construct a spreadsheet in the format of Figure 2.4.5. Setting $a = 0.1, b = 0.0005$, and $x_1 = 100$, estimate the value of a number d_0 such that the population persists for 60 years whenever $d < d_0$ but dies out within 60 years whenever $d > d_0$.

Problem 2.4.8 What is the effect of refinements $\Delta t \mapsto \Delta t / N$ on the number d_0 determined in Problem 2.4.7?

Such a provocative model raises many additional questions. By what mechanisms can nonrenewable resources inject delays into the damping process? What are the functional relationships between population size and the depletion of such nonrenewable resources? What other factors are related to the phenomenon of

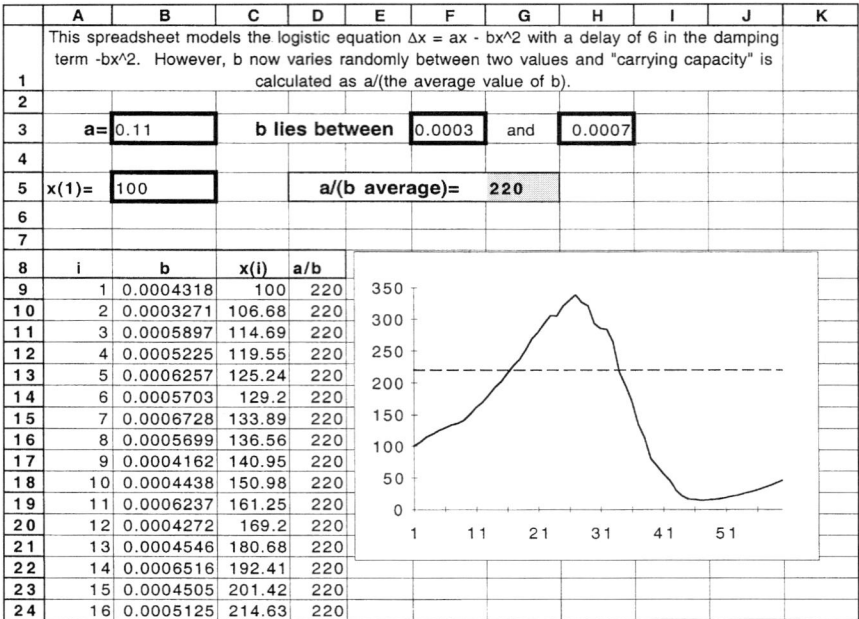

Figure 2.4.6 A Randomized Delay Logistic Equation

overshoot? As with many good models, Figure 2.4.5 motivates us to think about generalizations and improvements.

In the case at hand, we are led to questions about how a population interacts with the larger system that actually determines the size of delays and the carrying capacity of its environment. We will return to such questions in the context of *systems* of difference equations in Chapters 6 and 7.

As a further modification of these ideas, we can build a spreadsheet that fixes the delay at a specific value, say $d = 6$, and also builds an element of randomness into the associated delay logistic equation. Instead of taking the coefficient b to be a constant, the spreadsheet of Figure 2.4.6 programs b to assume *random values* between two constants. Successive runs of this model will generate different solutions, ones whose behavior depends on the random numbers being generated.

Problem 2.4.9 Construct a spreadsheet in the format of Figure 2.4.6. Setting $a = 0.11$, let b assume random values between 0.0003 and 0.0007. Do 10 or more successive runs of your model to estimate the likelihood that an initial population of 100 can sustain itself for 60 years.

Having related difference equations of the form $x_{i+1} - x_i = ax_i - bx_i^2$ to problems from population dynamics, there arises the question of whether such equations are related to other phenomena as well. As an example from economics, let us return to the Walrasian assumption considered in Section 2.3. Here the price of a commodity at time $i+1$ was determined by the excess demand E_i according to the rule

$$\Delta x_i = x_{i+1} - x_i = m\,E_i.$$

(Recall that $E_i = C_i - P_i$, whereas formulas for C_i and P_i reflect the responsiveness of consumers and producers to price.)

In Section 2.3, the linear rules (2.3.4) and (2.3.5) gave rise to a difference equation of the form $x_{i+1} - x_i = a - bx_i$. However, if these linear rules are modified to reflect *nonlinear* responses to price, more general difference equations will arise. For example, the rules

(2.4.8) $$P_i = c + d\,x_i^2,$$
(2.4.9) $$C_i = e - f\,x_i,$$

give rise to

(2.4.10) $$x_{i+1} - x_i = m\big[e - c - f\,x_i - d\,x_i^2\big].$$

Now the question "will a particular population approach its carrying capacity?" is replaced by "will the market for a particular commodity approach equilibrium?" The spreadsheet of Figure 2.4.7 recasts ideas associated with the logistic equation in an economic context.

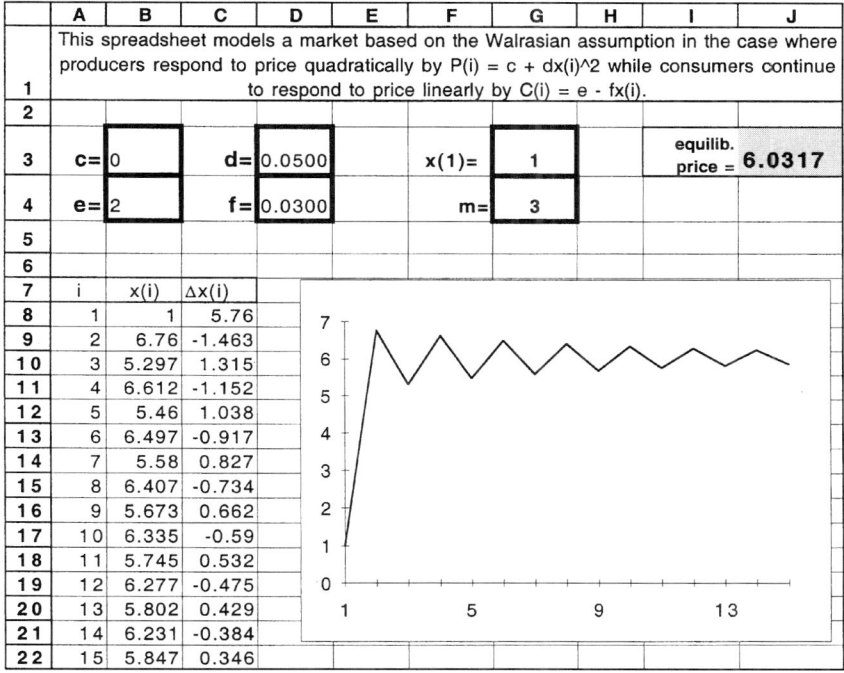

Figure 2.4.7 Producers Responding to Price Quadratically

Problem 2.4.10 Find the equilibrium values for (2.4.10). Use a spreadsheet such as the one above to give an example of a market that approaches equilibrium and of one that does not.

2.5. Higher-Order Equations   Throughout this chapter we have been studying difference equations of the form

$$\Delta x_i = f(x_i),$$

where Δx_i is used to represent the *first difference*

(2.5.1) $\Delta x_i = x_{i+1} - x_i.$

With Δ denoting "change," $\Delta x_i = f(x_i)$ prescribes the change that x_i undergoes during the ith iteration.

Having learned to interpret an application of Δ to x_i in this way, what happens if we apply Δ to Δx_i? Here $\Delta \Delta x_i = \Delta x_{i+1} - \Delta x_i$, and

(2.5.2) $\Delta \Delta x_i = (x_{i+2} - x_{i+1}) - (x_{i+1} - x_i) = x_{i+2} - 2x_{i+1} + x_i.$

In this way we can pose *second-order* difference equations of the form

(2.5.3) $\Delta \Delta x_i = f(x_i).$

Given initial values for x_1 and x_2, (2.5.3) enables us to determine x_3 by

$$x_3 = -x_1 + 2x_2 + f(x_1).$$

Having solved for x_3, we can then solve for x_4 as

$$x_4 = -x_2 + 2x_3 + f(x_2),$$

and so on. In other words, given initial values for x_1 and x_2, we are again able to solve (2.5.3) by iteration. In harnessing spreadsheets for this iterative task, it will be convenient to use the format shown in Figure 2.5.1.

i	x_i	Δx_i	$\Delta \Delta x_i = f(x_i)$
1	x_1 is given	$\Delta x_1 = x_2 - x_1$ is given	$f(x_1)$
2	$x_2 = x_1 + \Delta x_1$	$\Delta x_2 = \Delta x_1 + f(x_1)$	$f(x_2)$
3	$x_3 = x_2 + \Delta x_2$	$\Delta x_3 = \Delta x_2 + f(x_2)$	$f(x_3)$

etc.

Figure 2.5.1 Spreadsheet Format for $\Delta \Delta x_i = f(x_i)$

It turns out that even the simplest second-order difference equation

(2.5.4) $\Delta \Delta x_i = \text{constant}$

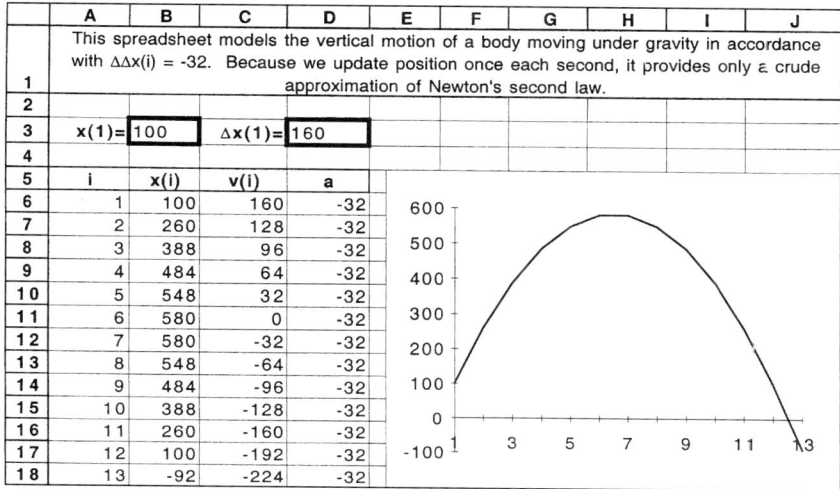

	A	B	C	D	E	F	G	H	I	J
1	This spreadsheet models the vertical motion of a body moving under gravity in accordance with $\Delta\Delta x(i) = -32$. Because we update position once each second, it provides only a crude approximation of Newton's second law.									
2										
3	x(1)=	100		Δx(1)=	160					
4										
5	i	x(i)	v(i)	a						
6	1	100	160	-32						
7	2	260	128	-32						
8	3	388	96	-32						
9	4	484	64	-32						
10	5	548	32	-32						
11	6	580	0	-32						
12	7	580	-32	-32						
13	8	548	-64	-32						
14	9	484	-96	-32						
15	10	388	-128	-32						
16	11	260	-160	-32						
17	12	100	-192	-32						
18	13	-92	-224	-32						

Figure 2.5.2 Height of a Body Moving Under Gravity

is of considerable interest. According to Newton's laws of motion, a body subject to a constant force (such as gravity near the surface of the earth) satisfies an equation[1] closely related to (2.5.4). Assuming that x is measured in feet and that the time between iterations corresponds to seconds, the difference equation

$$(2.5.5) \qquad\qquad \Delta\Delta x_i = -32$$

provides a good approximation to the motion of a falling body near the earth's surface. Here $\Delta\Delta x_i$ corresponds to the body's constant acceleration $a = -32$. Knowing the body's initial height x_1 measured in feet and approximating its initial velocity by $\Delta x_1 = x_2 - x_1$, one can approximate the body's subsequent height by a spreadsheet such as that of Figure 2.5.2.

Problem 2.5.1 Construct a spreadsheet in the format of Figure 2.5.2. Use this spreadsheet to model 12 seconds of the vertical motion of a ball thrown upward from a rooftop 100 feet high with an initial velocity of 160 feet/second.

Relying on (2.5.5) to represent the vertical motion of a body can be criticized on three grounds:

1. It neglects air resistance, which can profoundly affect the motion of a body near the earth's surface.
2. The force of gravity is only approximately constant near the surface of the earth.
3. Newton's law is based on a *differential* equation, rather than a difference equation.

[1] Newton's equation is actually a "differential equation," one that is closely related to (2.5.5). We will be able to approximate solutions of this differential equation by refinement of our difference equation.

The first two of these are a criticism of the assumptions underlying the model. A falling body is in fact subject to forces other than gravity. The force of gravity depends on the distance to the earth's center, and moreover, especially at high speeds, air resistance cannot be ignored. Responding to this latter fact will require modification of the right side of (2.5.5), as will be addressed later.

The third criticism deals with the use of (2.5.5) to approximate the differential equation

$$(2.5.6) \qquad\qquad \frac{d^2 x}{dt^2} = f(x)$$

that actually embodies Newton's second law. While iterative algebra does not enable us to solve (2.5.6) directly, it does provide a basis for *approximating* its solutions by means of the difference equation (2.5.5). However, achieving a *good* approximation may call for the technique of refinement encountered in modeling compound interest.

In solving the difference equation (2.5.5) with $\Delta t = 1$, we are updating the body's position once each second. Moving toward the solution of (2.5.6) requires more frequent updating. In the context of refinement, this calls for replacing $\Delta t = 1$ second by $\Delta t = 1/N$ seconds (as denoted by $\Delta t \mapsto \Delta t/N$).

In the context of Figure 2.5.2, this calls for introducing a column "Time" whose entries are $(i - 1)/N$. Then, continuing to set

$$f(x_i) = a(i) = -32,$$

we update our entries under $v(i)$ and $x(i)$ by the rules

$$v(i + 1) = v(i) + a(i)/N,$$
$$x(i + 1) = x(i) + v(i)/N.$$

By embodying these modifications, Figure 2.5.3 provides a refinement of our previous model for a falling body. With increasing values of N, solutions based on this refinement provide increasingly close approximations to the calculus-based solution of (2.5.6).

There is, however, a practical problem in implementing such refinements on a spreadsheet. While Figure 2.5.2 requires 10 iterations to simulate 10 seconds of motion, Figure 2.5.3 requires $10N$ iterations. In other words, implementing such refinements will also call for increasing the length of the corresponding spreadsheets.

Problem 2.5.2 Repeat Problem 2.5.1 with Δt replaced by $\Delta t/4$.

What degree of refinement is required to achieve a good approximation of the underlying differential equation? This can be a complicated question, one that varies from model to model and eventually calls for techniques other than those

This spreadsheet models the vertical motion of a body moving under gravity in accordance with $\Delta\Delta x(i) = -32$. By increasing N (and the length of the spreadsheet!) we obtain improved approximations to Newton's second law.

x(1)= 100   v(1)= 160   N= 4

i	Time	x(i)	v(i)	a(i)
1	0	100	160	-32
2	0.25	140	152	-32
3	0.5	178	144	-32
4	0.75	214	136	-32
5	1	248	128	-32
6	1.25	280	120	-32
7	1.5	310	112	-32
8	1.75	338	104	-32
9	2	364	96	-32
10	2.25	388	88	-32
11	2.5	410	80	-32
12	2.75	430	72	-32
13	3	448	64	-32
14	3.25	464	56	-32
15	3.5	478	48	-32
16	3.75	490	40	-32
17	4	500	32	-32
18	4.25	508	24	-32
19	4.5	514	16	-32
20	4.75	518	8	-32

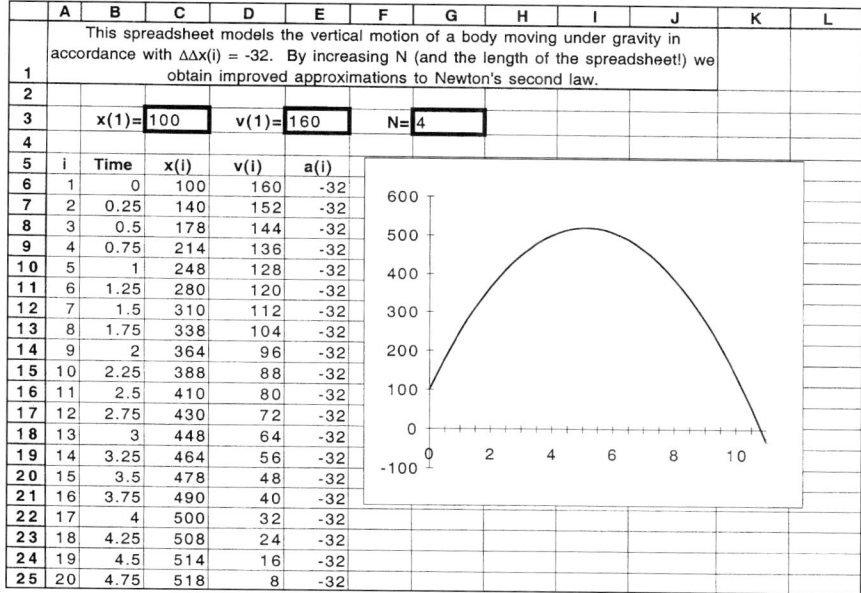

Figure 2.5.3 Refinement of a Falling Body Model

discussed here. However, a good rule of thumb is to ask whether a refinement $\Delta t \mapsto \Delta t/2$ produces significant change during the time period under consideration. If not, the solution of the difference equation is likely to be a good approximation to the underlying differential equation as well. However, like all "rules of thumb," this criterion should be applied with caution.

Having developed an ability to refine second-order difference equations, let us return to the possibility that air resistance might affect motion of a falling body. Such resistance depends both on the speed at which a body is moving and the viscosity of the medium through which it moves. Let us assume that the resisting force is proportional to the speed of the body and that the viscosity of the air is reflected by a constant of proportionality c. Such assumptions would lead us to replace (2.5.5) by

$$(2.5.7) \qquad \Delta\Delta x_i = -32 - cv(i).$$

Equation (2.5.7) is modeled in Figure 2.5.4. This spreadsheet format also allows us to test the effect of refinements of the form $\Delta t \mapsto \Delta t/N$ in order to determine an appropriate length for the spreadsheet, as in Figure 2.5.3.

Problem 2.5.3 In the format of Figure 2.5.4, model the vertical motion of a body encountering air resistance that is equal to one-tenth of the *square* of its velocity. [Hint: You may want to make use of Excel's built-in function ABS(·) that returns the absolute value of the quantity inside the parentheses.]

	A	B	C	D	E	F	G	H	I	J
1	This spreadsheet models the vertical motion of a body moving under gravity and air resistance that is proportional to the velocity. This leads to $\Delta\Delta x(i) = -32 -cv(i)$.									
2										
3		x(1)=	100		Δx(1)=	160	N=	4	c=	0.2
4										
5	i	Time	x(i)	v(i)	a(i)					

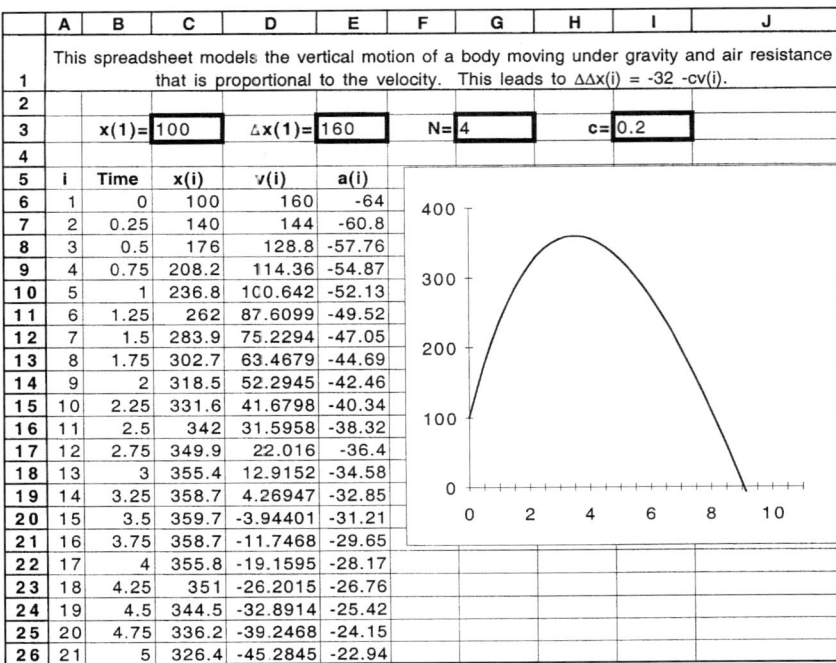

i	Time	x(i)	v(i)	a(i)
1	0	100	160	-64
2	0.25	140	144	-60.8
3	0.5	176	128.8	-57.76
4	0.75	208.2	114.36	-54.87
5	1	236.8	100.642	-52.13
6	1.25	262	87.6099	-49.52
7	1.5	283.9	75.2294	-47.05
8	1.75	302.7	63.4679	-44.69
9	2	318.5	52.2945	-42.46
10	2.25	331.6	41.6798	-40.34
11	2.5	342	31.5958	-38.32
12	2.75	349.9	22.016	-36.4
13	3	355.4	12.9152	-34.58
14	3.25	358.7	4.26947	-32.85
15	3.5	359.7	-3.94401	-31.21
16	3.75	358.7	-11.7468	-29.65
17	4	355.8	-19.1595	-28.17
18	4.25	351	-26.2015	-26.76
19	4.5	344.5	-32.8914	-25.42
20	4.75	336.2	-39.2468	-24.15
21	5	326.4	-45.2845	-22.94

Figure 2.5.4 Vertical Motion of a Body Subject to Air Resistance

By way of another example that gives rise to a second-order difference equation, we return to the example from microeconomics considered in Section 2.3. Here the behavior of producers of broccoli was determined by price, and the rule $P_i = c + d x_i$ reflected a linear relationship between price during the ith period and the amount of broccoli produced. However, noting the lag time between the decision to produce and the actual production, this rule was later modified to read

$$(2.5.8) \qquad\qquad P_i = c + d x_{i-1}.$$

Here (2.5.8) can be interpreted as reflecting the behavior of a producer who bases P_i on the price that is *anticipated* to exist during the ith period. However, in the absence of a better projection, this producer simply assumes that x_i will be the same as x_{i-1}.

This raises the question of whether, on the basis of additional information available during the $(i-1)$st period, a producer might make a more enlightened guess regarding the value of x_i. Denoting the anticipated price by y_i, one possibility is to replace the rather crude guess $y_i = x_{i-1}$ by

$$(2.5.9) \qquad y_i = x_{i-1} + m(x_{i-1} - x_{i-2}) = (1 + m)x_{i-1} - mx_{i-2},$$

where m is a constant yet to be chosen. This corresponds to a producer who bases y_i partly on x_{i-1} but also projects the prior *rate* of price change to arrive at y_i.

Based on this projected price, production during the ith period would be given by

(2.5.10) $P_i = c + d y_i = c + d(1 + m)x_{i-1} - d m x_{i-2}.$

Assuming that we continue to describe the behavior of consumers by

(2.5.11) $C_i = e - f x_i,$

a commitment to sell all available broccoli during the ith year will again lead to

$$E_i = C_i - P_i = 0.$$

Equating the right sides of (2.5.10) and (2.5.11), followed by some algebraic simplifying, now leads to

(2.5.12) $f \Delta \Delta x_i = e - c - (d + 2f + dm) \Delta x_i - (d + f)x_i.$

Problem 2.5.4 Derive (2.5.12) from the assumptions stated above.

While it is possible to use the format of Figure 2.5.1 to model (2.5.12) directly, it is easier to use a format that reflects the relationships leading up to this second-order difference equation. This calls for:

1. Specifying x_1 and x_2.
2. Using $y_3 = x_2 + m[x_2 - x_1]$ as the projected price that determines P_3 according to the relationship $P_3 = c + d y_3$.
3. Using $E_3 = 0$ to determine x_3 from the relationship $P_3 = e - f x_3$.
4. Determining y_4 from x_2 and x_3, etc.

It is these relationships that underlie the format for the spreadsheet model for (2.5.12) appearing as Figure 2.5.5.

Problem 2.5.5 Create a spreadsheet for (2.5.12) in the format of Figure 2.5.5. For $c = 200, d = 5, e = 0, f = 3, x_1 = 100,$ and $x_2 = 80$, find a value of m for which x_{10} is within a dollar of the equilibrium price.

A central question raised by such examples is, "To what extent do economic models for change provide meaningful insights into actual economic behavior?" Surely, the real world is more complicated than the difference equations of this chapter would suggest. Especially in situations involving human behavior, there are issues that mathematical models are unlikely to reflect.

One way of addressing such questions is to regard models as the bases for *scenarios*. In this context, economic models assert that, "If producers and consumers behave in a particular way, then the economy will respond accordingly." Such an interpretation allows us to construct models based on a variety of assumptions, and given their solutions, we can visualize a range of outcomes. While the models themselves may not be conclusive, their scenarios can provide us with more

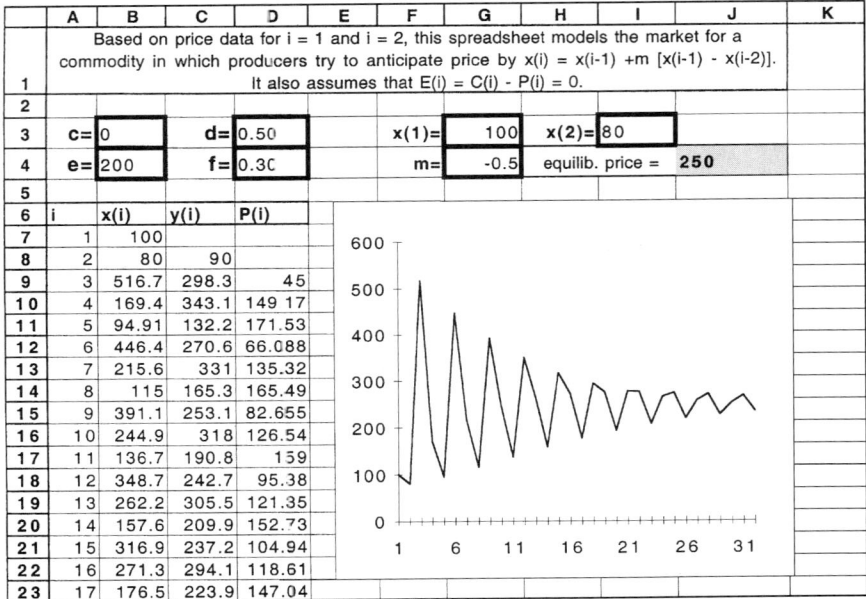

Figure 2.5.5 Producers Seeking to Anticipate Price Change

incisive ways of thinking about the real world than relying on intuition or wishful thinking.

The tools we have brought to bear in Chapter 2 also enabled us to build into our scenarios a wide range of phenomena. While it is tempting to start with simple rules for change (e.g., ones based on linear functional relationships), meaningful modeling requires an ability to deal with more general functions and with phenomena such as randomness and delays. Given a technology such as Excel to carry out the required computations, it was relatively easy to elaborate on the simple models with which our thinking began.

Continuing in this direction will require an ability to deal with several phenomena at the same time. Accordingly, in Chapters 6 and 7 we will return to "rules for change" in the context of *systems*.

2.6. Problems, Exercises, and Projects   In this chapter we have noted the close connection between iterative schemes of the form $x_{i+1} = F(x_i)$ and difference equations of the form $x_{i+1} - x_i = f(x_i)$. This connection has enabled us to develop rules for change in physics, demography, and economics, using iteration to study phenomena that are usually regarded as "calculus-based."

The problems that follow elaborate on such rules for change and on the mathematical ideas that underlie them. For those familiar with calculus, there are also some starred problems that illustrate the transition from our discrete methods to calculus-based techniques.

Solving $\Delta x_i = r x_i$ in Closed Form

Problem 2.6.1 By rewriting $x_{i+1} - x_i = r x_i$ in the form $x_{i+1} = (1+r)x_i$:

(a) Show that $x_{n+1} = (1+r)^n x_1, n = 0, 1, 2, \ldots$.

(b) At 5% annual interest, calculate the number of years required for an initial deposit x_1 to (at least) double in value.

(c) Determine the value of r at which an initial deposit x_1 doubles in exactly 40 years.

Limits to Compounding

Problem 2.6.2 Recalling Problem 2.2.2:

(a) Show that the difference equation $x_{i+1} - x_i = r x_i / N$ has the closed-form solution $x_{n+1} = (1 + r/N)^n x_1, n = 0, 1, 2, \ldots$.

(b) Explain why this x_{n+1} represents the value after n/N years of an initial deposit x_1 earning $100r\%$ interest compounded N times a year.

(c) At year's end, an account earning 5% compounded quarterly will have earned more than 5%. Find the value of this "annual percentage yield," or APY.

Problem 2.6.3 Answer the last question in Problem 2.2.7 by showing that for all $N = 1, 2, \ldots$,

$$\left(1 + \frac{1}{N}\right)^N \leq 1 + 1 + \frac{1}{2!} + \frac{1}{3!} + \cdots + \frac{1}{N!} \leq 1 + 1 + \frac{1}{2} + \frac{1}{2^2} + \cdots + \frac{1}{2^N} < 3.$$

[Hint: The first inequality can be established via the binomial theorem.]

Problem 2.6.4 Consider a bank that pays $100r\%$ interest compounded N times/year, but only credits the interest to your account at the end of each year. This is equivalent to annual compounding at the APY corresponding to r and N and means that the "doubling time" is a whole number.

(a) For $r = 0.05$, what is the smallest positive integer N that leads to a "doubling time" less than that obtained in Problem 2.6.1(b)?

(b) For $r = 0.06$, what is the smallest positive integer N that gives a doubling time less than that obtained when $N = 1$?

(c) Show that for any percentage rate r, $0 < r < 2$, the doubling time is at least equal to $\ln 2/\ln(\frac{2+r}{2-r})$. [Hint: Show that $(1 + \frac{r}{N})^N \le 1 + r +$

$$\frac{r^2}{2!} + \cdots + \frac{r^N}{N!} \le 1 + r + \frac{r^2}{2} + \frac{r^3}{2^2} + \cdots + \frac{r^N}{2^{N-1}} = 1 + r(1 + \frac{r}{2} +$$

$$(\frac{r}{2})^2 + \cdots + (\frac{r}{2})^{N-1}) < \frac{2+r}{2-r}].$$

Solving $\Delta x_i = a x_i + b$ in Closed Form

Problem 2.6.5 In connection with Problem 2.3.8:

(a) Show that the substitution $y_i = x_i + b/a$ converts $x_{i+1} - x_i = a x_i + b$ to $y_{i+1} - y_i = a y_i$.

(b) Use part (a) to show that

$$x_n = \begin{cases} (1 + a)^{n-1} x_1 + ((1 + a)^{n-1} - 1)(b/a) & \text{if } a \ne 0, \\ x_1 + b((n - 1) & \text{if } a = 0. \end{cases}$$

(c) Use part (b) to show that x_n approaches $-b/a$ whenever $-2 < a < 0$.

The Exponential Differential Equation

*Problem 2.6.6 As mentioned in Section 2.2, the difference equation $x_{i+1} - x_i = r x_i$ is an iteration-based analogue of the differential equation $dx/dt = r x$. The latter can be written as $x'(t) = r x(t)$, where $x'(t) = dx/dt$.

(a) Show that if $x'(t) = r x(t)$, then $g(t) = e^{-rt} x(t)$ satisfies $g'(t) = 0$. [Hint: Use the "product rule" for differentiation to calculate $g'(t)$.]

(b) Deduce from part (a) that $g(t)$ is constant and that $x(t) = x(0)e^{rt}$.

(c) If $r < 0$, use part (b) to show that the half-life of a substance governed by $x'(t) = r x(t)$ is $(\ln 2)/(-r)$.

(d) For $r = -0.05$, compare the answer of part (c) with the half-life value $(\ln 2)/\ln(1 + r)$ obtained from the difference equation $x_{i+1} - x_i = r x_i$. Repeat for $r = -0.1$ and $r = -0.5$.

A Different Kind of Broccoli Consumer

Problem 2.6.7 Consider a broccoli market in which producers continue to respond to price according to $P_i = c + d x_i$ but consumers follow the rule $C_i = e + f x_i^{-q}$, where $c \ge 0$ and d, e, f, and q are positive constants.

(a) Write down an equation that determines the market's equilibrium price.

(b) For $P_i = 2 x_i$ and $C_i = 1 + x_i^{-2}$, determine the equilibrium price.

(c) Construct a spreadsheet analogous to that in Figure 2.3.3 and confirm that for appropriate values of c, d, e, f, q, and m, the market does approach the equilibrium price determined in part (b).

Problem 2.6.8 Consider a broccoli market in which $P_i = c + dx_i^r$ and $C_i = e + fx_i^{-q}$, where $c \geq 0$ and d, e, f, r, and q are positive constants.

(a) Explain why $c < e$ assures that such a market has an equilibrium price.

(b) For $P_i = 1 + x_i^2$ and $C_i = 3 + 2x_i^{-1/2}$, find the value of the equilibrium price.

(c) Construct a spreadsheet analogous to that in Figure 2.3.3 and confirm that for appropriate values of c, d, e, f, r, q, and m, the market does approach the equilibrium price determined in part (b).

Refining $x_{i+1} - x_i = f(i, x_i)$

In Section 2.3 we approached the concept of refinement in terms of the difference equation $x_{i+1} - x_i = r x_i$ used to calculate the annual compounding of interest. As suggested by Problem 2.2.8, this concept extends to difference equations of the form

$$x_{i+1} - x_i = f(i, x_i).$$

Here it is useful to think of a bank that expresses its rule for calculating interest in the functional form "$f(t, x(t))$," where t corresponds to *time measured in years* and $x(t)$ is the balance at time t.

Problem 2.6.9 Consider a bank that pays interest annually according to the rule $x_{i+1} - x_i = f(i, x_i)$, where i corresponds to years and $f(i, x_i)$ is the amount of interest accrued by an account with balance x_i during the ith year.

(a) Suppose such a bank decides to switch to quarterly compounding. Letting $y_1 = x_1$, it denotes by $y_2, y_3, y_4, \ldots$ the value of such an initial deposit after 3 months, 6 months, 9 months, etc. Confirm that y_j is the balance after $(j - 1)/4$ years.

(b) Recalling that this bank expresses its rule for calculating interest in the form $f(t, x(t))$, justify the "quarterly compounding rule"

$$y_{j+1} - y_j = \frac{f\left(\frac{j}{4}, y_j\right)}{4}.$$

(c) Generalize part (b) to compounding N times per year and the rule

$$y_{j+1} - y_j = \frac{f\left(\frac{j}{N}, y_j\right)}{N}.$$

Note. Problem 2.6.9 provides a rationale for defining the *N-fold refinement* of the difference equation $x_{i+1} - x_i = f(i, x_i)$ to be

$$x_{i+1} - x_i = \frac{f\left(\dfrac{i}{N}, x_i\right)}{N}.$$

Refinement and the Definite Integral

Problem 2.6.10 A special case of $x_{i+1} - x_i = f(i, x_i)$ is the difference equation

$$x_{i+1} - x_i = f(i), \qquad i = 1, 2, \ldots.$$

(a) Show that the above difference equation has the closed-form solution

$$x_{n+1} = x_1 + [f(1) + f(2) + \cdots + f(n)].$$

(b) Show that in this case, our definition of N-fold refinement leads to

$$x_{n+1} = x_1 + \frac{1}{N}\left[f\left(\frac{1}{N}\right) + f\left(\frac{2}{N}\right) + \cdots + f\left(\frac{n}{N}\right)\right].$$

(c) Confirm that in setting $n = rN$, the N-fold refinement of $x_{i+1} - x_i = f(i)$ becomes

$$x_{n+1} = x_1 + \frac{r}{n}\left[f\left(\frac{r}{n}\right) + f\left(\frac{2r}{n}\right) + \cdots + f(r)\right].$$

(d) Assuming $f(t) > 0$ for $0 \le t \le r$, reconcile the cumulative area of the rectangles in the figure below with the expression

$$\frac{r}{n}\left[f\left(\frac{r}{n}\right) + f\left(\frac{2r}{n}\right) + \cdots + f(r)\right]$$

appearing in the answer to part (c).

*(e) In integral calculus, the expression

$$\frac{r}{n}\left[f\left(\frac{r}{n}\right) + f\left(\frac{2r}{n}\right) + \cdots + f(r)\right]$$

is used to approximate the integral $\int_0^r f(t)\,dt$. Use this fact to reconcile the closed-form solution to part (a) with the calculus-based solution of

$$\frac{dx}{dt} = f(t).$$

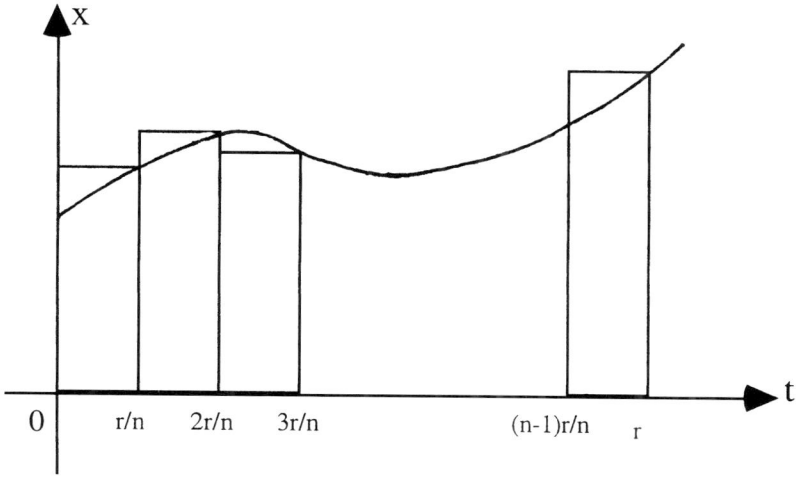

Some Modified Logistic Equations

Problem 2.6.11 In Section 2.4 we saw that the logistic equation

$$x_{i+1} - x_i = ax_i - bx_i^2$$

can be interpreted as modeling a population subject to a carrying capacity a/b.

(a) Find the carrying capacity that corresponds to the modified logistic equation $x_{i+1} - x_i = ax_i - bx_i^k$, where a, b, and k are positive constants.

(b) Find the carrying capacity that corresponds to the modified logistic equation $x_{i+1} - x_i = ax_i - bx_i^3 + c$, where a, b, and c are positive constants.

Malthus and the Pernicious Amoebae

Problem 2.6.12 While we tend to describe population growth in terms of birth rates and death rates, Malthus focused on "the ratio of births to deaths" as well as the fraction of the population that dies each year.

(a) Letting q denote the ratio of births to deaths and letting d denote the fraction of the population that dies each year, derive the rule for change

$$x_{i+1} - x_i = d(q - 1)x_i.$$

(b) Find a formula for the doubling time of a population in terms of d and q. Use this formula to confirm the result of Problem 2.4.2: If $d = 1/36$ and $q = 3$, the doubling time is slightly more than 12.8 years.

(c) The assumption $d = 1/36$ corresponds to a population in which life expectancy is about 36 years. Find the doubling time for a population whose life expectancy corresponds to the biblical "three score and ten" and in which there are two births to every death.

(d) Given Malthus's assumption of $q = 3$, for what value of d would the population have a doubling time of 1 year? (Calculate the answer using part (a), and also give a direct argument that doesn't use part (a)).

Problem 2.6.13 A population of superamoebae is such that each fissioning results in 4 (rather than 2) new amoebae. If the doubling time of this population is 1 year, what fraction of the population fissions each year? (Keep in mind that the old amoebae do not die, they just fission away.)

Problem 2.6.14 Continuing our single-celled fission expedition, assume that each fissioning of a superamoeba gives rise to q new amoebae. If the population's doubling time is 1 year, find the fraction of the population that fissions each year.

3

REFINEMENT, SLOPPINESS, AND CHAOS

3.1. Introduction In our study of difference equations, great importance was attached to the process of refinement. Here $\Delta t \mapsto \Delta t/N$ corresponded to an increase in the rate of iteration, one that was realized by setting $j - 1 = N(i - 1)$. As a result,

$$(3.1.1) \qquad x_{i+1} - x_i = f(x_i), \qquad i = 1, 2, \ldots, n,$$

was transformed into

$$(3.1.2) \qquad x_{j+1} - x_j = \frac{1}{N} f(x_j), \qquad j = 1, 2, \ldots, 2n, \ldots, Nn,$$

calling for an N-fold increase in the length of the corresponding spreadsheet.

For the special equation

$$(3.1.3) \qquad x_{i+1} - x_i = r x_i$$

with $r > 0$, refinement was given a monetary interpretation in terms of compound interest. This example was used to illustrate an important link between iterative algebra and calculus: As N increases and (3.1.3) undergoes the corresponding degree of refinement, solutions of the refined difference equation approach the solution of the corresponding "differential equation"

$$(3.1.4) \qquad\qquad \frac{dx}{dt} = rx.$$

But what about going the other way? Just as $\Delta t \mapsto \Delta t/N$ can be used to represent an N-fold increase in the rate of iteration so can $\Delta t \mapsto M(\Delta t)$ be used for $M > 1$ to represent an M-fold decrease. For example, if

$$x_{i+1} - x_i = 0.1x_i$$

represents a bank paying 10% interest with annual compounding, then

$$x_{j+1} - x_j = 0.1Mx_j$$

would correspond to a bank paying 10% annual interest, but only updating its books every M years. If $M = 5$, then the balance generated by an initial deposit of \$100 would remain at \$100 for five years! Only at the end of five years would the balance jump to \$150. While we know of no banks that engage in such questionable practices, this form of "sloppy bookkeeping" does provide a useful interpretation of "$\Delta t \mapsto M(\Delta t)$."

In this chapter we explore the implications of $\Delta t \mapsto M(\Delta t)$, first for (3.1.3), then for the logistic equation $x_{i+1} - x_i = ax_i - bx_i^2$, and then for more general equations of the form (3.1.1). For sufficiently large values of M, difference equations whose solutions had rather ordinary behavior in Chapter 2 will be shown to take on a very different character. In some cases their solutions will become quite unpredictable, and the term "chaos" is often used to describe their gyrations.

While there is a tendency to regard this form of chaos as a mysterious or even sinister phenomenon, our banking analogy will provide a more mundane interpretation:

> "Sloppy bookkeeping," as represented by infrequent updating of data, can transform an orderly iterative process into a radically different one.

This interpretation will apply not only to bank balances governed by (3.1.3), but to the solutions of more general difference equations as well.

3.2. Exponential Growth and Decay   In Chapter 2 we came to associate the difference equation

$$(3.2.1) \qquad\qquad x_{i+1} - x_i = rx_i$$

with various forms of exponential behavior. For $r > 0$, solutions of (3.2.1) were shown to grow exponentially, corresponding to a bank account growing at $100\,r\,\%/$year. For $-1 < r < 0$, its solutions were shown to decline exponentially toward an equilibrium value $x_0 = 0$. Rather than money deposited in a bank, the case $r < 0$ was associated with the decay of a radioactive substance.

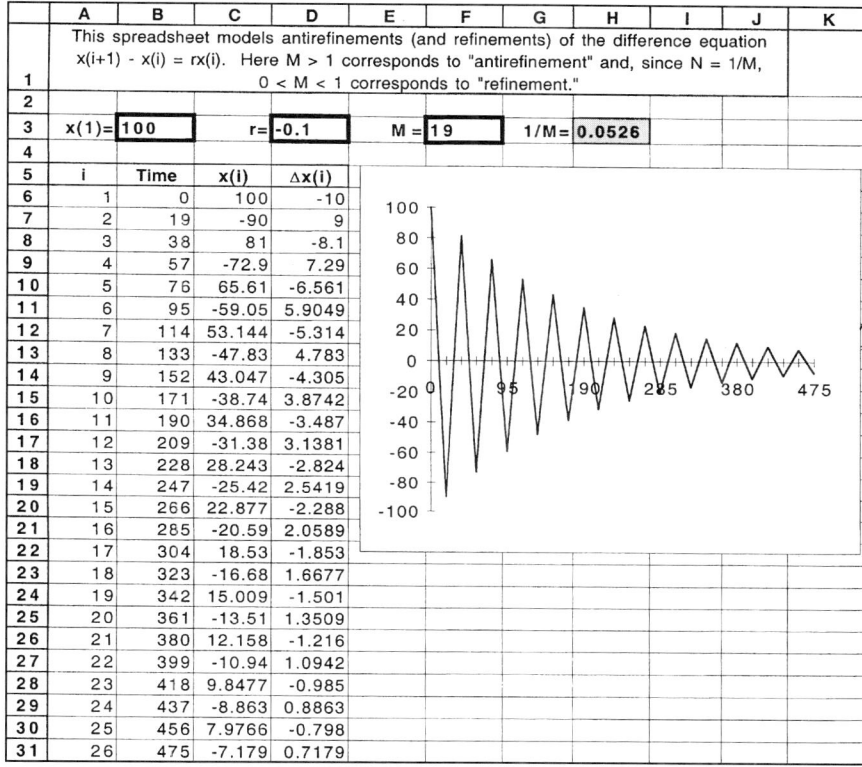

	A	B	C	D	E	F	G	H	I	J	K
	This spreadsheet models antirefinements (and refinements) of the difference equation										
1	x(i+1) - x(i) = rx(i). Here M > 1 corresponds to "antirefinement" and, since N = 1/M, 0 < M < 1 corresponds to "refinement."										
2											
3	x(1)=100			r=-0.1		M =19			1/M=0.0526		
4											
5	i	Time	x(i)	Δx(i)							
6	1	0	100	-10							
7	2	19	-90	9							
8	3	38	81	-8.1							
9	4	57	-72.9	7.29							
10	5	76	65.61	-6.561							
11	6	95	-59.05	5.9049							
12	7	114	53.144	-5.314							
13	8	133	-47.83	4.783							
14	9	152	43.047	-4.305							
15	10	171	-38.74	3.8742							
16	11	190	34.868	-3.487							
17	12	209	-31.38	3.1381							
18	13	228	28.243	-2.824							
19	14	247	-25.42	2.5419							
20	15	266	22.877	-2.288							
21	16	285	-20.59	2.0589							
22	17	304	18.53	-1.853							
23	18	323	-16.68	1.6677							
24	19	342	15.009	-1.501							
25	20	361	-13.51	1.3509							
26	21	380	12.158	-1.216							
27	22	399	-10.94	1.0942							
28	23	418	9.8477	-0.985							
29	24	437	-8.863	0.8863							
30	25	456	7.9766	-0.798							
31	26	475	-7.179	0.7179							

Figure 3.2.1 Antirefinement of $x_{i+1} - x_i = r x_i$

In the case of (3.2.1) and $N > 1$, refinements of the form $\Delta t \mapsto \Delta t/N$ tended to

*in*crease the rate at which x_i grows in the case $r > 0$, and

*de*crease the rate at which x_i decays in the case $-1 < r < 0$.

On this basis, it is reasonable to conjecture that for $M > 1$, "antirefinements" of the form $\Delta t \mapsto M\Delta t$ will serve to

*de*crease the rate at which x_i grows in the case $r > 0$, and

*in*crease the rate at which x_i decays in the case $r < 0$.

Further insight into these phenomena can be obtained via a spreadsheet[1] that models antirefinements of (3.2.1) while allowing us to alter the values of x_1, r, and M.

Problem 3.2.1 Create a spreadsheet in the format of Figure 3.2.1. Describe the effect of antirefinement ($M > 1$) on the behavior of solutions of (3.2.1) in the

---

[1] Actually, we could study this question by setting $N = 1/M$ in one of the spreadsheets used to study refinement, but creating a new one won't take long.

case:

(a) $r > 0$.   (d) $-2 < Mr < -1$.

(b) $-1 < r < 0$.   (e) $r = -2$.

(c) $r = -1$.   (f) $r < -2$.

While experimentation with such a spreadsheet is likely to reinforce our expectations for $r > 0$, the case $r < 0$ may involve some surprises. As M increases beyond a critical value, antirefinement of (3.2.1) leads to solutions $x_1, x_2, x_3, \ldots$ whose exponential decay is accompanied by alternating signs. As M increases beyond a second critical value, antirefinement leads not only to alternating signs but also to solutions for which $|x_i|$ grows without bound.

To see why antirefinement leads to such behavior for $r < 0$, we need only implement $\Delta t \mapsto M(\Delta t)$, transforming (3.2.1) into

(3.2.2) $x_{j+1} - x_j = Mr\,x_j$.

Going on to write (3.2.2) in the iterative format

(3.2.3) $x_{j+1} = (1 + Mr)x_j$,

we see that x_{j+1} will have the opposite sign of x_j whenever $1 + Mr < 0$, or

$$M > -\frac{1}{r}.$$

Although negative values of x may not make physical sense in modeling radioactive decay, they do arise in literal interpretations of the difference equation (3.2.1). Large antirefinements generate "decays greater than 100% of the amount present," and the resulting overshoot of the equilibrium value $x_0 = 0$ leads to an alternation in the sign of x_j. When M exceeds $-1/r$ by a small amount, such overshoots can be self-correcting, leading to a solution $x_1, x_2, x_3, \ldots$ with $|x_1| > |x_2| > |x_3| > \cdots$. However, for sufficiently large values of M, the overshoot is too large to self-correct. Rather than approaching $x_0 = 0$, the x_j now satisfy $|x_1| < |x_2| < |x_3| < \cdots$, exhibiting increasingly large oscillations about this equilibrium value.

As in Chapter 1, we can represent solutions of (3.2.3) by means of a staircase diagram (Figure 3.2.2) corresponding to the iterator

(3.2.4) $F(x) = (1 + Mr)x$.

Problem 3.2.2 With $r = -0.1$ and $x_1 = 1$, use a staircase diagram for $F(x) = (1 + Mr)x$ to represent solutions of the difference equation $x_{j+1} - x_j = Mr\,x_j$ in the case:

(a) $M = 8$.   (b) $M = 12$.   (c) $M = 25$.

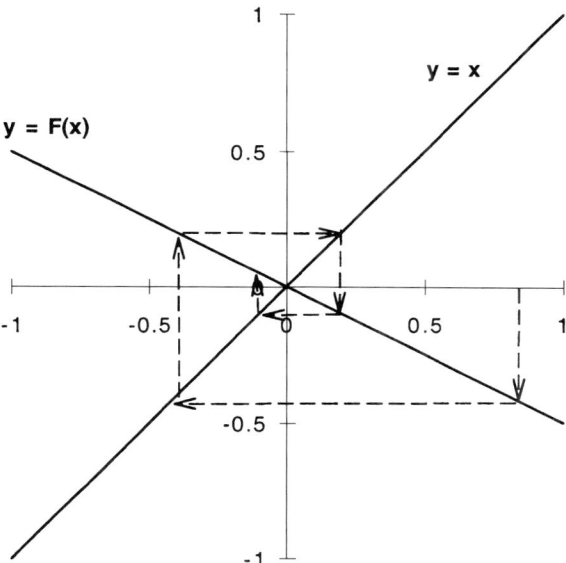

Figure 3.2.2 Staircase Diagrams for $F(x) = (1 + Mr)x$

Problem 3.2.3 In Problem 3.2.2, determine the range of values of M for which solutions of the difference equation $x_{j+1} - x_j = Mr x_j$:

(a) Satisfy $x_1 > x_2 > x_3 > \cdots > 0$.
(b) Satisfy $|x_1| > |x_2| > |x_3| > \cdots$.
(c) Satisfy $|x_1| < |x_2| < |x_3| < \cdots$.

As already noted in Chapter 2, many of the concepts we have developed in modeling radioactive decay extend to the more general difference equation

(3.2.5) $$x_{i+1} - x_i = ax_i + b$$

when $a < 0$. Here $\Delta x_i = x_{i+1} - x_i$ is zero whenever $ax_i + b = 0$, or, equivalently, whenever $x_i = -b/a$. With $a < 0$ we again have reason to expect that solutions of (3.2.5) will approach an equilibrium value, one that is now given by $x_0 = -b/a$.

Problem 3.2.4 This problem deals with a generalization of Problem 3.2.2 to the difference equation $x_{i+1} - x_i = ax_i + b$, where $a < 0$.

(a) Create a spreadsheet that models $x_{i+1} - x_i = ax_i + b$.
(b) For $a = -3, b = 2$, and $x_1 = 1$, find x_2, x_3, and x_4.

(c) Determine a refinement $\Delta t \mapsto \Delta t / N$ for which the corresponding x_j approach a constant value x_0.

(d) Can the value of x_0 be predicted? If so, how?

As Problem 3.2.4(c) suggests, we are always able to find a refinement $\Delta t \mapsto \Delta t / N$ for which the sequence $x_1, x_2, x_3, \ldots$ generated by (3.2.5) approaches $-b/a$. It is, however, also possible to find *antirefinements* for which the solution alternates about $x_0 = -b/a$, perhaps with ever growing amplitudes. Such antirefinements can be realized by sufficiently small values of $N = 1/M < 1$ in the spreadsheet of Problem 3.2.4. General conditions for achieving these forms of behavior can be established by considering a refinement $\Delta t \mapsto \Delta t / N$ that transforms (3.2.5) into the form

$$(3.2.6) \qquad x_{j+1} - x_j = \frac{a}{N} x_j + \frac{b}{N}.$$

Examining the corresponding iterator

$$F(x) = \left(1 + \frac{a}{N} \right) x + \frac{b}{N},$$

we find (not unexpectedly) that $F(x)$ has the equilibrium value $x_0 = -b/a$ as a fixed point. Since the graph of F is a straight line with slope $m = 1 + a/N$, a staircase diagram will show that the sequence $x_1, x_2, x_3, \ldots$ approaches $-b/a$ whenever

$$(3.2.7) \qquad \left| 1 + \frac{a}{N} \right| < 1.$$

Problem 3.2.5 Recalling that $a < 0$, show that (3.2.7) is satisfied whenever $N > -a/2$.

If $|1 + a| < 1$, solutions of (3.2.5) will approach $-b/a$ without further refinement, i.e., (3.2.7) is satisfied with $N = 1$. However, in this case there are values of M for which *antirefinements* of the form $\Delta t \mapsto M \Delta t$ destroy such an approach to equilibrium. To see this we recall that $\Delta t \mapsto M \Delta t$ transforms (3.2.5) into

$$(3.2.8) \qquad x_{j+1} - x_j = Max_j + Mb$$

and examine the corresponding iterator

$$(3.2.9) \qquad \bar{F}(x) = (1 + Ma)x + Mb.$$

Recalling the assumption $a < 0$, we find that

$M < -\frac{2}{a}$ leaves intact the solution's approach to equilibrium, and

$M > -\frac{2}{a}$ leads to increasing oscillations about $-\frac{b}{a}$.

Problem 3.2.6 Derive the above conditions by examining the slope of the iterator (3.2.9).

Problem 3.2.7 Keeping in mind that $a < 0$, show that:

(a) If $M < -1/a$, then (3.2.8) leads to a solution that approaches $-b/a$ either from above or from below.
(b) If $-1/a < M < -2/a$, then (3.2.8) leads to a solution that oscillates about $-b/a$ as it approaches this equilibrium value.
(c) What happens if $M = -1/a$?

It now remains to explore similar questions for more general difference equations of the form

$$\text{(3.2.10)} \qquad x_{i+1} - x_i = f(x_i).$$

Here values of x_0 for which $f(x_0) = 0$ are again termed "equilibrium values" for (3.2.10), and there arises the question of whether their solutions will in fact approach such x_0. This question can be addressed by examining the corresponding iterative process

$$\text{(3.2.11)} \qquad x_{i+1} = x_i + f(x_i)$$

and determining whether x_0 is an attractor or repeller for the iterator $F(x) = x + f(x)$.

In Section 3.3 we address these questions for $f(x) = ax - bx^2$. This corresponds to the logistic equation studied in Chapter 2, and as illustrated by Figure 2.4.2, there *are* cases in which solutions of the logistic equation $x_{i+1} - x_i = ax_i + bx_i^2$ tend to the equilibrium value $x_0 = a/b$.

What is the effect of "sloppy bookkeeping" for the logistic equation and, more generally, for (3.2.10)? As in the case of radioactive decay, we will find that an infrequent updating of data in the iterative process causes solutions to overshoot their (attracting) equilibrium values. Now, however, rather than generating the simple kinds of behavior encountered in the case of $x_{i+1} - x_i = ax_i + bx_i^2$, such overshoot will take on more varied and interesting forms. Insights into the reasons for such "interesting behavior" are provided in Section 3.3 to follow.

3.3. The Logistic Equation Revisited

To determine the impact of refinement and antirefinement on the logistic equation

$$\text{(3.3.1)} \qquad x_{i+1} - x_i = ax_i - bx_i^2$$

it will be useful to begin by constructing a spreadsheet (Figure 3.3.1) corresponding to (3.3.1), one in which we are able to vary the values of x_1, a, b, and $M = 1/N$. In keeping with the phenomena[1] associated with (3.3.1) in Chapter 2, we assume throughout that $a > 0$ and $b > 0$.

---

[1] In population dynamics, the constant a would represent a normal growth rate, while b corresponds to a superlinear damping of growth

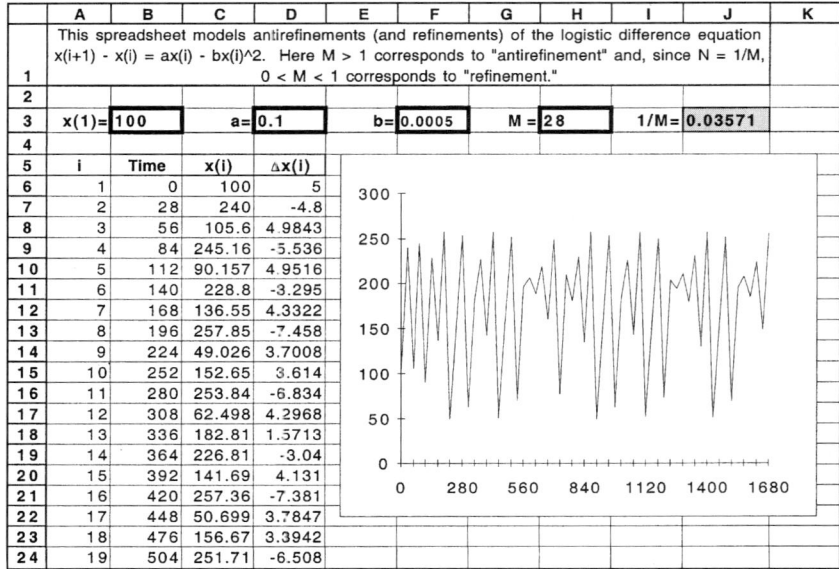

The table portion:

	A	B	C	D	E	F	G	H	I	J	K
1	This spreadsheet models antirefinements (and refinements) of the logistic difference equation x(i+1) - x(i) = ax(i) - bx(i)^2. Here M > 1 corresponds to "antirefinement" and, since N = 1/M, 0 < M < 1 corresponds to "refinement."										
2											
3	x(1)=	100		a=	0.1		b=	0.0005	M =	28	1/M= 0.03571
4											
5	i	Time	x(i)	Δx(i)							
6	1	0	100	5							
7	2	28	240	-4.8							
8	3	56	105.6	4.9843							
9	4	84	245.16	-5.536							
10	5	112	90.157	4.9516							
11	6	140	228.8	-3.295							
12	7	168	136.55	4.3322							
13	8	196	257.85	-7.458							
14	9	224	49.026	3.7008							
15	10	252	152.65	3.614							
16	11	280	253.84	-6.834							
17	12	308	62.498	4.2968							
18	13	336	182.81	1.5713							
19	14	364	226.81	-3.04							
20	15	392	141.69	4.131							
21	16	420	257.36	-7.381							
22	17	448	50.699	3.7847							
23	18	476	156.67	3.3942							
24	19	504	251.71	-6.508							

Figure 3.3.1 The Logistic Equation with Antirefinement

Writing (3.3.1) in the form

$$(3.3.2) \qquad x_{i+1} - x_i = bx_i \left(\frac{a}{b} - x_i \right)$$

we see that the logistic equation has *two* equilibrium values, namely $x_0 = 0$ and $x_0 = a/b$. Antirefinements of the form $\Delta t \mapsto M \Delta t$ now transform (3.3.2) into

$$(3.3.3) \qquad x_{j+1} - x_j = Mbx_j \left(\frac{a}{b} - x_j \right),$$

which has the same equilibrium values as the original equation.

In order to analyze the solutions generated by Figure 3.3.1, it is useful to write (3.3.3) in the iterative form

$$(3.3.4) \qquad x_{j+1} = (1 + Ma)x_j - Mbx_j^2$$

and to create a spreadsheet that graphs both the parabola $F(x) = (1 + Ma)x - Mbx^2$ and the $45°$ line $y = x$ for varying values of a, b, and M. (See Figure 3.3.2.)

The question of whether the fixed points $x_0 = 0$ and $x_0 = a/b$ are attractors or repellers depends on the slope of the parabola $y = (1 + Ma)x - Mbx^2$ at $(0, 0)$ and at $(a/b, a/b)$. Here it can be shown (see Problem 3.7.1 at the end of the chapter) that at $x = 0$ the term $-Mbx^2$ does not contribute to the slope of the graph of $F(x)$. That is, the slope of the parabola at $(0, 0)$ is the same as that of

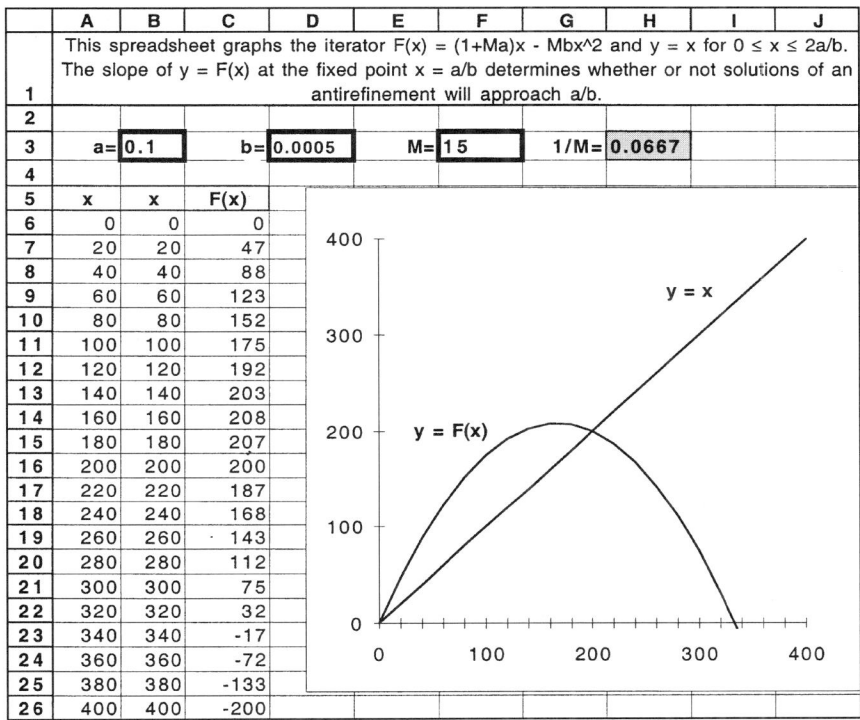

	A	B	C	D	E	F	G	H	I	J
1	This spreadsheet graphs the iterator F(x) = (1+Ma)x - Mbx^2 and y = x for 0 ≤ x ≤ 2a/b. The slope of y = F(x) at the fixed point x = a/b determines whether or not solutions of an antirefinement will approach a/b.									
2										
3	a=|0.1		b=|0.0005		M=|15		1/M=|0.0667	
4										
5	x	x	F(x)							
6	0	0	0							
7	20	20	47							
8	40	40	88							
9	60	60	123							
10	80	80	152							
11	100	100	175							
12	120	120	192							
13	140	140	203							
14	160	160	208							
15	180	180	207							
16	200	200	200							
17	220	220	187							
18	240	240	168							
19	260	260	143							
20	280	280	112							
21	300	300	75							
22	320	320	32							
23	340	340	-17							
24	360	360	-72							
25	380	380	-133							
26	400	400	-200							

Figure 3.3.2 Iterators Associated with Antirefinements

the straight line $y = (1 + Ma)x$. Since $a > 0$ and $M > 0$, this line always has slope $m > 1$. As a result,

$x_0 = 0$ will always be a repeller for $F(x) = (1 + Ma)x - Mbx^2$.

This is the case for any refinement corresponding to $0 < M < 1$ as well as for any antirefinement corresponding to $M > 1$.

At $(a/b, a/b)$, however, the parabola can have a wide range of slopes, and determining the relationship between the sizes of a and b, the size of M, and the slope of the graph of $F(x) = (1 + Ma)x - Mbx^2$ at $x = a/b$ is part of the problem before us. Because the graph of $F(x)$ is a parabola, it is possible to bypass calculus and study its slope by techniques rooted in Greek geometry. Even "calculus cognoscenti" are invited to join us in addressing this question by means of classical geometric tools.

Problem 3.3.1

(a) Use Problem 1.10.2 to show that the focus of the parabola
$y = (1 + Ma)x - Mbx^2$ is at (h, k), where $h = (1 + Ma)/(2Mb)$ and
$k = (2a + Ma^2)/(4b)$.

(b) Recalling that the latus rectum of a parabola is the chord perpendicular to its axis of symmetry and passing through the focus, show that the inequality

$$\frac{a}{2b} \le x \le \frac{2 + Ma}{2Mb}$$

determines the values of x for which the parabola in part (a) lies above the latus rectum.

(c) Problem 1.11.22(f) at the end of Chapter I shows that the slope m of the parabola $y = ax^2 + bx + c$ satisfies $|m| \le 1$ on the arc joining the ends of the latus rectum and $m > 1$ at other points of the curve. Use this to confirm (3.3.5) below:[2]

If $x < a/(2b)$ or if $x > (2 + Ma)/(2Mb)$, then $|m| > 1$.

(3.3.5)   If $a/(2b) < x < (1 + Ma)/(2Mb)$, then $0 < m < 1$.

If $(1 + Ma)/(2Mb) < x < (2 + Ma)/(2Mb)$, then

$$0 > m > -1.$$

With (3.3.5) at hand, we can explain some of the phenomena encountered in experimentation with Figure 3.3.1. The next three problems use (3.3.5) to study the slope of $F(x)$ at the fixed point $x_0 = a/b$.

Problem 3.3.2 Whenever $0 < M < 1/a$, the slope of $F(x)$ at a/b satisfies $0 < m < 1$.

Problem 3.3.3 Whenever $1/a < M < 2/a$, the slope of $F(x)$ at a/b satisfies $-1 < m < 0$.

Problem 3.3.4 Whenever $M > 2/a$, the slope of $F(x)$ at a/b satisfies $m < -1$.

These results can now be combined to conclude that

whenever $0 < M < 2/a$, $x_0 = a/b$ is an attractor for $F(x) = (1 + Ma)x - Mbx^2$.

Based on Figure 3.3.2, we can now use staircase diagrams to gain geometric insight into the implications of this statement. As illustrated in Figure 3.3.3, if $0 < M < 2/a$, then for an initial x_1 satisfying $0 < x_1 < (1 + Ma)/(Mb)$, the iterates $x_2, x_3, x_4, \ldots$ will approach a/b.

Problem 3.3.5 Create a spreadsheet in the format of Figure 3.3.2 and set $a = 0.1$ and $b = 0.0005$. Use this spreadsheet to generate accurate graphs of $F(x)$ for:

(a) $M = 8$.

---

[2] For readers familiar with calculus, (3.3.5) is also established in the starred Problem 3.7.2 at the end of this chapter.

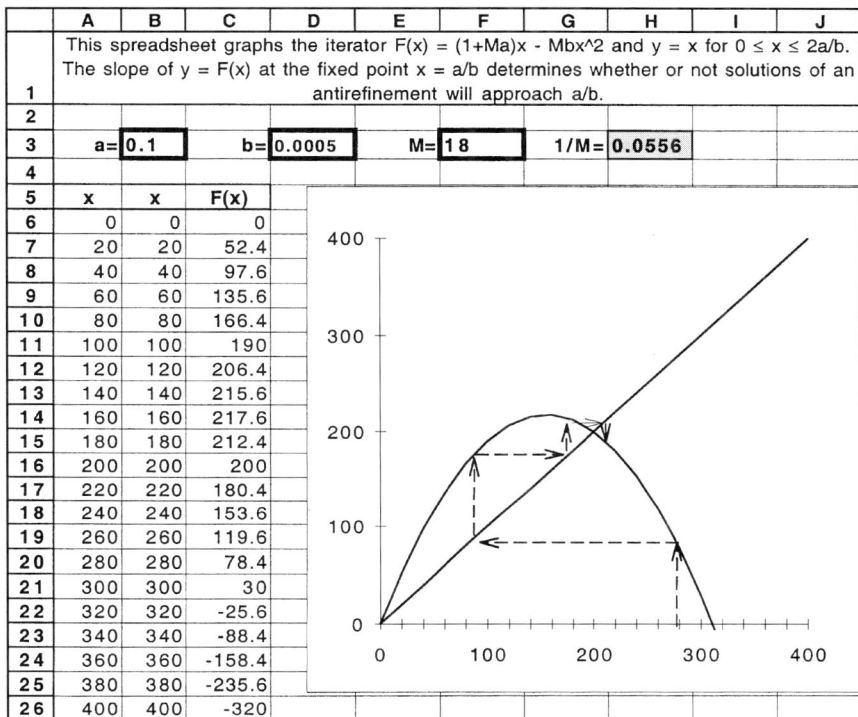

Figure 3.3.3 Antirefinements with Attractor at *a*/*b*

(b) $M = 10$.

(c) $M = 12$.

[Scale the graphs so that the line $y = x$ makes a $45°$ angle with the x-axis.] In each case, draw a staircase diagram for an initial value satisfying $0 < x_1 < (1 + Ma)/(Mb)$ and for an initial value slightly larger than $(1 + Ma)/(Mb)$.

While Problem 3.3.5 provides a fairly complete picture of what happens when $0 < M < 2/a$, it does not deal with the phenomena encountered when M exceeds $2/a$. Accordingly, it will be desirable to return to Figure 3.3.1 and to examine with care the solutions of antirefinements of the logistic equation for a range of values of $M > 1/a$.

Problem 3.3.6 For $x_1 = 50, a = 0.1, b = 0.0005$, give qualitative descriptions of the graph generated by Figure 3.3.1 when $M = 18, M = 22, M = 25, M = 28, M = 30, M = 32$, and two other values of M of your choice. [You may want to use a bar graph in place of the line graph used in Figure 3.3.1.]

How does one explain the phenomena encountered in Problem 3.3.6? Answering this important question is at the heart of Section 3.4 to follow. For now, let us simply observe that when $M > 2/a$, *both* fixed points $x_0 = 0$ and $x_0 = a/b$

are repellers for the iterator $F(x) = (1 + Ma)x - Mbx^2$. What seems to be happening here is that values of x_i lying between $x = 0$ and $x = (1 + Ma)/(Mb)$ become trapped between these two repellers, much like a Ping-Pong ball between two championship players. For some values of M the values of x_i fall into a cyclic pattern, while for others these values become quite chaotic. Furthermore, when M becomes sufficiently large, the values of x_i tend to grow without bound.

In Section 3.4 we will employ Excel's graphing capability to gain a deeper understanding of these remarkable phenomena.

3.4. Iteration of Functions

In an effort to understand the effect of antirefinement on the logistic equation, we have been led to study iterators of the form

$$(3.4.1) \qquad F(x) = (1 + Ma)x - Mbx^2.$$

Our analysis of (3.4.1) was based on the fact that the graph of $F(x)$ is a parabola having a vertical axis of symmetry and intersecting the line $y = x$ at fixed points that are readily calculated. Furthermore, as indicated by the spreadsheet in Figure 3.4.1, increases in M tend to raise the parabola's vertex (the highest point

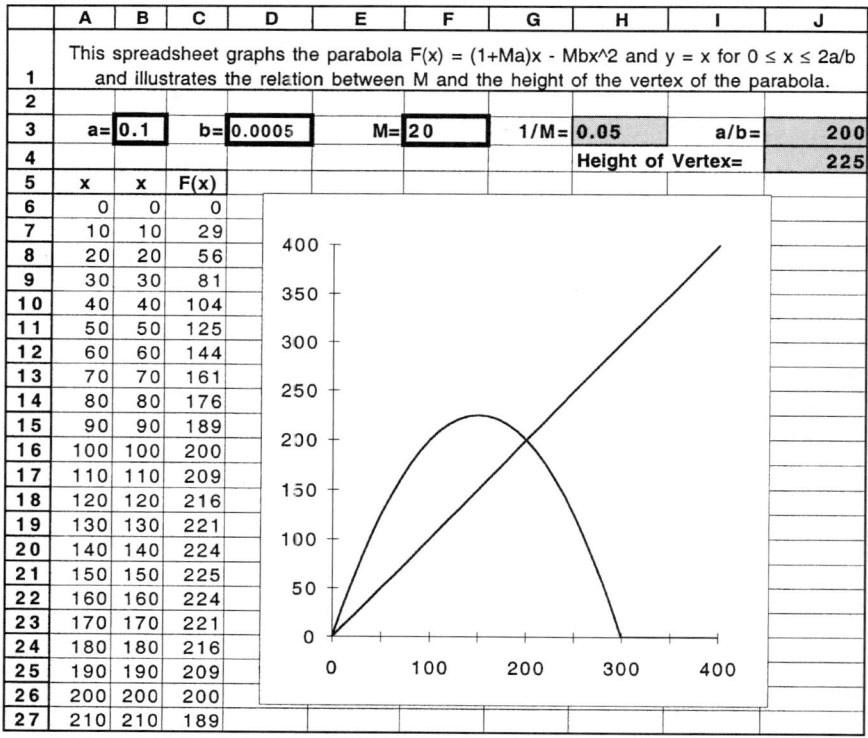

Figure 3.4.1 The Geometry of Antirefinement

in the graph) and result in increasingly steep slopes at the two fixed points $(0, 0)$ and $(a/b, a/b)$.

Problem 3.4.1 Assuming $a > 0, b > 0$, and $M > 0$, find the coordinates of the highest point on the graph of the parabola (3.4.1) in the (x, y)-plane. Show that for $M \geq 1/a$, further increases in M increase the y coordinate of this vertex.

Our reason for describing antirefinements of the logistic equation in such general terms is to set the stage for a recasting of this problem into a form that makes it easier to analyze. In place of (3.4.1), we will consider the simpler parabola

(3.4.2) $$y = x^2 - c.$$

Now, instead of having three parameters a, b, and M to contend with, we have a single parameter c. To be sure, the parabola (3.4.2) opens upward rather than downward, and determining its intersections with the line $y = x$ requires some additional calculations. But it is now easy to see that changes in c simply result in vertical translation of this parabola. Putting old adages aside, it will be advantageous to change equations in midstream and to approach the topic of chaos while riding (3.4.2) rather than (3.4.1).

Problem 3.4.2 For $c \geq -\frac{1}{4}$, find the fixed points of the function $F(x) = x^2 - c$. What happens if $c < -\frac{1}{4}$?

But before exploring the geometry of (3.4.2) with a spreadsheet, let us recall from Problem 1.10.2(b) that:

(i) The latus rectum of the parabola $y = x^2 - c$ lies on the line $y = \frac{1}{2} - c$.
(ii) The slope of $y = x^2 - c$ satisfies satisfies $|m| < 1$ below the latus rectum and $|m| > 1$ above the latus rectum.

These observations will enable us to determine whether the fixed points of $F(x) = x^2 - c$ are repellers or attractors.

Problem 3.4.3 Show that for $c > -\frac{1}{4}$, the fixed point $x = \frac{1}{2} + \sqrt{c + \frac{1}{4}}$ lies above the latus rectum $y = \frac{1}{4} - c$ and therefore is a repeller for $F(x) = x^2 - c$.

The spreadsheet in Figure 3.4.2 provides further insight into the effect of changes in c by constructing the graphs of $y = x^2 - c$, $y = x$, and the latus rectum $y = \frac{1}{4} - c$ for varying values of c.

Problem 3.4.4 By locating the fixed point at $x = \frac{1}{2} - \sqrt{c + \frac{1}{4}}$ relative to the latus rectum, show that this fixed point is:

(a) an attractor if $-\frac{1}{4} < c < \frac{3}{4}$.
(b) a repeller if $c > \frac{3}{4}$.

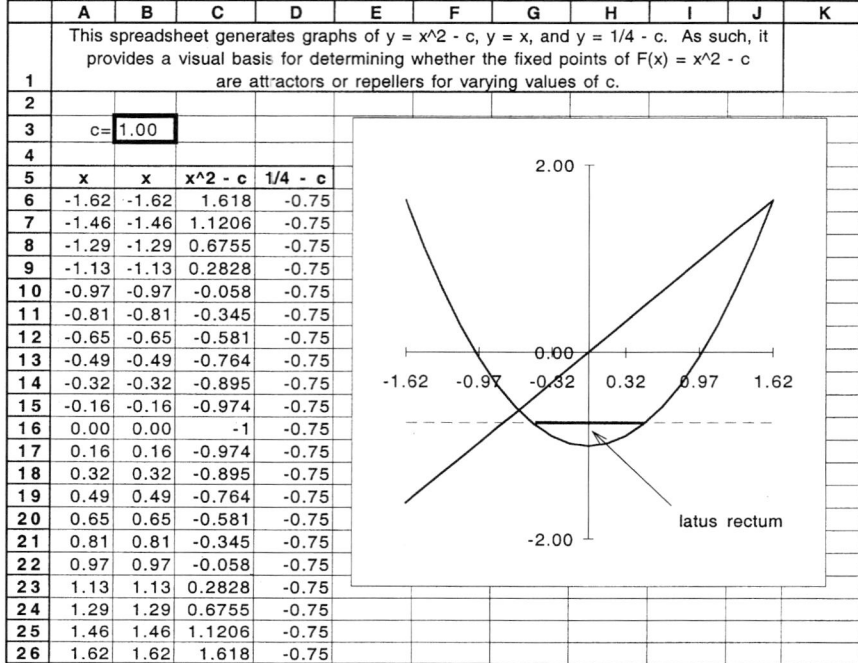

	A	B	C	D	E	F	G	H	I	J	K
1	This spreadsheet generates graphs of y = x^2 - c, y = x, and y = 1/4 - c. As such, it provides a visual basis for determining whether the fixed points of F(x) = x^2 - c are attractors or repellers for varying values of c.										
2											
3	c=	1.00									
4											
5	x	x	x^2 - c	1/4 - c							
6	-1.62	-1.62	1.618	-0.75							
7	-1.46	-1.46	1.1206	-0.75							
8	-1.29	-1.29	0.6755	-0.75							
9	-1.13	-1.13	0.2828	-0.75							
10	-0.97	-0.97	-0.058	-0.75							
11	-0.81	-0.81	-0.345	-0.75							
12	-0.65	-0.65	-0.581	-0.75							
13	-0.49	-0.49	-0.764	-0.75							
14	-0.32	-0.32	-0.895	-0.75							
15	-0.16	-0.16	-0.974	-0.75							
16	0.00	0.00	-1	-0.75							
17	0.16	0.16	-0.974	-0.75							
18	0.32	0.32	-0.895	-0.75							
19	0.49	0.49	-0.764	-0.75							
20	0.65	0.65	-0.581	-0.75							
21	0.81	0.81	-0.345	-0.75							
22	0.97	0.97	-0.058	-0.75							
23	1.13	1.13	0.2828	-0.75							
24	1.29	1.29	0.6755	-0.75							
25	1.46	1.46	1.1206	-0.75							
26	1.62	1.62	1.618	-0.75							

Figure 3.4.2 Locating Fixed Points of $x^2 - c$ Relative to the Latus Rectum

Having developed an ability to locate and classify the fixed points of $F(x) = x^2 - c$, we are also in a position to analyze the iterative process

$$(3.4.3) \qquad\qquad x_{i+1} = x_i^2 - c,$$

or equivalently,

$$(3.4.4) \qquad\qquad x_{i+1} = F(x_i) \quad \text{with} \quad F(x) = x^2 - c.$$

From Problems 3.4.3 and 3.4.4 we know that for $-\frac{1}{4} < c < \frac{3}{4}$

$\frac{1}{2} - \sqrt{c+\frac{1}{4}}$ is an attractor for $F(x)$,
$\frac{1}{2} + \sqrt{c+\frac{1}{4}}$ is a repeller for $F(x)$.

As in Figure 3.4.3, staircase diagrams enable us to illustrate how for any initial value x_1 satisfying

$$-\frac{1}{2} - \sqrt{c + \frac{1}{4}} < x_1 < \frac{1}{2} + \sqrt{c + \frac{1}{4}}$$

the sequence $x_2, x_3, x_4, \ldots$ defined by (3.4.3) approaches $\frac{1}{2} - \sqrt{c+\frac{1}{4}}$.

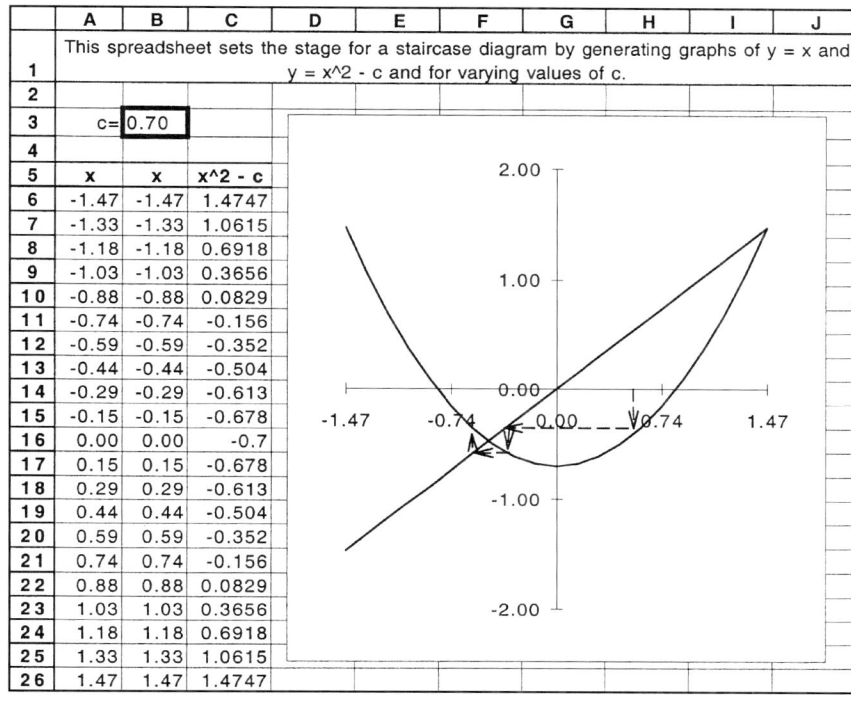

	A	B	C	D	E	F	G	H	I	J
1	This spreadsheet sets the stage for a staircase diagram by generating graphs of y = x and y = x^2 - c and for varying values of c.									
2										
3	c=	0.70								
4										
5	x	x	x^2 - c							
6	-1.47	-1.47	1.4747							
7	-1.33	-1.33	1.0615							
8	-1.18	-1.18	0.6918							
9	-1.03	-1.03	0.3656							
10	-0.88	-0.88	0.0829							
11	-0.74	-0.74	-0.156							
12	-0.59	-0.59	-0.352							
13	-0.44	-0.44	-0.504							
14	-0.29	-0.29	-0.613							
15	-0.15	-0.15	-0.678							
16	0.00	0.00	-0.7							
17	0.15	0.15	-0.678							
18	0.29	0.29	-0.613							
19	0.44	0.44	-0.504							
20	0.59	0.59	-0.352							
21	0.74	0.74	-0.156							
22	0.88	0.88	0.0829							
23	1.03	1.03	0.3656							
24	1.18	1.18	0.6918							
25	1.33	1.33	1.0615							
26	1.47	1.47	1.4747							

Figure 3.4.3   A Staircase Diagram for Iteration by $F(x) = x^2 - c$

Problem 3.4.5 Using a spreadsheet if you wish, draw an accurate graph of $y = x^2 - \frac{1}{2}$. Illustrate how the iterates of any x_1 satisfying $|x_1| < 1.3660254$ approach the fixed point at $x_0 \approx -0.3660254$.

Problem 3.4.6 Construct a spreadsheet that calculates iterations of x_1 by $F(x) = x^2 - c$ while enabling you to vary the values of x_1 and c.

(a) Use this spreadsheet to confirm the results of Problem 3.4.5.

(b) Describe what happens when $c = \frac{1}{4}$. In particular, for what range of initial values x_1 do the iterates approach a fixed point? What is the value of that fixed point?

Well, so far so good. But what happens when c becomes greater than $\frac{3}{4}$? The spreadsheet in Figure 3.4.4 (much like the one you constructed for Problem 3.4.6) enables you to explore this question. However, staircase diagrams such as the one in Figure 3.4.3 are unlikely to provide much insight into the phenomena you observe.

The reason things change so dramatically at $c = \frac{3}{4}$ was already hinted at in Section 3.3. For $c > \frac{3}{4}$, the interval defined by the two fixed points,

$$(3.4.5) \qquad\qquad \tfrac{1}{2} - \sqrt{c + \tfrac{1}{4}} < x < \tfrac{1}{2} + \sqrt{c + \tfrac{1}{4}}$$

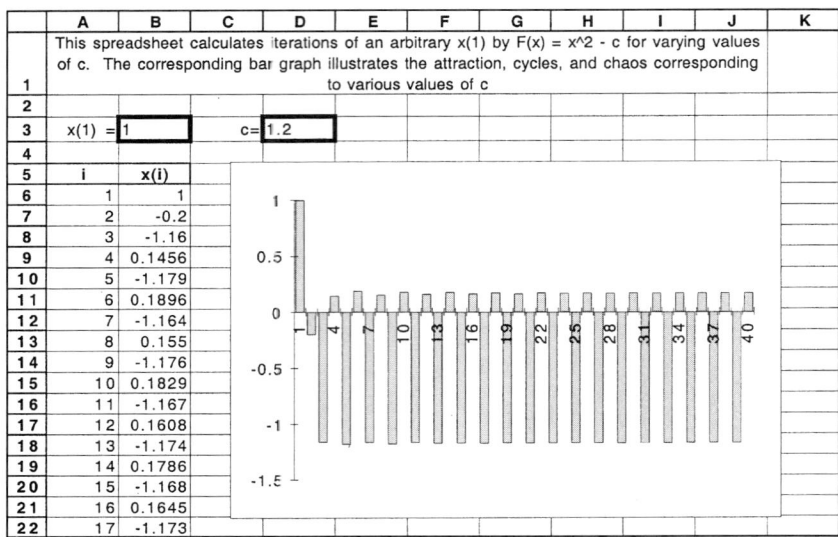

Figure 3.4.4 Iteration by $F(x) = x^2 - c$

has repellers at both ends. This leads to a situation like that encountered with antirefinements of the logistic equation when $M > 2/a$. If x_1 satisfies

$$|x_1| < \tfrac{1}{2} + \sqrt{c + \tfrac{1}{4}},$$

then the iterates $x_2, x_3, x_4, \ldots$ will be trapped within the interval defined by (3.4.5), and this sets the stage for both cyclic and chaotic behavior.

To understand the origin of the "cycles of length two" that appear as c surpasses $\tfrac{3}{4}$, it will be useful to construct a staircase diagram for both $F(x) = x^2 - c$ and for

(3.4.6) $$\begin{aligned} F(F(x)) &= F(x^2 - c) \\ &= (x^2 - c)^2 - c \\ &= x^4 - 2cx^2 + c^2 - c. \end{aligned}$$

Accordingly, the spreadsheet of Figure 3.4.5 generates graphs of x, $F(x)$, and $F(F(x))$ for varying values of c.

To see what happens as c passes through the value $\tfrac{3}{4}$, we consider the graphs generated in Figure 3.4.5 when $c = 0.65$, $c = 0.75$, and $c = 0.9$. (See Figure 3.4.6.)

For $c = 0.65$, both $F(x)$ and $F(F(x))$ have an attracting fixed point at $\tfrac{1}{2} - \sqrt{c + \tfrac{1}{4}}$. However, as we "pass through $c = \tfrac{3}{4}$" this attractor turns into a repeller for both $F(x)$ and $F(F(x))$. While $F(x)$ now has two fixed points, both of which are repellers, $F(F(x))$ has given birth to *two new attracting fixed points*. These new attractors for $F(F(x))$ are solutions of the quartic equation

(3.4.7) $$x^4 - 2cx^2 + c^2 - c = x.$$

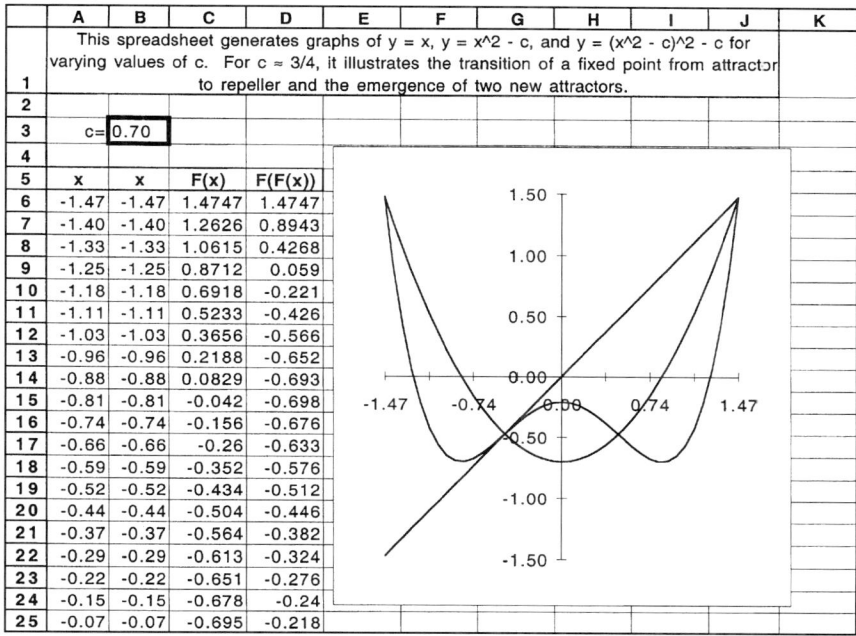

Figure 3.4.5 **Locating Fixed Points for** $F(x)$ **and** $F(F(x))$

Problem 3.4.7 Confirm that $x = 0$ and $x = -1$ are solutions of $x^4 - 2x^2 - x = 0$. Setting $c = 1$ and choosing x_1 in the interval $-0.6181 < x < 1.618$, use the spreadsheet of Figure 3.4.4 to confirm that the resulting sequence of iterates does approach a cycle alternating between these two solutions.

Note that if $F(F(x_1)) = x_1$, then iteration of x_1 by F leads to the sequence $x_1, F(x_1), x_1, F(x_1), \ldots$.

Problem 3.4.8 Recalling the iterator used to solve the general quartic in Problem 1.9.3, set $c = 1.1$ in (3.4.7) and find approximate values for the four real solutions of $x^4 - 2.2x^2 - x + 0.11 = 0$. Use the spreadsheet of Figure 3.4.4 to

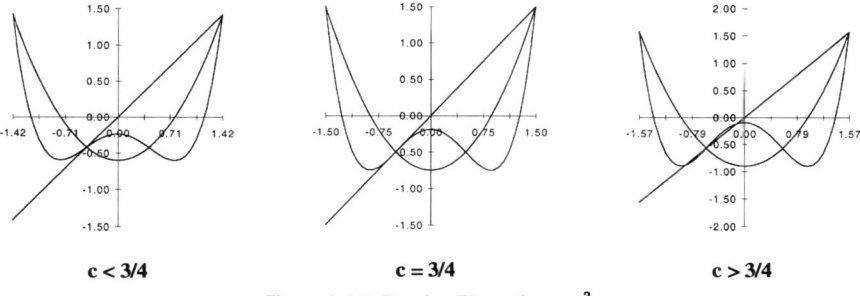

Figure 3.4.6 **Passing Through** $c = \frac{3}{4}$

confirm that when $c = 1.1$ and $|x_1| < 1.66$, the resulting sequence of iterates does approach a cycle alternating between two of these solutions.

The insight provided by Figures 3.4.4 and 3.4.5 provides an explanation for the cyclic behavior of the iterates

$$x_{i+1} = x_i^2 - c$$

from $c = \frac{3}{4}$ until $c \approx 1.3$. At this point, however, the spreadsheet of Figure 3.4.4 shows that cycles of length two are replaced by cycles of length four. To explain this new turn of events the reader is invited to generalize Figure 3.4.5, locating fixed points not only for $F(x)$ and $F(F(x))$, but also for

$$\begin{aligned}
F(F(F(x))) &= F(x^4 - 2cx^2 + c^2 - c) \\
&= (x^4 - 2cx^2 + c^2 - c)^2 - c \\
&= x^8 - 4cx^6 + (6c^2 - 2c)x^4 - 4(c^3 - c^2)x^2 + c^4 - 2c^3 + c^2 - c.
\end{aligned}$$

With formulas for iterates becoming so complicated, a single (spreadsheet) picture becomes worth a thousand words. On this basis, we leave it to Problem 3.7.4 at the end of this chapter to explore the graphs of $F(x)$, $F(F(x))$, $F(F(F(x)))$, ... These graphs will provide insight into why in the range $1.3 < c < 2$ further increases in c lead to other cyclic patterns and a subsequent "descent into chaos."

At $c = 2$, however, there is a new and interesting turn of events that is discussed in the following section.

3.5. Escape From Chaos   In studying the iterative scheme

$$(3.5.1) \qquad\qquad x_{i+1} = x_i^2 - c$$

we have assumed that the initial x_1 belongs to the interval

$$(3.5.2) \qquad\qquad -\tfrac{1}{2} - \sqrt{\tfrac{1}{4} + c} < x < \tfrac{1}{2} + \sqrt{\tfrac{1}{4} + c}.$$

The reason for this assumption is that if

$$(3.5.3) \qquad\qquad |x_1| > \tfrac{1}{2} + \sqrt{\tfrac{1}{4} + c},$$

then the process is completely predictable.

Problem 3.5.1 Use a staircase diagram to represent (3.5.1) when $c = \frac{1}{2}$ and $x_1 = \frac{3}{2}$. Repeat for $c = 0$ and $x_1 = -\frac{5}{4}$.

Problem 3.5.2 For $c > -\frac{1}{4}$, show that if $|x_i| > \frac{1}{2} + \sqrt{\frac{1}{4} + c}$, then $|x_{i+1}| > |x_i|$.

In fact, if we ever have

$$(3.5.4) \qquad\qquad |x_i| > \frac{1}{2} + \sqrt{\frac{1}{4} + c}$$

for some $i \geq 1$, then the iterates of this x_i will grow without bound (see Problem 3.7.3).

For the values of c considered in Section 3.4, these observations were not of central importance. As long as $-\frac{1}{4} \leq c < 2$, choosing x_1 inside the interval defined by (3.5.2) assures that the x_{i+1} will stay inside this interval. The reason for this is illustrated by a spreadsheet that graphs $y = x^2 - c$ together with a "critical square" that is centered at the origin, and has $(-\frac{1}{2} - \sqrt{\frac{1}{4}+c}, -\frac{1}{2} - \sqrt{\frac{1}{4}+c})$ and $(\frac{1}{2} + \sqrt{\frac{1}{4}+c}, \frac{1}{2} + \sqrt{\frac{1}{4}+c})$ as diagonally opposite vertices.

As shown in Figure 3.5.1, for $-\frac{1}{4} < c < 2$ the parabola $y = x^2 - c$ is fully contained within the critical square. This implies that whenever x_i satisfies (3.5.2), so does x_{i+1}. As a consequence, the iterates of any x_1 satisfying (3.5.2) will remain in that interval. [Note: The spreadsheet can be programmed to draw the horizontal sides of the critical square by graphing $y = \frac{1}{2} + \sqrt{\frac{1}{4}+c}$ and $y = -\frac{1}{2} - \sqrt{\frac{1}{4}+c}$. However, the vertical sides have to be entered manually—or else left to the imagination.]

Problem 3.5.3 Construct a spreadsheet in the format of Figure 3.5.1 and confirm that whenever $-\frac{1}{4} < c < 2$, the vertex of $y = x^2 - c$ lies inside the critical square. Construct a staircase diagram for $c = 1.5$ and $x_1 = 0.5$ that illustrates the assertion that $\frac{1}{2} - \sqrt{\frac{1}{4}+c} < x_i < \frac{1}{2} + \sqrt{\frac{1}{4}+c}$ for $i = 2, 3. \ldots.$.

So what happens as c grows larger than 2? As readily illustrated in Figure 3.5.1, if $c > 2$, the vertex of the parabola $y = x^2 - c$ dips below the line $y = -\frac{1}{2} - \sqrt{c + \frac{1}{4}}$.

Problem 3.5.4 For $c = 9/4$, determine the interval of values of x for which the parabola $y = x^2 - c$ lies below the line $y = -\frac{1}{2} - \sqrt{c + \frac{1}{4}}$.

The implications of such a "breaching of the boundary" of the critical square are substantial. If x_1 lies within the interval determined in Problem 3.5.4, then

$$x_2 < -\frac{1}{2} - \sqrt{\frac{1}{4} + c},$$

$$x_3 > \frac{1}{2} + \sqrt{\frac{1}{4} + c},$$

$$x_4 > x_3,$$

$$x_5 > x_4, \quad \text{etc.}$$

In other words, the interval determined in Problem 3.5.4 provides an "escape window" through which iterates of an x_1 lying between the two repelling fixed points can escape the interval defined by (3.5.2). Given such a possibility, it

	A	B	C	D	E	F	G	H	I	J	K
1	This spreadsheet plots y = x^2 - c for -1/2-√(c + 1/4) ≤ x ≤ 1/2 + √(c + 1/4) and y = x. It illustrates the relationship between the vertex of the parabola y = x^2 - c and the line y = -1/2-√(c + 1/4) for varying values of c.										
2											
3	c=	2.2					L=	-2.065	R=	2.0652	
4											
5	x	x	x^2-c	Bottom	Top						
6	-2.07	-2.07	2.065	-2.07	2.1						
7	-1.93	-1.93	1.516	-2.07	2.1						
8	-1.79	-1.79	1.004								
9	-1.65	-1.65	0.53								
10	-1.51	-1.51	0.094								
11	-1.38	-1.38	-0.304								
12	-1.24	-1.24	-0.665								
13	-1.10	-1.10	-0.987								
14	-0.96	-0.96	-1.271								
15	-0.83	-0.83	-1.518								
16	-0.69	-0.69	-1.726								
17	-0.55	-0.55	-1.897								
18	-0.41	-0.41	-2.029								
19	-0.28	-0.28	-2.124								
20	-0.14	-0.14	-2.181								
21	0.00	0.00	-2.2								
22	0.14	0.14	-2.181								
23	0.28	0.28	-2.124								
24	0.41	0.41	-2.029								
25	0.55	0.55	-1.897								
26	0.69	0.69	-1.726								
27	0.83	0.83	-1.518								
28	0.96	0.96	-1.271								
29	1.10	1.10	-0.987								
30	1.24	1.24	-0.665	-2.07	2.1						
31	1.38	1.38	-0.304	-2.07	2.1						
32	1.51	1.51	0.094	-2.07	2.1						
33	1.65	1.65	0.53	-2.07	2.1						
34	1.79	1.79	1.004	-2.07	2.1						
35	1.93	1.93	1.516	-2.07	2.1						
36	2.07	2.07	2.065	-2.07	2.1						

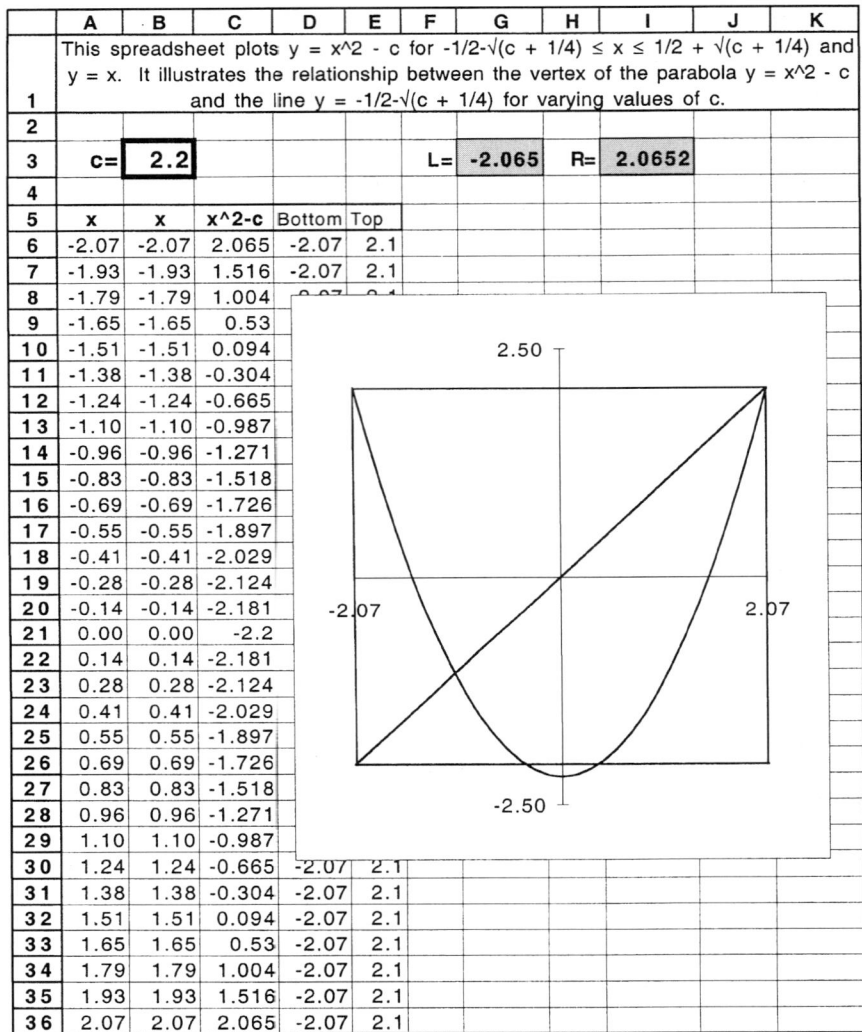

Figure 3.5.1 The Parabola $y = x^2 - c$ Relative to the Critical Square

seems desirable to return to Figure 3.4.4 and try to obtain some insights into the behavior of $x_{i+1} = x_i^2 - c$ for $c > 2$.

Problem 3.5.5 Use the spreadsheet of Figure 3.4.4 to investigate the iterative scheme $x_{i+1} = x_i^2 - c$ for $c = \frac{9}{4}$ and for two other values of $c > 2$. Try to find values of x_1 whose iterates do *not* grow without bound.

Problem 3.5.6 Find the ratio of the length of the escape window determined in Problem 3.5.4 to that of the base of the corresponding critical square.

Problem 3.5.7 Confirm that for $c = 45/16$, the critical square corresponding to $x_{i+1} = x_i^2 - c$ is given by $-\frac{9}{4} < x, y < \frac{9}{4}$. Use a staircase diagram to find a value of x_1, not itself a fixed point, for which iterations by $F(x) = x^2 - 45/16$ do not grow without bound. [Hint: Try the ends of the escape window.]

Given the relatively small size of the escape window found in Problem 3.5.6, the results of Problem 3.5.5 may seem surprising. Even though the escape window is less than half the base of the critical square, the iterates of virtually every x_1 eventually escape from between the two repellers.

To explain this phenomenon, we note that the window determined in Problem 3.5.4 enables iterates of an initial x_1 to escape (3.5.2) *in a single step*. If we are also interested in values of x_1 whose iterates escape in *two steps*, then we would have to look at the points where the graph of

$$F(F(x)) = (x^2 - c)^2 - c$$

breaches the boundary of the critical square. This is readily done in Figure 3.5.2, which is a modification of the spreadsheet of Figure 3.5.1.

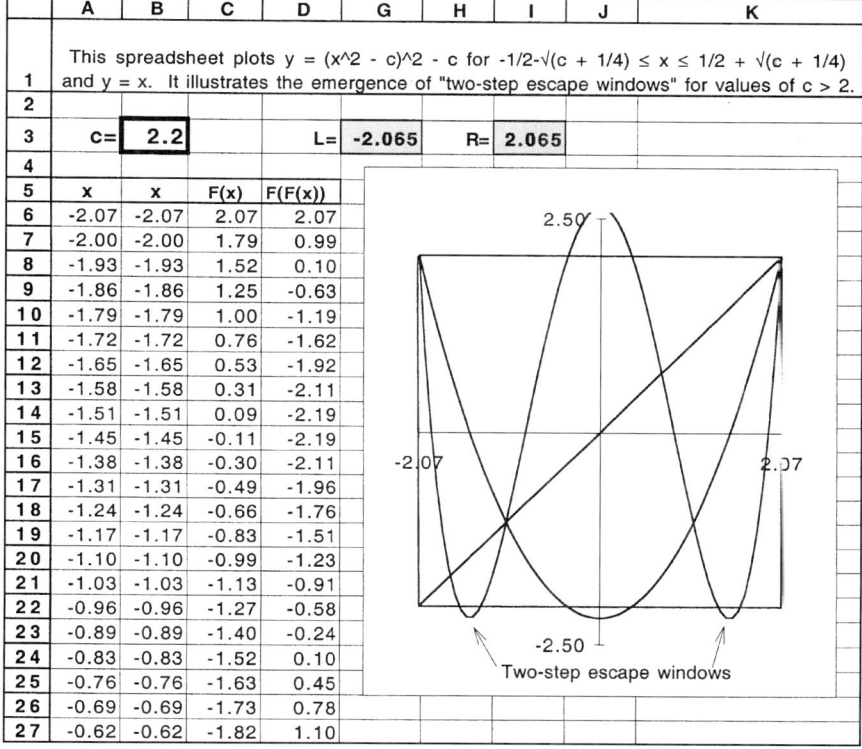

	A	B	C	D	G	H	I	J	K
1	\multicolumn{9}{	c	}{This spreadsheet plots y = (x^2 - c)^2 - c for -1/2-√(c + 1/4) ≤ x ≤ 1/2 + √(c + 1/4) and y = x. It illustrates the emergence of "two-step escape windows" for values of c > 2.}
2									
3	c=	**2.2**			L=	**-2.065**	R=	**2.065**	
4									
5	x	x	F(x)	F(F(x))					
6	-2.07	-2.07	2.07	2.07					
7	-2.00	-2.00	1.79	0.99					
8	-1.93	-1.93	1.52	0.10					
9	-1.86	-1.86	1.25	-0.63					
10	-1.79	-1.79	1.00	-1.19					
11	-1.72	-1.72	0.76	-1.62					
12	-1.65	-1.65	0.53	-1.92					
13	-1.58	-1.58	0.31	-2.11					
14	-1.51	-1.51	0.09	-2.19					
15	-1.45	-1.45	-0.11	-2.19					
16	-1.38	-1.38	-0.30	-2.11					
17	-1.31	-1.31	-0.49	-1.96					
18	-1.24	-1.24	-0.66	-1.76					
19	-1.17	-1.17	-0.83	-1.51					
20	-1.10	-1.10	-0.99	-1.23					
21	-1.03	-1.03	-1.13	-0.91					
22	-0.96	-0.96	-1.27	-0.58					
23	-0.89	-0.89	-1.40	-0.24					
24	-0.83	-0.83	-1.52	0.10					
25	-0.76	-0.76	-1.63	0.45					
26	-0.69	-0.69	-1.73	0.78					
27	-0.62	-0.62	-1.82	1.10					

Figure 3.5.2 More Escape Windows in the Boundary of the Critical Square

Problem 3.5.8 Create a spreadsheet whose graph determines the values of x_1 for which iterates escape the critical square in *three* steps.

As you may have guessed by now, if we are willing to be patient in allowing the iterates of an initial x_1 to escape, the number of escape windows will grow without bound. To what extent do these windows fill the interval between the two repellers? This intriguing question is explored in problems at the end of this chapter and again in Chapter 4.

3.6. Restoring Order   Having explored the phenomena of cycles and chaos for iterators of the form

$$(3.6.1) \qquad\qquad F(x) = x^2 - c,$$

let us retrace some of the steps that led to this juncture. It was our study of antirefinements of the logistic difference equation

$$(3.6.2) \qquad\qquad x_{i+1} - x_i = a x_i - b x_i^2$$

with $b > 0$ that led us to consider iterators of the form

$$(3.6.3) \qquad\qquad G(x) = (1 + Ma) x - Mb x^2.$$

For sufficiently small values of $M > 0$, solutions of (3.6.3) approached the equilibrium value $x_0 = a/b$. However, for sufficiently large values of M, iteration by (3.6.3) began to produce the phenomena of cycles and chaos. It was with the aim of trying to understand such behavior that we replaced the parabola $y = (1 + Ma)x - Mbx^2$ with the simpler (one-parameter) parabola $y = x^2 - c$.

Having achieved such understanding, let us return to the notion that chaos, as produced by "sloppy bookkeeping," can be reversed by refinement. In this vein, can the process of refinement applied to (3.6.1) eliminate the cyclic and chaotic behavior encountered for $c > \frac{3}{4}$?

To answer this question in the affirmative, we need only write (3.6.1) as an equivalent difference equation

$$(3.6.4) \qquad\qquad x_{i+1} - x_i = x_i^2 - x_i - c$$

whose equilibrium values are the fixed points of F, i.e., $x = \frac{1}{2} \pm \sqrt{c + \frac{1}{4}}$. For refinements of the form $\Delta t \mapsto \Delta t / N$, (3.6.4) becomes

$$(3.6.5) \qquad\qquad x_{i+1} - x_i = \frac{x_i^2 - x_i - c}{N},$$

giving rise to the equivalent iterative process

$$(3.6.6) \qquad\qquad x_{i+1} = \left(1 - \frac{1}{N}\right) x_i + \frac{1}{N} x_i^2 - \frac{c}{N}.$$

By way of showing that "refinement creates order" for (3.6.1), we shall show that for sufficiently large values of N, the smaller fixed point of

(3.6.7)
$$H(x) = \left(1 - \frac{1}{N}\right)x + \frac{1}{N}x^2 - \frac{c}{N},$$

viz. $\frac{1}{2} - \sqrt{c+\frac{1}{4}}$, becomes an attractor. This will return us to the situation $c < \frac{3}{4}$ in (3.6.1).

A visual way of getting insight into the fixed points of (3.6.7) is to construct a spreadsheet (Figure 3.6.1) that graphs $y = x$ and $y = H(x)$ for variable choices of c and N.

Problem 3.6.1 Create a spreadsheet for the difference equation $x_{i+1} - x_i = x_i^2 - x_i - c$ that allows for refinement. Confirm that for each of the values

(a) $c = 0.8$,   (c) $c = 1.45$,

(b) $c = 1.35$,   (d) $c = 2.2$,

there is a refinement for which solutions approach the equilibrium value $\frac{1}{2} - \sqrt{c+\frac{1}{4}}$.

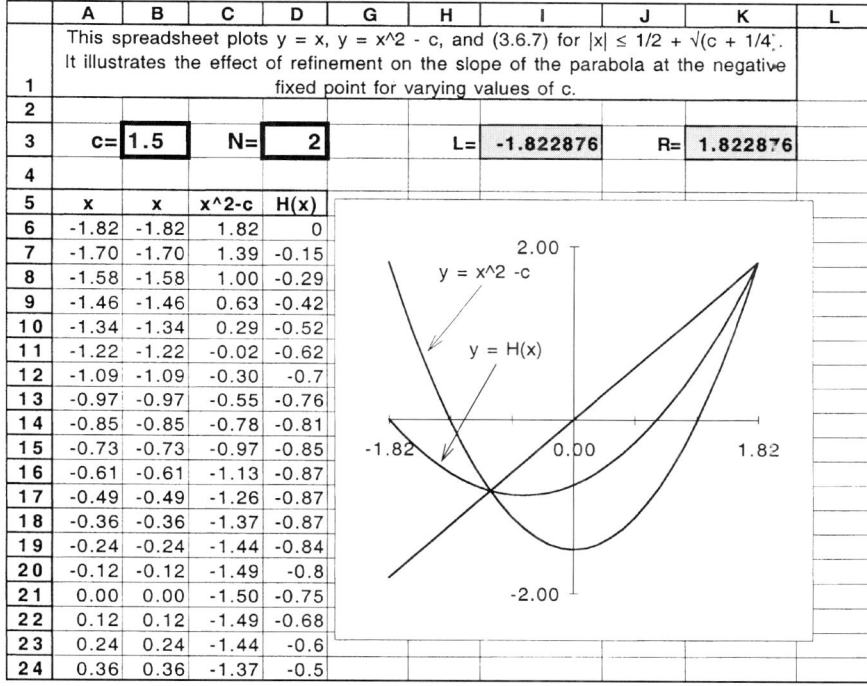

	A	B	C	D	G	H	I	J	K	L
	This spreadsheet plots y = x, y = x^2 - c, and (3.6.7) for |x| ≤ 1/2 + √(c + 1/4).									
	It illustrates the effect of refinement on the slope of the parabola at the negative									
1	fixed point for varying values of c.									
2										
3	c=	1.5		N=	2		L=	-1.822876		R=	1.822876
4										
5	x	x	x^2-c	H(x)						
6	-1.82	-1.82	1.82	0						
7	-1.70	-1.70	1.39	-0.15						
8	-1.58	-1.58	1.00	-0.29						
9	-1.46	-1.46	0.63	-0.42						
10	-1.34	-1.34	0.29	-0.52						
11	-1.22	-1.22	-0.02	-0.62						
12	-1.09	-1.09	-0.30	-0.7						
13	-0.97	-0.97	-0.55	-0.76						
14	-0.85	-0.85	-0.78	-0.81						
15	-0.73	-0.73	-0.97	-0.85						
16	-0.61	-0.61	-1.13	-0.87						
17	-0.49	-0.49	-1.26	-0.87						
18	-0.36	-0.36	-1.37	-0.87						
19	-0.24	-0.24	-1.44	-0.84						
20	-0.12	-0.12	-1.49	-0.8						
21	0.00	0.00	-1.50	-0.75						
22	0.12	0.12	-1.49	-0.68						
23	0.24	0.24	-1.44	-0.6						
24	0.36	0.36	-1.37	-0.5						

Figure 3.6.1 Refinement Corresponding to $F(x) = x^2 - c$

Problem 3.6.2 By way of formal confirmation of the results of Problem 3.6.1:

(a) Use Problem 1.10.2 to show that the latus rectum of the parabola

$$y = \left(1 - \frac{1}{N}\right)x + \frac{1}{N}x^2 - \frac{c}{N}$$

lies on the line $y = \frac{1}{2} - (c + \frac{1}{4})/N$.

(b) Recalling that the slope m of this parabola satisfies $|m| \leq 1$ on the arc below the latus rectum, show that the smaller fixed point of (3.6.7) is an attractor when $N > \sqrt{c+\frac{1}{4}}$.

(c) Can the fixed point at $\frac{1}{2} + \sqrt{\frac{1}{4}+c}$ ever be an attractor? Justify your answer.

As we can see from the above discussion, the iterative processes that give rise to cycles and chaos are closely related to difference equations. These difference equations are similar to those that arise in population dynamics and theoretical economics. Indeed, it is possible to return to applications considered in Chapter 2 and to discover the phenomena of cycles and chaos embedded within them. However, since one sometimes encounters suggestions to the effect that "the new science of chaos will revolutionize the fields of ecology and economics," it is important to keep the fundamental message of this chapter in mind: In many instances, cyclic and chaotic behavior are simply a form of overshoot resulting from "sloppy bookkeeping."

Having said this, there *are* instances in which difference equations give rise to cycles and chaos but where these phenomena *cannot* be eliminated by refinements of the form $\Delta t \mapsto \Delta t/N$. Some examples of this more pervasive form of chaos will be given in the context of *systems* of difference equations, as addressed in Chapter 6.

3.7. Problems, Exercises, and Projects
This chapter has focused on the nonlinear function $F(x) = x^2 - c$ as the source of phenomena associated with the term "chaos." Because the graph of $F(x)$ is a parabola, classical ideas from Greek geometry can help us analyze the iterative scheme $x_{i+1} = F(x_i)$.

The Slope of a Parabola (Without Calculus)

Problem 3.7.1 In Problem 1.11.22 we established the following construction for the tangent line to a parabola at a point P. Let Q be the foot of the perpendicular dropped from P to the parabola's directrix. Then the tangent line t is the perpendicular bisector of the line segment joining Q to the focus F of the parabola.

(a) Given that the slope of the line segment QF is m, what is the slope of the tangent line t?

(b) Given a parabola $y = Ax - Bx^2$, where $A, B > 0$, use Problem 1.10.2 to show that the focus F has coordinates $(A/(2B), (A^2 - 1)/(4B))$ and that the directrix $y = (A^2 + 1)/(4B)$ intersects the y-axis at $Q = (0, (A^2 + 1)/(4B))$.

(c) Deduce from the above that the slope of the tangent to the parabola $y = Ax - Bx^2$ at $P = (0, 0)$ is the same as that of the line $y = Ax$, i.e., that "the term $-Bx^2$ does not contribute to the slope at $x = 0$."

(d) Deduce from part (c) that the line tangent to the parabola $y = (1 + Ma)x - Mbx^2$ at $(0, 0)$ has slope $m = 1 + Ma$.

The Slope of a Parabola (with Calculus)

Problem 3.7.2 As in the problem above, consider the parabola $y = Ax - Bx^2$ with $A > 0$ and $B > 0$.

(a) Using the fact that the derivative of a function is the slope of its graph, confirm that the slope m of the parabola satisfies.

$$|m| > 1 \quad \text{if} \quad x < \frac{A-1}{2B} \quad \text{or} \quad x > \frac{A+1}{2B},$$

$$0 < m < 1 \quad \text{if} \quad \frac{A-1}{2B} < x < \frac{A}{2B},$$

$$-1 < m < 0 \quad \text{if} \quad \frac{A}{2B} < x < \frac{A+1}{2B}.$$

(b) Use part (a) to confirm (3.3.5) for the parabola with equation

$$y = (1 + Ma)x - Mbx^2,$$

where M, a, and b are positive.

Iterates Marching to Infinity

Problem 3.7.3 Let $F(x) = x^2 - c$ and $x_{i+1} = F(x_{i+1})$, $i = 1, 2, \ldots$. If $c > -\frac{1}{4}$ and $|x_i| > \frac{1}{2} + \sqrt{\frac{1}{4} + c}$, then we have seen in Problem 3.5.2 that $|x_{i+1}| > |x_i|$. But this inequality does not by itself imply what our spreadsheet experiments strongly indicate, namely that the x_i grow without bound. This problem shows that this stronger conclusion is indeed true.

(a) Assume $c > -\frac{1}{4}$ and $x_0 = \frac{1}{2} + \sqrt{\frac{1}{4} + c}$. Let $y = G(x)$ be the equation of the line tangent to the parabola $y = x^2 - c$ at (x_0, x_0). Use Problem 3.4.3 to show that this tangent line has slope $m > 1$, i.e., is of the form $y = ax + b$ where $a > 1$.

(b) Choosing $x_1 > x_0$ and $z_1 = x_1$, consider the iterates generated by F and G, respectively, i.e., consider the sequences

$$x_1, x_2, \ldots, \quad \text{where} \quad x_{i+1} = F(x_i) \quad \text{and} \quad x_1 > x_0,$$

$$z_1, z_2, \ldots, \quad \text{where} \quad z_{i+1} = G(z_i) \quad \text{and} \quad z_1 = x_1.$$

Use the fact that the parabola lies *above* the tangent line and part (a) to show that $x_i > z_i$ for $i = 2, 3, \ldots$. [Hint: If for some i, $x_i > z_i$, then $x_{i+1} = F(x_i) > G(x_i) > G(z_i) = z_{i+1}$.]

(c) Recalling the closed-form solution of $z_{i+1} = G(z_i)$ when $G(x) = ax + b$ (see Problem 1.11.11), show that the z_i grow without bound for increasing values of i. [Hint: Show that $z_n = a^n x_1 + b(\frac{a^n - 1}{a - 1}) = a^n(x_1 + \frac{b}{a-1}) - \frac{b}{a-1}$. Why is $x_1 + \frac{b}{a-1} > 0$?]

(d) Use parts (a)–(c) to show that the x_i grow without bound whenever $x_1 > x_0$. If $x_1 < -x_0$, show that $x_2 > x_0$ and conclude again that the x_i grow without bound.

Problem 3.7.4 Generalize the spreadsheet in Figure 3.4.5 by adding a column under E5 that calculates $F(F(F(F(x))))$. Create a line graph of the data under B5, C5, D5, and E5.

(a) Confirm that for $c < 1.1$ all fixed points of $F(F(F(F(x))))$ are repellers.

(b) Confirm that for at $c = 1.2$, $F(F(F(F(x))))$ has four fixed points that are attractors.

(c) As c increases from 1.1 to 1.2, determine the value at which these attractors first appear with an accuracy of ± 0.01.

Feigenbaum's "Descent into Chaos"

The diagram below provides a graphical representation of the "descent into chaos" that was observed in the spreadsheet of Figure 3.4.4. As c increases along the horizontal axis from $c = -0.25$ to $c = 2$, the diagram locates first the attracting fixed point, then the cyles of length 2, length 4, and so forth. To make these connections more explicit, the following problem asks the reader to use a spreadsheet to construct an approximation of the "Feigenbaum diagram" as given below.

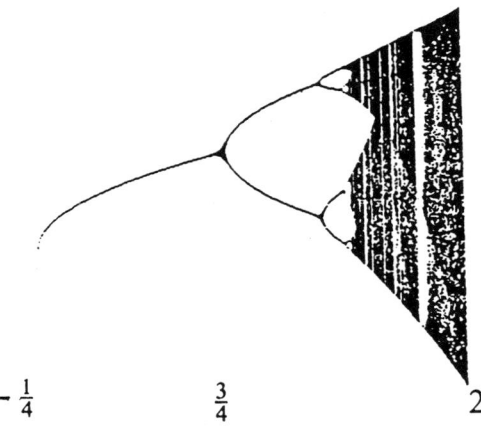

$$-\frac{1}{4} \qquad\qquad \frac{3}{4} \qquad\qquad 2$$

Problem 3.7.5 Create a spreadsheet with the values $c = -0.25, -0.23, -0.21,$ $\ldots 0.73, 0.75$ in cells A1, B1, C1, $\ldots$, AX1, AY1. Use this spreadsheet to compute 50 iterations of $x_1 = 0.5$ by $x_{i+1} = x_i^2 - c$ when $c = -0.25, c = -0.23, \ldots,$ $c = 0.75$.

(a) Explain why the values in cells A51, B51, C51, $\ldots$, AY51 are very close to $\frac{1}{2} - \sqrt{c - \frac{1}{4}}$, where c is the value in A1, B1, C1, $\ldots$, AY1.

(b) Extend the spreadsheet to the right so that the values $c = 0.77, 0.79, \ldots,$ 1.11 appear in cells AZ1, BA1, $\ldots$, BS1 and compute 50 iterations of $x_1 = 0.5$ by $x_{i+1} = x_i^2 - c$ when $c = -0.25, c = -0.23, \ldots, c = 1.09,$ $c = 1.11$. Create a line graph of the data in rows 50 and 51 and explain the "forked behavior" (bifurcation) at $c = 0.75$.

(c) Extend the spreadsheet to the right so that the values 0.77, 0.79, $\ldots,$ 1.11, $\ldots,$ 1.97, 1.99 appear in cells AZ1, BA1, $\ldots,$ BS1, $\ldots,$ DH1, DI1 and compute 100 iterations of $x_1 = 0.5$ by $x_{i+1} = x_i^2 - c$ when $c = -0.25, c = -0.23, \ldots, c = 1.09, c = 1.11, \ldots, c = 1.99$. Create a line graph of the data in rows 50, 51, $\ldots,$ 100 and reconcile the resulting pattern with the "descent into chaos" referred to at the end of Section 3.4.

Chaotic Behavior Associated with $F(x) = x^4 - c$

Until now we have focused on the iterator $F(x) = x^2 - c$ as the source of chaotic behavior. While the phenomena encountered in studying this iterator are typical for nonlinear functions $F(x)$, the analysis of the transition from escape to attraction, from attraction to cycles, from cycles to chaos, and from chaos to escape tends to be quite challenging for more general functions. The problems that follow pursue such an analysis in a somewhat tractable case, namely that of $F(x) = x^4 - c$.

Problem 3.7.6 Determining the critical square for the iterator $F(x) = x^4 - c$ requires finding the positive root of $x^4 - x - c = 0$.

(a) Construct a spreadsheet that for specified values of c gives a good approximation for the positive root of $x^4 - x - c = 0$ (e.g., recall Problem 1.9.3 and use the 30th iteration).

(b) Within the spreadsheet in part (a), construct the critical square together with the graph associated with $y = x^4 - c$ (as Figure 3.5.1 does for the parabola $y = x^2 - c$.)

(c) Use the above spreadsheet to check that the "vertex" of $y = x^4 - c$ (i.e., the point with coordinates $(0, -c)$) lies inside the critical square only if $c < 2^{1/3}$.

(d) Give an algebraic argument showing that the "vertex" of $y = x^4 - c$ belongs to the bottom edge of the critical square when $c = 2^{1/3}$. [Hint: If α is the positive root of $x^4 - x - c = 0$, then $y = -\alpha$ is the line determined by the base of the critical square.]

Ferrari's Method Revisited

Problem 3.7.7 In the problem above, we used a spreadsheet to approximate the positive solution of

(∗) $$x^4 - x - c = 0.$$

An alternative approach is to apply Ferrari's method for solving quartic equations, as was developed in Chapter 1.

(a) Show that if x is a solution of (∗), then for any t,

$$(x^2 + t)^2 = 2tx^2 + x + (t^2 + c).$$

(b) Show that if t satisfies the cubic $8t^3 + 8ct - 1 = 0$, then the equation in part (a) takes the form

$$(x^2 + t)^2 = 2t\left(x + \frac{1}{4t}\right)^2$$

and deduce that $x = \frac{\sqrt{2}}{2}(\sqrt{t} + \sqrt{\frac{1}{\sqrt{2t}} - t})$ is a root of (∗).

(c) Use Cardano's method to solve the cubic $8t^3 + 8ct - 1 = 0$ appearing in part (b). Confirm that in case $c = 2^{1/3}$, the solution of this cubic leads to the root $x = 2^{1/3}$ for (∗).

Problem 3.7.8 Referring to Problem 3.7.5, construct a bifurcation diagram corresponding to the iterator $F(x) = x^4 - c$. [Hint: Be sure to restrict c to values that keep the graph of the iterator within the critical square.]

Finding Bifurcation Values by Long Division

We have seen that the iterator $F(x) = x^2 - c$ has no fixed points when $c < -\frac{1}{4}$, since in that case the parabola $y = x^2 - c$ lies above the line $y = x$. But as c increases, the parabola descends, and at $c = -\frac{1}{4}$ the parabola becomes tangent to $y = x$. As c increases from $c < -\frac{1}{4}$ to $c > -\frac{1}{4}$, the corresponding iterator makes a transition from "no fixed points" to "two fixed points" (one of which is an attracting fixed point). The following problems show that the usual division algorithm for polynomials can be used to determine such "bifurcation values" for iterators.

Problem 3.7.9 Suppose the parabola $y = x^2 - c$ is tangent to $y = x$ for some $x = a$. Observing that $(x^2 - c) - x > 0$ when $x < a$ and also when $x > a$, while $(x^2 - c) - x = 0$ when $x = a$, we conclude that $x = a$ must be a multiple root of the polynomial $x^2 - x - c$. This means that $x^2 - x - c$ is *divisible* by $(x - a)^2 = x^2 - 2ax + a^2$.

(a) Solve the long-division problem $(x^2 - x - c) \div (x^2 - 2ax + a^2)$.
(b) Show that the remainder in part (a) is $(2a - 1)x - (a^2 + c)$.

(c) Show that $x^2 - x - c$ is divisible by $(x - a)^2$ if and only if $2a - 1 = 0$ and $a^2 + c = 0$.

(d) Use part (c) to determine the "first bifurcation value" for the iterator $F(x) = x^2 - c$ and the value of x at which this bifurcation occurs.

(e) In Section 3.3 we saw that as c passes through the value $\frac{3}{4}$, the iterator $F(x) = x^2 - c$ gives rise to cycles of length two. As illustrated in Figure 3.4.6, this occurs when the line $y = x$ becomes tangent to the graph of $y = F(F(x))$. After confirming the identity $F(F(x)) - x = x^4 - 2cx^2 - x + c^2 - c$, solve the long-division problem $(x^4 - 2cx^2 - x + c^2 - c) \div (x^2 - 2ax + a^2)$ to conclude that:

  (i) bifurcation occurs at $c = \frac{3}{4}$.

  (ii) the graph of $x^4 - \frac{3}{2}x^2 - \frac{3}{16}$ is tangent to $y = x$ at $x = -\frac{1}{2}$.

Problem 3.7.10 In this problem we extend the reasoning from Problem 3.7.9 to $F(x) = x^4 - c$.

(a) For the iterator $F(x) = x^4 - c$, use long division of polynomials to show that the transition from "no fixed points" to "two fixed points" occurs for $c = -\frac{3}{4^{4/3}}$ at $x = \frac{1}{4^{1/3}}$. [Hint: Divide $x^4 - x - c$ by $x^2 - 2ax + a^2$.]

*(b) Use calculus methods to give an independent verification that $y = x^4 + \frac{3}{4^{4/3}}$ is tangent to $y = x$ at $x = \frac{1}{4^{1/3}}$.

(c) For the iterator $F(x) = x^n - c$, show that the transition from "no fixed points" to "two fixed points" occurs for $c = -(n - 1)/n^{n/(n-1)}$.

(d) The ambitious reader may now extend the method of Problem 3.7.9(e) to determine a pair of equations whose solution yields values of c and and x at which the iterator $F(x) = x^4 - c$ first encounters cycles of length 2.

The Size of the Escape Window

Problem 3.7.11 Let $F(x) = x^2 - c$, with $c \geq 2$.

(a) Find a value of c such that the length of the escape window in Figure 3.5.1 is exactly $\frac{1}{3}$ the base of the critical square.

(b) Show that if r is the ratio of the length of the escape window to the base of the critical square, then $r^2 = \frac{c + 1 - \sqrt{4c+1}}{c}$ and $c = \frac{2(1+r^2)}{(1-r^2)^2}$.

How to Escape in Several Steps

Problem 3.7.12 This problem deals with escape windows for the iterator $F(x) = x^2 - 6$.

(a) Show that the critical square for $F(x) = x^2 - 6$ has its vertices at the points $(\pm 3, \pm 3)$ and that if $-\sqrt{3} < x < \sqrt{3}$, then $F(x)$ lies below the base of the critical square.

(b) In the context of a graph of $F(x)$ for $-3 < x < 3$, use a staircase diagram to describe the "one-step escape" of $x = -1$ under the iterative scheme $x_{i+1} = F(x_i)$.

(c) In order that x escape the critical square in *two* steps, $F(x)$ must belong to the *one-step* escape window, so we must have

$$-\sqrt{3} < F(x) < \sqrt{3}.$$

Show that this condition on x is equivalent to

$$-\sqrt{6 + \sqrt{3}} < x < -\sqrt{6 - \sqrt{3}} \quad \text{or} \quad \sqrt{6 - \sqrt{3}} < x < \sqrt{6 + \sqrt{3}}.$$

[Hint: First show that if $0 < a < b$, then the inequality $a < x^2 < b$ is equivalent to $-\sqrt{b} < x < -\sqrt{a}$ or $\sqrt{a} < x < \sqrt{b}$.]

(d) In the context of graphs of $F(x)$ and $F(F(x))$ for $-3 < x < 3$, use a staircase diagram to describe the "two-step escape" of $x = 2.5$ under the iterative scheme $x_{i+1} = F(x_i)$.

(e) Show that the "three-step escape window" for $F(x) = x^2 - 6$ consists of the *four* intervals

$$\left(-\sqrt{6 + \sqrt{6 + \sqrt{3}}}, \ -\sqrt{6 + \sqrt{6 - \sqrt{3}}} \right),$$

$$\left(-\sqrt{6 - \sqrt{6 - \sqrt{3}}}, \ -\sqrt{6 - \sqrt{6 + \sqrt{3}}} \right),$$

$$\left(\sqrt{6 - \sqrt{6 + \sqrt{3}}}, \ \sqrt{6 - \sqrt{6 - \sqrt{3}}} \right),$$

$$\left(\sqrt{6 + \sqrt{6 - \sqrt{3}}}, \ \sqrt{6 + \sqrt{6 + \sqrt{3}}} \right).$$

[Hint: Use the result from part (c), replacing the condition on x by the same condition on $F(x) = x^2 - 6$.]

(f) Use a spreadsheet such as that of Problem 3.5.8 to show that the graph of $y = F(F(F(x)))$ breaches the base of the critical square in four intervals, corresponding to those in part (e).

(g) Show that the "four-step escape window" for $F(x) = x^2 - 6$ consists of *eight* intervals whose endpoints have the form $\pm\sqrt{6 \pm \sqrt{6 \pm \sqrt{6 \pm 3}}}$.

Problem 3.7.13 The ambitious reader is now invited to explore the pattern arising in Problem 3.7.12.

(a) Considering the iterator $F(x) = x^2 - c$, where $-\frac{1}{4} < c < 2$, and $k = \frac{1}{2} + \sqrt{c + \frac{1}{4}}$, show that the "$n$-step escape window" consists of 2^{n-1}

disjoint intervals whose 2^n endpoints are of the form

$$\pm\sqrt{c \pm \sqrt{c \pm \sqrt{c \pm \cdots \pm \sqrt{c - k}}}}.$$

[Remark: This problem will require a rule for pairing off the above 2^n endpoints to yield the 2^{n-1} intervals comprising the escape window.]

(b) Show that the iteration of any of the 2^n *endpoints* in part (a) eventually ends up at the fixed point k.

Problem 3.7.14 Continuing in the vein of the preceding problem, we now consider the case $c = 2$, where the graph of the iterator $F(x) = x^2 - 2$ is *tangent* to the base of the critical square.

(a) Show that the graph of $y = F(F(x))$ touches the base of the critical square at the two points $(\pm\sqrt{2}, -2)$.
(b) Show that the graph of $y = F(F(F(x)))$ touches the base of the critical square at the four points $(\pm\sqrt{2 \pm \sqrt{2}}, -2)$.
(c) Denoting $F(x)$ by $F^{[1]}(x)$, $F(F(x))$ by $F^{[2]}(x)$, $F(F(F(x)))$ by $F^{[3]}(x)$, and so forth, show that the graph of $F^{[n]}(x)$ touches the base of the critical square at 2^{n-1} points of the form

$$\left(\pm\sqrt{2 \pm \sqrt{2 \pm \cdots \pm \sqrt{2}}}, -2\right),$$

where we have $n - 1$ "nested radical signs."

(d) Investigate the distribution of points of the type $\pm\sqrt{2 \pm \sqrt{2 \pm \cdots \pm \sqrt{2}}}$ in the interval $[-2, 2]$. For example, by alternating between $+$ and $-$ signs, we obtain the sequence

$$\sqrt{2}, \sqrt{2 - \sqrt{2}}, \sqrt{2 - \sqrt{2 + \sqrt{2}}}, \sqrt{2 - \sqrt{2 + \sqrt{2 - \sqrt{2}}}},$$

$$\sqrt{2 - \sqrt{2 + \sqrt{2 - \sqrt{2 + \sqrt{2}}}}}, \ldots,$$

which appears to approach a solution of $x^2 + x - 1 = 0$. Establish similar associations for

$$\sqrt{2 + \sqrt{2}}, \sqrt{2 + \sqrt{2 - \sqrt{2}}}, \sqrt{2 + \sqrt{2 - \sqrt{2 + \sqrt{2}}}},$$

$$\sqrt{2 + \sqrt{2 - \sqrt{2 + \sqrt{2 - \sqrt{2}}}}}, \ldots$$

and for

$$\sqrt{2}, \sqrt{2+\sqrt{2}}, \sqrt{2+\sqrt{2+\sqrt{2}}}, \sqrt{2+\sqrt{2+\sqrt{2+\sqrt{2}}}},$$

$$\sqrt{2+\sqrt{2+\sqrt{2+\sqrt{2+\sqrt{2}}}}}, \dots.$$

(e) Show that iteration of any point of the type $\pm\sqrt{2\pm\sqrt{2\pm\cdots\pm\sqrt{2}}}$ eventually ends up at the fixed point $k = 2$.

(f) Consider now the indefinite continuation of the first sequence considered in part (d). That is, let x_1 be given by the "infinite nested radicals"

$$x_1 = \sqrt{2-\sqrt{2+\sqrt{2-\sqrt{2+\sqrt{2}-\cdots}}}}$$

where the $\pm$ signs alternate indefinitely. Explain why you would expect that $F(F(x_1)) = x_1$ and why the iterates of x_1 should fall into a cycle of length 2.

(g) Why would you expect numbers represented by $\pm\sqrt{2\pm\sqrt{2\pm\sqrt{2\pm\cdots}}}$ and having a *periodic repeating* pattern of $\pm$ signs to lead to cycles under iteration by F?

(h) What would you conjecture concerning the iteration of numbers represented by $\pm\sqrt{2\pm\sqrt{2\pm\sqrt{2\pm\cdots}}}$ but without a repeating pattern of $\pm$ signs? [For example, the sequence $+ - + + - + + + - + + + + - + + + + + - \cdots$ has no repeating pattern].

4
COMPLEX NUMBERS, POLYNOMIAL EQUATIONS, AND FRACTALS

4.1. Introduction Until now, our iterative methods have dealt exclusively with *real* numbers. In this context, efforts to solve the quadratic equation

$$(4.1.1) \qquad x^2 + 4x + 5 = 0$$

by means of the iterator

$$(4.1.2) \qquad F(x) = \frac{x^2 - 5}{2x + 4}$$

would not be successful. Indeed, applying the "quadratic equation solver" of Figure 1.7.3 when $a = 1$, $b = 4$, and $c = 5$ yields a sequence of iterates whose behavior is more suggestive of chaos than of an orderly approach to a root of $x^2 + 4x + 5$.

The fact that some quadratic equations have no real solutions was known in medieval times. Moreover, Cardano (1501–1576) observed that his now famous formula for solving cubic equations requires the introduction of a new class of "numbers," even when all three roots of the underlying cubic are real (see Problems 4.6.18 and 4.6.19). In fact, such a system of "complex numbers" is needed to provide solutions for even the simplest of quadratic equations, such as

(4.1.3) $x^2 + 1 = 0.$

That is, no real number x can satisfy (4.1.3) because $x^2 + 1 \geq 1$ for any real x.

Problem 4.1.1 Without using the quadratic formula, show that there is no real number x such that $x^2 + 4x + 5 = 0$. [Hint: $x^2 + 4x + 5 = (x^2 + 4x + 4) + 1$.]

Problem 4.1.2 For $a \neq 0$, verify the identity $ax^2 + bx + c = a(x + \frac{b}{2a})^2 - \frac{b^2 - 4ac}{4a}$. Use this identity to show that the quadratic equation $ax^2 + bx + c = 0$ has no real solutions if $b^2 - 4ac < 0$.

As the reader is probably aware, this lack of solutions leads us to introduce a new number i having the property that $i^2 = -1$. With this definition, $i = \sqrt{-1}$ provides a "solution" to (4.1.3). Similarly, applying the quadratic formula to (4.1.1) leads to

$$x = \frac{-4 \pm \sqrt{16 - 20}}{2} = \frac{-4 \pm \sqrt{-4}}{2}.$$

Here we have been taught to write

$$\sqrt{-4} = \sqrt{4(-1)} = \sqrt{4} \cdot \sqrt{-1} = 2i$$

in order to arrive at the solutions $-2 \pm i$.

However, thoughtless application of such "rules for calculation" can lead to unpleasant paradoxes. For example, by assuming that $\sqrt{a} \cdot \sqrt{b} = \sqrt{ab}$ holds for all real numbers a and b, we may be led to conclude that

$$-1 = \sqrt{-1} \cdot \sqrt{-1} = \sqrt{(-1)(-1)} = \sqrt{1} = 1.$$

Even the great Leonard Euler (1707–1783) fell into such a trap before the complex number system was clearly understood (see Problem 4.2.14 for further discussion of this "proof" that $-1 = 1$).

Chastened by such an example, we would do well to confirm that $x = -2 + i$ is in fact a solution of $x^2 + 4x + 5 = 0$. Such a confirmation is based on

$$\begin{aligned}
x^2 + 4x + 5 &= (-2 + i)^2 + 4(-2 + i) + 5 \\
&= (-2)^2 + 2(-2)(i) + i^2 + 4(-2) + 4(i) + 5 \\
&= 4 - 4i - 1 - 8 + 4i + 5 = 0,
\end{aligned}$$

from which $-2+i$ does emerge as a solution of our equation. In these calculations we have applied *only* the commutative, associative, and distributive (CAD) laws of arithmetic,[1] supplemented by the rule $i^2 = -1$.

Problem 4.1.3 Confirm in the above manner that $x = -2 - i$ is also a solution of $x^2 + 4x + 5 = 0$.

Raphael Bombelli (1526–1572) was probably the first mathematician to use formally correct operations with complex numbers to determine the roots of polynomials. He seems to have had an intuitive understanding of the fact that algebra could be enlarged to a system of numbers of the form $a + bi$, where, in accordance with the "CAD rules" below, the real numbers a and b interact with a number "i" satisfying $i^2 = -1$.

Commutative Laws

$$a + b = b + a \quad \text{and} \quad a \times b = b \times a$$

Associative Laws

$$(a + b) + c = a + (b + c) \quad \text{and} \quad (a \times b) \times c = a \times (b \times c)$$

Distributive Laws

$$a \times (b + c) = a \times b + a \times c \quad \text{and} \quad (a + b) \times c = a \times c + b \times c$$

Problem 4.1.4 Bombelli came to the seemingly peculiar conclusion that

$$(2 + 11i)^{1/3} + (2 - 11i)^{1/3} = 4.$$

Show how he might have done this by first using the CAD laws to calculate $(2 + i)^3$ and $(2 - i)^3$.

In the context of Bombelli's arithmetic for the complex numbers, the quadratic formula

(4.1.4)
$$x = \frac{-b \pm \sqrt{b^2 - 4ac}}{2a}$$

assures us that every quadratic equation $ax^2 + bx + c = 0$ has a solution (indeed two solutions if $b^2 - 4ac \neq 0$). Computation aside, the importance of Cardano's and Ferrari's methods (see Chapter 1) is that they extend this conclusion to equations of third and fourth degree. Even in an age where "iteration is cheap and easy," a theoretical proof of the *existence* of solutions can be a crucially important mathematical tool.

---

[1] It is these three "CAD" rules that will guide us in our subsequent development of the system of complex numbers.

Given a paradox-free definition of $\sqrt{z}$ for any complex number z, (4.1.4) even asserts the existence of solutions to $ax^2 + bx + c = 0$ when a, b, and c are themselves allowed to be complex. Later in the history of mathematics this led mathematicians to ask whether *every* polynomial equation with real *or* complex coefficients has at least one solution among the complex numbers. The resolution of this question, in the form of the *fundamental theorem of algebra*, represents an important milestone in the history of mathematics. This "FTA" asserts that

every nonconstant polynomial with complex coefficients has at least one complex root.[1]

Problem 4.1.5 (a) Verify by direct substitution that $1 + 2i$ and $-4 - 2i$ are solutions of the quadratic equation $x^2 + 3x - 10i = 0$. (b) Derive this solution from (4.1.4).

To extend the iterative techniques from Chapter 1 to include complex numbers (e.g., to finding the solutions of $x^2 + 3x - 10i = 0$ by iteration), we will need to be adept in Bombelli's "complex arithmetic" while also relying on a paradox-free definition of concepts such as "the square root of z." Since application of the CAD rules to complex numbers is a familiar topic, we relegate this aspect of complex arithmetic to the problems at the end of this chapter. However, the somewhat more delicate issues surrounding concepts such as $\sqrt{z}$ and $z^{1/3}$ are addressed in Section 4.2. Here we use a geometric representation of complex numbers as points in the (x, y)-plane as the basis for extending the laws of exponents to numbers of the form $z = a + ib$.

With such matters firmly under control, we will be able to build on the Babylonian iterators developed in Chapter 1 to solve a host of interesting problems. In particular, the solution of polynomial equations by iterations of the form

$$(4.1.5) \qquad\qquad z_{n+1} = F(z_n)$$

(where $F(z_n)$ is itself complex-valued) will lead us to the important concept of a "basin of attraction" for such iterators.

The highlight of this excursion appears in Section 4.4, where we extend the use of iteration to include the complex solutions of cubic equations. Here we will find that basins of attraction can have boundaries with an extraordinary structure. In particular, efforts to solve

$$z^3 - 1 = 0$$

in the form (4.1.5) will lead to an encounter with a *Julia set*, one of the remarkable denizens of the teeming world of "fractals." The chapter ends with a brief discussion of fractals in Section 4.5.

---

[1] Once the fundamental theorem of algebra is proved, it is relatively easy to show that every polynomial of degree n has exactly n complex roots (some of which may be "multiple roots," which are counted with their multiplicity).

4.2. Does $\sqrt{-1}$ Exist?

To set the stage for our discussion of the complex numbers, we recall that for $k > 0$, the "Babylonian iterator"

$$(4.2.1) \qquad F(x) = \frac{x + k/x}{2}$$

enabled us to approximate $\sqrt{k}$ in spreadsheet format. As observed in Chapter 1, this iterator serves to approximate both $\sqrt{k}$ and $-\sqrt{k}$ as the solutions to the equation $x^2 - k = 0$ (see Figure 1.5.4).

Problem 4.2.1 Recreate the spreadsheet "square root machine" of Figure 1.5.4. Observe that if $x_1 > 0$, then the iterates x_1, x_2, x_3, ... approach $\sqrt{k}$, whereas if $x_1 < 0$, they approach $-\sqrt{k}$.

Problem 4.2.2 For $k = 2$, graph the iterator (4.2.1) for $-4 \le x \le 4$. Use a staircase diagram to explain why iterations based on $x_1 > 0$ approach $\sqrt{2}$, whereas iterations based on $x_1 < 0$ approach $-\sqrt{2}$.

Had Cardano, Bombelli, and their compatriots had access to such square root machines, they might have wondered if meaning could be attached to $\sqrt{-1}$ by choosing $k = -1$ in (4.2.1). This would mean applying the iterative process

$$(4.2.2) \qquad x_{n+1} = \frac{x_n - 1/x_n}{2}$$

to x_1 with the aim of approximating a solution of $x^2 + 1 = 0$.

Problem 4.2.3 Using the spreadsheet from Problem 4.2.1, describe the behavior of iterates x_1, x_2, x_3, ... produced by (4.2.2) when $x_1 > 0$. Repeat when $x_1 < 0$.

In light of Problem 4.2.3, and recalling the iterative processes considered in Chapter 3, it may be tempting to associate the iterator (4.2.2) with "chaos" In this context, it would be natural to try refinement as a means of obtaining a more orderly sequence of iterates. This would call for writing (4.2.2) as a difference equation

$$x_{n+1} - x_n = \frac{-x_n - 1/x_n}{2}$$

and then dividing the right side by N. This leads to the iterative process

$$(4.2.4) \qquad x_{n+1} = F(x_n) = \frac{(2N - 1)x_n - 1/x_n}{2N}.$$

Problem 4.2.4 Modify the spreadsheet of Problem 4.2.3 to allow for refinement. Does refinement lead to a more orderly sequence of iterates?

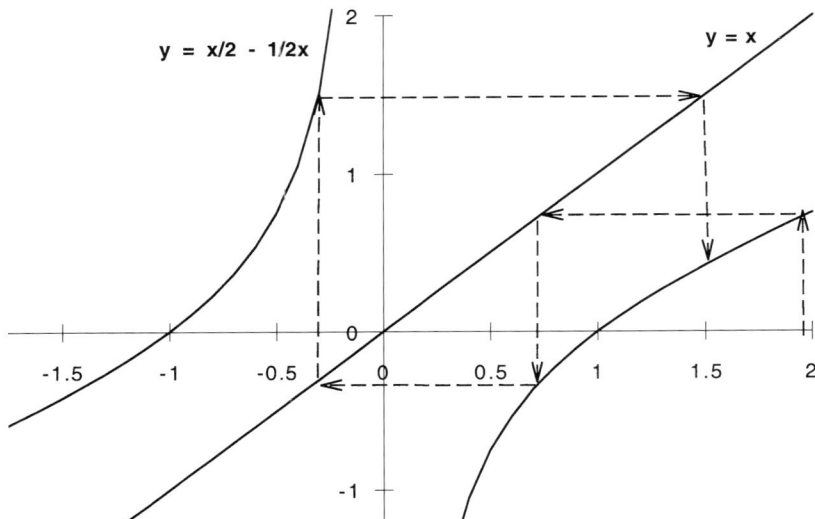

Figure 4.2.1 A Staircase Diagram for $F(x) = x/2 - 1/(2x)$

The failure of refinement in this case is due to the fact that regardless of the size of N, the iterator of (4.2.4) fails to have a (real) fixed point. As indicated by Figure 4.2.1, there is no "repeller" to be transformed into an "attractor," and this explains why the erratic behavior encounted in Problem 4.2.3 is not chaos in the sense of Chapter 3.

Problem 4.6.25 at the end of this chapter shows that there is a simple but subtle pattern underlying the behavior of the iterates produced by this staircase diagram. However, this pattern will not spell success for this effort to determine $\sqrt{-1}$ by iteration.

A more promising approach to solving $x^2 + 1 = 0$ by iteration can be based on a representation of complex numbers in the (x, y)-plane, one in which $z = a + ib$ corresponds to the point $(x, y) = (a, b)$. As indicated by problems at the end of this chapter, the arithmetic operations governed by the CAD rules allow for interesting geometric interpretations in this context. Here we can also resolve the paradoxes associated with a definition of $\sqrt{z}$ by introducing a *polar representation* for complex numbers.

In Figure 4.2.2, $r = \sqrt{a^2 + b^2}$ and $\tan\theta = b/a$. The right triangle shows that

$$\text{Re}(z) = \text{the real part of } z = a = r\cos\theta$$

and

$$\text{Im}(z) = \text{the imaginary part of } z = b = r\sin\theta.$$

Here $r = \sqrt{a^2 + b^2}$ is called the *modulus* or *absolute value* of z and denoted by

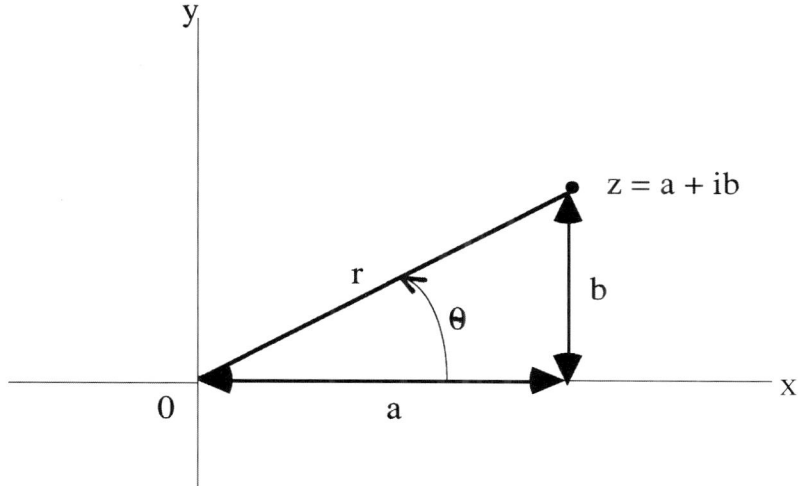

Figure 4.2.2 The Polar Representation of $z = a + ib$

$|z|$. The angle $\theta = \arctan b/a$ is called the *argument* of z and denoted by $\text{Arg}(z)$. For $z = a + ib$, we define the *complex conjugate* as $\bar{z} = a - ib$ and observe that $z\bar{z} = a^2 + b^2 = |z|^2$. These definitions underlie the polar representation

$$(4.2.3) \qquad z = r \cos\theta + ir \sin\theta = r(\cos\theta + i \sin\theta),$$

in which $r = |z|$ and $\theta = \text{Arg}(z)$.

Having decided to give the CAD rules absolute sway in our arithmetic of complex numbers, we now apply CAD to complex numbers in polar form. Writing

$$z_1 = r_1 \cos\theta_1 + ir_1 \sin\theta_1 \quad \text{and} \quad z_2 = r_2 \cos\theta_2 + ir_2 \sin\theta_2$$

we begin with multiplication.

Problem 4.2.5 Recalling the trigonometric identities

$$\sin(A + B) = \sin A \cos B + \cos A \sin B,$$
$$\cos(A + B) = \cos A \cos B - \sin A \sin B,$$

show that the polar form of $z_1 \cdot z_2$ is given by

$$z_1 \cdot z_2 = r_1 \cdot r_2[\cos(\theta_1 + \theta_2) + i \sin(\theta_1 + \theta_2)].$$

Problem 4.2.6 Given z in polar form (4.2.3), show that

$$z^2 = r^2(\cos 2\theta + i \sin 2\theta).$$

In other words, the multiplication of two complex numbers z_1 and z_2 in polar form corresponds to *multiplying* their absolute values while *adding* their arguments.

Problem 4.2.7 Sketch polar representations for $z_1 = 1 + 2i$, $z_2 = 3 + i$. Use this sketch to indicate the polar form of multiplication for z_1 and z_2. Does your sketch correspond to the fact that $z_1 z_2 = 1 + 7i$?

Conversely, the division of two complex numbers z_1 and z_2 in polar form corresponds to *dividing* their absolute values while *subtracting* their arguments.

Problem 4.2.8 Recalling that $z_1/z_2 = z$ if and only if $z_1 = z z_2$, confirm that (for $z_2 \neq 0$),

$$\frac{z_1}{z_2} = \frac{r_1}{r_2}[\cos(\theta_1 - \theta_2) + i \sin(\theta_1 - \theta_2)].$$

Problem 4.2.9 Assuming that $\sqrt{a}$ is characterized by the property that $\sqrt{a} \cdot \sqrt{a} = a$, show that $\sqrt{r}(\cos(\theta/2) + i \sin(\theta/2))$ is a square root of $z = r(\cos\theta + i \sin\theta)$.

Problem 4.2.10 Confirm that for the special case where z is a real number (i.e., $z = a + ib$ with $b = 0$), Problems 4.2.5, 4.2.8, and 4.2.9 reduce to the usual rules for multiplication, division, and square roots among the real numbers.

Problem 4.2.11 Given that $z^3 = z z^2$, use Problems 4.2.5 and 4.2.6 to show that

$$z^3 = r^3(\cos 3\theta + i \sin 3\theta)$$

and, more generally, that

$$z^n = r^n(\cos n\theta + i \sin n\theta), \qquad n = 1, 2, 3, \ldots.$$

On this basis derive De Moivre's theorem:

$$(\cos\theta + i \sin\theta)^n = \cos n\theta + i \sin n\theta, \qquad n = 1, 2, 3, \ldots.$$

Problem 4.2.9 provides the basis for defining the square root of a complex number. Writing $z = a + ib = r(\cos\theta + i \sin\theta)$ in polar form, we now define

$$(4.2.4) \qquad z^{1/2} = \sqrt{r}\left(\cos\frac{\theta}{2} + i \sin\frac{\theta}{2}\right),$$

where $\sqrt{r}$ represents the positive square root of $a^2 + b^2$. More generally, we define

$$z^{1/n} = \sqrt[n]{r}\left(\cos\frac{\theta}{n} + i \sin\frac{\theta}{n}\right).$$

Finally, there remains the question of the meaning to be assigned to $\sqrt{k}$ when k is not positive. Recall that for $k > 0$, $\sqrt{k}$ denotes the *positive* square root of k. To extend such specificity to the complex case, we note that z can be written in polar form as

$$r(\cos\theta + i \sin\theta)$$

or as

$$r(\cos(2n\pi + \theta) + i \sin(2n\pi + \theta))$$

for any integer n. By first specifying that $0 \le \theta < 2\pi$ and *then* defining

$$\sqrt{z} = \sqrt{r}\left(\cos\frac{\theta}{2} + i \sin\frac{\theta}{2}\right)$$

we obtain the so-called *principal value* of the square root of z. It is this principal value with which the symbol "$\sqrt{}$" is associated in the complex domain. Now, as in the case $k > 0$, we obtain *two* square roots for any $z \neq 0$. They are $\sqrt{z}$ and $-\sqrt{z}$.

Problem 4.2.12 Confirm that for any z, $0 \le \mathrm{Arg}(\sqrt{z}) < \pi$.

Problem 4.2.13 Confirm that for $k > 0$, the principal value of the square root of k is positive.

Problem 4.2.14 Show that if $0 \le \mathrm{Arg}\, z_1 < \pi$ and $0 \le \mathrm{Arg}\, z_2 < \pi$, then

$$\sqrt{z_1 \cdot z_2} = \sqrt{z_1} \cdot \sqrt{z_1},$$

where $\sqrt{z}$ represents the principal value of the square root of z. What goes wrong if we try to extend this to $z_1 = z_2 = -1$?

4.3. Quadratic Equations Revisited

Having reestablished a basis for confidence in our complex arithmetic, let us return to the problem of solving $z^2 + 1 = 0$ by iteration (here we have anticipated a complex solution by replacing the variable x by z). Our digression into the representation of complex numbers by points in the (x, y)-plane will serve us well in this regard. For now the "imaginary number" i has a concrete representation as the point with coordinates $(0, 1)$ in the (x, y)-plane, while $-i$ is similarly represented by $(0, -1)$. In this context, the problem of solving $z^2 + 1 = 0$ by iteration becomes one of finding a sequence of points

$$(x_1, y_1), \quad (x_2, y_2), \quad (x_3, y_3), \ldots$$

that approaches the point $(0, 1)$ or $(0, -1)$ in the (x, y)-plane.

While applying the iterative scheme

$$x_{n+1} = \frac{1}{2}\left(x_n - \frac{1}{x_n}\right)$$

to *real* numbers x_n produced disappointing results, let us now consider the iterator

(4.3.1) $$F(z) = \frac{1}{2}\left(z - \frac{1}{z}\right)$$

in the context of *complex* numbers.

The key to success in this area is to consider the real and imaginary parts of (4.3.1). Recalling that $\bar{z} = x - iy$ is the complex conjugate of $z = x + iy$, we have

$$F(z) = \frac{z}{2} + \frac{1}{2z} = \frac{z}{2} - \frac{\bar{z}}{2z\bar{z}} = \frac{x+iy}{2} - \frac{x-iy}{2(x^2+y^2)}.$$

Separating the last expression into its real and imaginary parts, we obtain

(4.3.2) $$F(z) = \frac{1}{2}\left(x - \frac{x}{x^2+y^2}\right) + \frac{i}{2}\left(y + \frac{y}{x^2+y^2}\right).$$

Now writing $F(z) = u(x, y) + iv(x, y)$ and equating real and imaginary parts, we obtain

(4.3.3) $$u(x, y) = \frac{1}{2}\left(x - \frac{x}{x^2+y^2}\right) \quad \text{and} \quad v(x, y) = \frac{1}{2}\left(y + \frac{y}{x^2+y^2}\right).$$

From (4.3.3) it follows that the single complex iterative scheme

$$z_{n+1} = F(z_n)$$

corresponds to *two real* iterative schemes, namely

(4.3.4)
$$x_{n+1} = \frac{1}{2}\left(x_n - \frac{x_n}{x_n^2 + y_n^2}\right),$$

$$y_{n+1} = \frac{1}{2}\left(y_n + \frac{y_n}{x_n^2 + y_n^2}\right).$$

These iteration formulas generate two real sequences $x_1, x_2, x_3, \ldots$ and $y_1, y_2, y_3, \ldots$, from which we can reconstruct $z_1, z_2, z_3, \ldots$ via $z_n = x_n + iy_n$, $n = 1, 2, \ldots$.

We can now implement (4.3.4) with a spreadsheet (Figure 4.3.1), keeping in mind that these two iterators are interdependent. That is, the value of x_{n+1} depends not only on x_n but also on y_n. Similarly, y_{n+1} depends on both x_n and y_n.

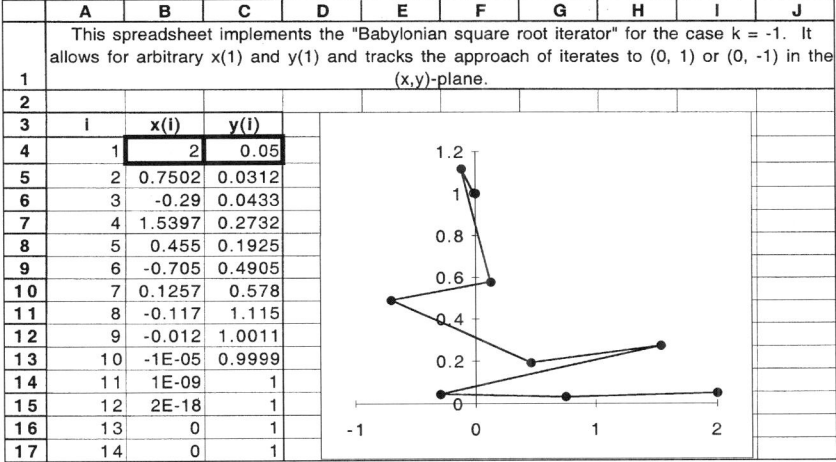

	A	B	C	D	E	F	G	H	I	J
1	This spreadsheet implements the "Babylonian square root iterator" for the case k = -1. It allows for arbitrary x(1) and y(1) and tracks the approach of iterates to (0, 1) or (0, -1) in the (x,y)-plane.									
2										
3	i	x(i)	y(i)							
4	1	2	0.05							
5	2	0.7502	0.0312							
6	3	-0.29	0.0433							
7	4	1.5397	0.2732							
8	5	0.455	0.1925							
9	6	-0.705	0.4905							
10	7	0.1257	0.578							
11	8	-0.117	1.115							
12	9	-0.012	1.0011							
13	10	-1E-05	0.9999							
14	11	1E-09	1							
15	12	2E-18	1							
16	13	0	1							
17	14	0	1							

Figure 4.3.1 The Babylonian Iterator for $k = -1$ in the Complex Plane

Problem 4.3.1 In the format of Figure 4.3.1, construct a spreadsheet that implements (4.3.4) for arbitrary values of x_1 and y_1. Represent the values of x_n and y_n so obtained on a "scatter graph." What determines whether the (x_n, y_n) approach $(0, 1)$ or $(0, -1)$?

Noting that i and $-i$ are the solutions of

(4.3.5) $$z^2 - k = 0$$

when $k = -1$, it is of interest to develop iterative schemes for more general case of

(4.3.6) $$F(z) = \frac{1}{2}\left(z + \frac{k}{z}\right).$$

In the case $k > 0$, (4.3.6) corresponds to the Babylonian iterator $F(x) = \frac{1}{2}(x + \frac{k}{x})$ used to approximate $\sqrt{k}$.

Problem 4.3.2 Given a complex number number $k = a + ib$, decompose $F(z) = \frac{1}{2}(z + \frac{k}{z})$ into its real and imaginary parts. Confirm that the pair of real iterative schemes

(4.3.7) $$x_{n+1} = \frac{1}{2}\left(x_n + \frac{a x_n + b y_n}{x_n^2 + y_n^2}\right); \quad y_{n+1} = \frac{1}{2}\left(y_n + \frac{b x_n - a y_n}{x_n^2 + y_n^2}\right)$$

is equivalent to the complex iterative scheme $z_{n+1} = F(z_n)$.

Problem 4.3.3 For $k = a + ib = \sqrt{a^2 + b^2}\,(\cos\theta + i\sin\theta)$, confirm that the square roots of k are given by

$$\pm\sqrt[4]{a^2 + b^2}\left(\cos\left(\frac{1}{2}\tan^{-1}\frac{b}{a}\right) + i\sin\left(\frac{1}{2}\tan^{-1}\frac{b}{a}\right)\right).$$

[Another formula for $\sqrt{k}$ is given in Problem 4.6.9 at the end of this chapter.]

The spreadsheet of Figure 4.3.2 allows us to assign arbitrary real values to a and b. Given initial values for x_1 and y_1, it generates sequences $x_1, x_2, x_3, \ldots$ and $y_1, y_2, y_3, \ldots$ for which $z_n = x_n + iy_n$ approaches a square root of $k = a + ib$. Making use of Problem 4.3.3, it also calculates $\pm(a + ib)^{1/2}$ from the polar form for k, namely $a + ib = \sqrt{a^2 + b^2}\,(\cos\theta + i\sin\theta)$, where $\tan\theta = b/a$.

While Figure 4.3.2 may provide compelling evidence of the effectiveness of such iterative schemes, we have not yet provided any mathematical justification for their validity. In Chapter 1 we used staircase diagrams to explain the success of such techniques, but extending these ideas to (4.3.7) would require an ability to represent graphs in four dimensions. Problem 4.3.7 will in fact provide a mathematical justification for using such iterative techniques to find $\pm k^{1/2}$. However, to set the stage for this proof, some additional preparatory work is desirable.

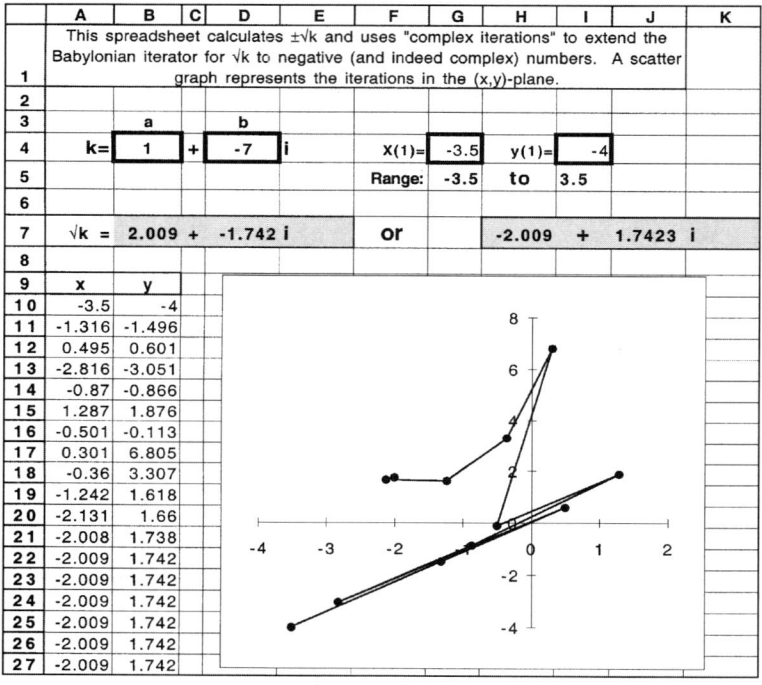

Figure 4.3.2 The Babylonian Iterator for $k = a + ib$

Problem 4.3.4 Construct a spreadsheet for approximating $\pm k^{1/2}$ in the format of Figure 4.3.2 above. Setting $b = 0$ and $k = a > 0$, confirm that starting values of $y_1 = 0$ and $x_1 \neq 0$ yield the same sequence of x-iterates as does the "real square root machine" of Figure 1.5.4.

Problem 4.3.4 reflects the fact that the Babylonian square root machine of Chapter 1 is embedded in the spreadsheet of Figure 4.3.2. Here it is interesting to recall that for $a > 0$, we have a simple criterion for determining whether the iterates generated by

$$x_{n+1} = \frac{1}{2}\left(x_n + \frac{a}{x_n}\right)$$

will approach $\sqrt{a}$ or $-\sqrt{a}$:

(4.3.8)

　　　　If $x_1 > 0$, then $x_n > 0$ for $n = 1, 2, \ldots$ and $x_1, x_2, \ldots$ approaches $\sqrt{a}$.

　　　　If $x_1 < 0$, then $x_n < 0$ for $n = 1, 2, \ldots$ and $x_1, x_2, \ldots$ approaches $-\sqrt{a}$.

But what if we allow $y_n \neq 0$?

Problem 4.3.5 Use the spreadsheet of Problem 4.3.4 to give credence to the assertion that (4.3.8) holds even if $y_1 \neq 0$.

On the basis of Problem 4.3.5 we will say that when $b = 0$, the right half-plane $x_1 > 0$ serves as a "basin of attraction" for the fixed point $\sqrt{a}$ of the iterator (4.3.7), whereas the left half-plane $x_1 < 0$ serves as a basin of attraction for $-\sqrt{a}$. This fact is illustrated in Figure 4.3.3.

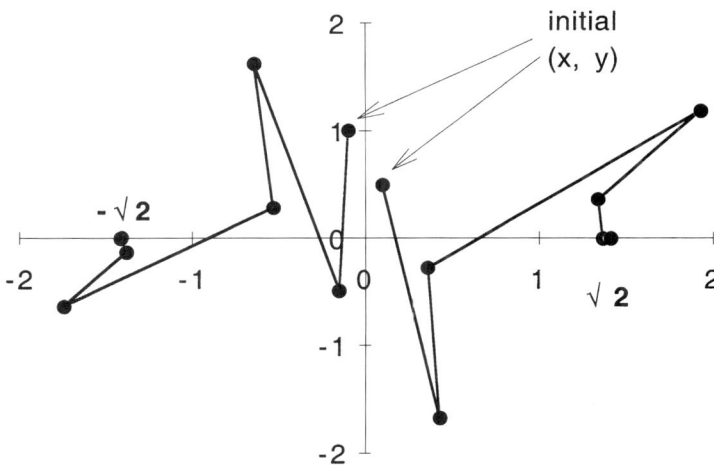

Figure 4.3.3 Half-Planes as Basins of Attraction for $\sqrt{a}$ and $-\sqrt{a}$

This brings us to some mundane observations, ones that will, however, assume epic proportions when extended to cube roots in Section 4.4. The boundary between the basins of attraction for $\sqrt{a}$ and $-\sqrt{a}$ is a straight line, namely the y-axis. Choosing z_1 near the y-axis (i.e., with $|x_1|$ small) tends to increase the number of iterations required to obtain a good approximation for $\sqrt{a}$ or $-\sqrt{a}$. Choosing z_1 *on* the y-axis (i.e., with $x_1 = 0$) leads to a sequence of iterates with $x_2 = x_3 = x_4 = \cdots = 0$. As a result, the z_n remain on the y-axis, which is the perpendicular bisector of the line segment joining the roots $(-\sqrt{a}, 0)$ and $(\sqrt{a}, 0)$. Equidistant from these two attracting fixed points, such iterates leap about frantically, unable to escape the confines of the y-axis in their quest to approach one of the square roots of a. (The reader will find a less romantic explanation in the problems at the end of this chapter.)

Problem 4.3.6 Use the spreadsheet of Problem 4.3.1 to determine the basins of attraction for the fixed points of the iterator $F(z) = \frac{1}{2}(z - \frac{1}{z})$.

We now return to the iterator $F(z) = \frac{1}{2}(z + \frac{k}{z})$ for the general case where $k = a + ib$ and the iterative scheme $z_{n+1} = F(z_n)$ gives rise to (4.3.7). The spreadsheet of Figure 4.3.4 provides a scatter diagram for the iterations produced by (4.3.7) *and* draws the perpendicular bisector of the line segment connecting $k^{1/2}$ and $-k^{1/2}$. It gives added credence to the following assertions:

1. The perpendicular bisector of the line segment connecting $k^{1/2}$ and $-k^{1/2}$ divides the (x, y)-plane into basins of attraction for the fixed points of $F(z) = \frac{1}{2}(z + \frac{k}{z})$ (which are $\pm k^{1/2}$).

2. If an initial z_1 is chosen on this perpendicular bisector, its iterates will also lie on this line.

3. If an initial z_1 is chosen inside a basin of attraction, its iterates will remain in that basin and approach the square root of k belonging to that basin.

It remains to *prove* that iterations of the form

$$(4.3.9) \qquad z_{n+1} = \frac{1}{2}\left(z_n + \frac{k}{z_n}\right)$$

do in fact approach $\pm k^{1/2}$ in accordance with the behavior observed in Figure 4.3.4. Here we will ask the reader to confirm the underlying algebraic calculations.

Problem 4.3.7 Assuming that $z_{n+1} = \frac{1}{2}(z_n + \frac{k}{z_n})$, show that

(a)
$$\left(\frac{z_n - \sqrt{k}}{z_n + \sqrt{k}}\right)^2 = \frac{z_{n+1} - \sqrt{k}}{z_{n+1} + \sqrt{k}}.$$

	A	B	C	D	E	F	G	H	I	J	K
1	This spreadsheet calculates "complex iterations" allowing one to extend the Babylonian iterator for √k to negative (and indeed complex) numbers. A scatter graph represents the iterations in the (x,y)-plane.										
2											
3	The graph also contains the line dividing the plane into two "basins of attraction" corresponding to the two square roots. The iterates approach the root belonging to the basin to which (x(1),y(1)) belongs.										
4											
5		a		b							
6	k=	2	+	-7	i	X(1)=	-3.4	y(1)=	-4.2		
7						Range:	-3.4	to	3.4		
8											
9	√k =	2.154 +	-1.625 i			or		-2.154	+	1.6248 i	
10											
11	x	y									
12	-3.4	-4.2									
13	-1.313	-1.549									
14	0.34	0.716									
15	-3.278	-2.674									
16	-1.299	-0.547									
17	-0.34	2.29									
18	-1.729	0.94									
19	-2.16	1.79									
20	-2.151	1.628									
21	-2.154	1.625									
22	-2.154	1.625									
23	-2.154	1.625									
24	-2.154	1.625									
25	-2.154	1.625									
26	-2.154	1.625									
27	-2.154	1.625									
28											
29	2.154	-1.625									
30											
31											
32											
33											
34											
35											
36	-3.4	-4.507									
37	0	0									
38	3.4	4.507									

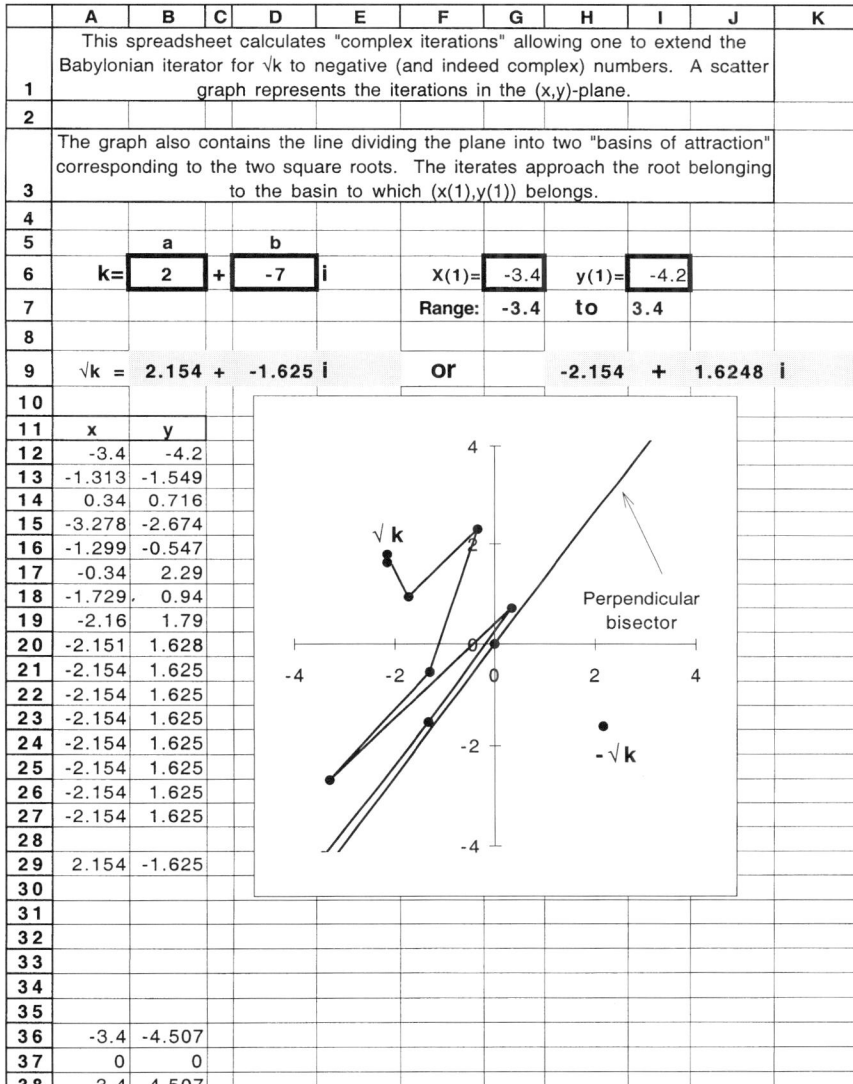

Figure 4.3.4 Basins of Attraction for $F(z) = \frac{1}{2}(z + \frac{k}{z})$

(b) If $z_1, z_2, z_3, \ldots$ is a sequence of iterates generated by (4.3.9), then

$$\frac{z_3 - \sqrt{k}}{z_3 + \sqrt{k}} = \left(\frac{z_1 - \sqrt{k}}{z_1 + \sqrt{k}}\right)^4,$$

and more generally,

$$\frac{z_n - \sqrt{k}}{z_n + \sqrt{k}} = \left(\frac{z_1 - \sqrt{k}}{z_1 + \sqrt{k}}\right)^{2^{n-1}}.$$

(c) Since the absolute value of a product [quotient] is the product [quotient] of the absolute values.

$$\frac{|z_n - \sqrt{k}|}{|z_n + \sqrt{k}|} = \left(\frac{|z_1 - \sqrt{k}|}{|z_1 + \sqrt{k}|}\right)^{2^{n-1}}.$$

(d) Let $d_1 = |z_1 - \sqrt{k}|$ and $d_2 = |z_1 + \sqrt{k}| = |z_1 - (-\sqrt{k})|$, so that d_1 is the distance from z_1 to $\sqrt{k}$ and d_2 is the distance from z_1 to $-\sqrt{k}$. Let L denote the perpendicular bisector of the line segment connecting $\sqrt{k}$ and $-\sqrt{k}$. Draw a diagram and give a geometric argument showing that

$$d_1 < d_2 \text{ if and only if } z_1 \text{ is on the same side of } L \text{ as } \sqrt{k};$$
$$d_1 > d_2 \text{ if and only if } z_1 \text{ is on the same side of } L \text{ as } -\sqrt{k};$$
$$d_1 = d_2 \text{ if and only if } z_1 \text{ belongs to } L.$$

(e) Use (c) to conclude that $z_1, z_2, z_3, \ldots$ approaches $\sqrt{k}$ if and only if z_1 is on the same side of L as $\sqrt{k}$, whereas this sequence approaches $-\sqrt{k}$ if and only if z_1 is on the same side of L as $-\sqrt{k}$.

(f) Show that if z_1 belongs to L, then all succeeding iterates $z_2, z_3, \ldots$ also belong to L.

Problem 4.3.7 shows that the basins of attraction for the iterator $F(z) = \frac{1}{2}(z + \frac{k}{z})$ are the two half-planes defined in part (d). Problem 4.6.13 at the end of this chapter shows that L is given by the equation

(4.3.10) $$(a - \sqrt{a^2 + b^2})x + by = 0,$$

where $a = \text{Re}(k)$ and $b = \text{Im}(k)$.

Having determined the basins of attraction for the iterator associated with $z^2 - k = 0$, it is natural to turn to the general quadratic equation $az^2 + bz + c = 0$, where we allow a, b, and c to assume complex values, $a \neq 0$. In Chapter 1 we successfully applied the "Newton iterator" $G(x) = (ax^2 - c)/(2ax + b)$ to approximate solutions to the quadratic equation $ax^2 + bx + c = 0$ with real coefficients. This impels us to try

(4.3.11) $$F(z) = \frac{az^2 - c}{2az + b}$$

as an iterator for generating sequences of complex numbers $z_1, z_2, z_3, \ldots$ that approximate

$$(4.3.12) \quad q_1 = \frac{-b + \sqrt{b^2 - 4ac}}{2a} \quad \text{and} \quad q_2 = \frac{-b - \sqrt{b^2 - 4ac}}{2a}.$$

Assuming that (4.3.9) does generate such iterates, there arises the question of whether a particular sequence of iterates will approach q_1 or q_2. Here we can build on Problem 4.3.7 to show that the basins of attraction for (4.3.9) are again half-planes defined by the perpendicular bisector of the line segment connecting q_1 and q_2 (see Problem 4.6.14 at the end of this chapter).

Finally, we return to the issue of staircase diagrams. In Chapter 1 we used staircase diagrams for $F(x) = \frac{1}{2}(x + \frac{k}{x})$ to explain why the iterates of an initial $x_1 > 0$ approach $\sqrt{k}$ and how the "flatness" of the graph of $F(x)$ at $x = \sqrt{k}$ contributes to the remarkable effectiveness of this procedure. In this chapter we have not been able to provide any comparable geometric explanation as to why iteration by $F(z) = \frac{1}{2}(z + \frac{k}{z})$ yields approximations to $\pm k^{1/2}$. Instead, we were forced to rely on spreadsheets to illustrate the success of our methods and on Problem 4.3.7 to demonstrate their validity.

While four-dimensional staircase diagrams continue to elude us, we are able to use Excel's remarkable graphing capability to describe some "stepping stones" that correspond to (4.3.7). For example, for $k = -1$ we can generate 3-D graphs for

$$(4.3.13) \quad u(x, y) = \frac{1}{2}\left(x - \frac{x}{x^2 + y^2} \right) \quad \text{and} \quad v(x, y) = \frac{1}{2}\left(y + \frac{y}{x^2 + y^2} \right).$$

(See Figure 4.3.5.) Without providing a full geometric representation of the iterative phenomena under discussion, we do note that $(0, 1)$ *is* a fixed point for (4.3.13) and these two graphs are "flat" near the fixed point $(0, 1)$ corresponding to the complex number i.

4.4. Cubic Equations and Julia Sets

We turn now to cubic equations and the use of iteration to find both real *and* complex solutions. Here our experience with quadratic equations will serve us well. To approximate solutions of the general cubic

$$(4.4.1) \qquad\qquad az^3 + bz^2 + cz + d = 0$$

we need only invoke the complex form of the iterator $F(x) = \frac{2ax^3 + bx^2 - d}{3ax^2 + 2bx + c}$ used in Chapter 1. As in Section 4.3, by expressing $F(z)$ in terms of its real and imaginary parts, one can transform the iterative scheme $z_{n+1} = F(z_n)$ into a *pair* of real

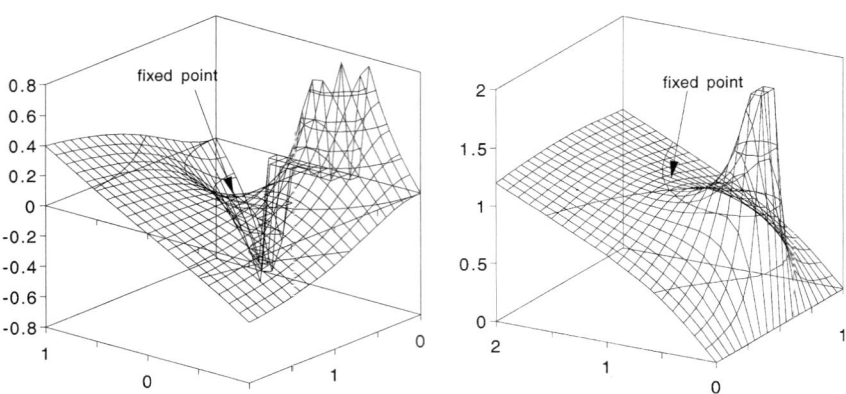

Figure 4.3.5 "Flat Stepping Stones" at $z = i$

iterative schemes

(4.4.2) $x_{n+1} = u(x_n, y_n)$ and $y_{n+1} = v(x_n, y_n)$.

For each pair of initial values x_1 and y_1, (4.4.2) defines a sequence of iterates in the (x, y)-plane. This process underlies the concept of "basins of attraction" associated with the attracting fixed points of the iterator $F(z)$.

Despite this close similarity with the quadratic case, the resulting basins of attraction lead to some startling new phenomena. Rather than consisting of easily described regions in the (x, y)-plane (they were half-planes in the quadratic case), the basins of attraction associated with (4.4.2) have an elaborate structure. The exquisite complexity of this structure is reflected not only in the basins of attraction themselves, but also by the boundary that separates them. In the quadratic case this boundary was a straight line separating two half-planes. In the cubic case, this boundary is an example of a geometric object called a *fractal*.

The pertinent features we want to study do not require consideration of the general cubic (4.4.1). They are exhibited by the simplest kinds of cubic equations, such as

(4.4.3) $z^3 - k = 0,$

where k is a complex number. Expressing k in polar form as $r(\cos\theta + i \sin\theta)$, we obtain the solution

(4.4.4) $z = \sqrt[3]{r}\left(\cos\frac{\theta}{3} + i \sin\frac{\theta}{3}\right).$

However, (4.4.4) yields just one solution of $z^3 - k = 0$, whereas the fundamental theorem of algebra leads us to expect three solutions (some or all of which may be complex). Where are the other two?

An answer can be based on the fact that if $k = r(\cos\theta + i \sin\theta)$, then it is also true that

(4.4.5) $$k = r(\cos(\theta + 2n\pi) + i \sin(\theta + 2n\pi))$$

where n is any integer, $n = 0, \pm 1, \pm 2, \ldots$. Therefore, (4.4.4) can be replaced by

(4.4.6) $$\sqrt[3]{r}\left(\cos\left(\frac{\theta + 2n\pi}{3}\right) + i \sin\left(\frac{\theta + 2n\pi}{3}\right)\right)$$

for $n = 0, \pm 1, \pm 2, \ldots$.

Although this may suggest that we now have *more* than three solutions, this is not the case. For given values of $r > 0$ and θ, there are only three different complex numbers of the form (4.4.6). These can be obtained by setting $n = 0, 1$, and 2 in (4.4.6).

Problem 4.4.1 Show that for any integer n, $\sqrt[3]{r}(\cos(\frac{\theta + 2n\pi}{3}) + i \sin(\frac{\theta + 2n\pi}{3}))$ must equal one of the following complex numbers:

$$\sqrt[3]{r}\left(\cos\frac{\theta}{3} + i \sin\frac{\theta}{3}\right),$$

$$\sqrt[3]{r}\left(\cos\left(\frac{\theta + 2\pi}{3}\right) + i \sin\left(\frac{\theta + 2\pi}{3}\right)\right)$$

or

$$\sqrt[3]{r}\left(\cos\left(\frac{\theta + 4\pi}{3}\right) + i \sin\left(\frac{\theta + 4\pi}{3}\right)\right).$$

Problem 4.4.2 Show that the equation $z^3 - 1 = 0$ has the three solutions

$$z = 1, \quad z = -\frac{1}{2} + i\frac{\sqrt{3}}{2}, \quad \text{and} \quad z = -\frac{1}{2} - i\frac{\sqrt{3}}{2}.$$

Sketch these three "cube roots of unity" as points on the unit circle $x^2 + y^2 = 1$ and show that they form the vertices of an equilateral triangle in the (x, y)-plane.

In order to determine the basins of attraction of these three points, we note that for $a = 1$, $b = 0$, $c = 0$, and $d = -1$, the iterator $F(x) = \frac{2ax^3 + bx^2 - d}{3ax^2 + 2bx + c}$ becomes $\frac{2x^3 + 1}{3x^2}$. However, since we are seeking complex as well as real solutions, we now consider

$$(4.4.7) \qquad\qquad F(z) = \frac{2z^3 + 1}{3z^2}$$

as the complex iterator for approximating $z = 1, -\frac{1}{2} + i\frac{\sqrt{3}}{2}$, and $-\frac{1}{2} - i\frac{\sqrt{3}}{2}$.

Problem 4.4.3 By breaking (4.4.7) into its real and imaginary parts, show that the iterative scheme $z_{i+1} = F(z_i)$ is equivalent to $x_{n+1} = u(x_n, y_n)$ and $y_{n+1} = v(x_n, y_n)$, where

$$u(x, y) = \frac{1}{3}\left(2x + \frac{x^2 - y^2}{(x^2 + y^2)^2}\right) \quad \text{and} \quad v(x, y) = \frac{1}{3}\left(2y - \frac{2xy}{(x^2 + y^2)^2}\right).$$

(4.4.8)

As in Section 4.3, we can create a spreadsheet to implement (4.4.8) in the (x, y)-plane (see Figure 4.4.1). After constructing such a spreadsheet, the reader is urged to tarry a while in reflecting on the problems that follow.

Problem 4.4.4 Modify the spreadsheet of Problem 4.3.4 to implement (4.4.8). Find at least three different values of $x_1 + iy_1$ for which the iterates $z_1, z_2, z_3, \ldots$ approach $z = 1$.

	A	B	C	D	E	F	G	H	I	J
1	This spreadsheet implements the complex iterator F(z) = (2z^3 + 1)/3z^2 for arbitrary initial values of z(1) = (x(1), y(1)). The cube root of unity approached depends on the choice of initial value.									
2										
3	x(1) =	-0.8		y(1) =	0.1					
4										
5	i	x(i)	y(i)							
6	1	-0.8	0.1							
7	2	-0.036	0.1929							
8	3	-8.085	3.2728							
9	4	-5.387	2.1849							
10	5	-3.584	1.4635							
11	6	-2.373	0.9972							
12	7	-1.547	0.6967							
13	8	-0.954	0.5512							
14	9	-0.499	0.6051							
15	10	-0.436	0.9353							
16	11	-0.492	0.8632							
17	12	-0.5	0.856							
18	13	-0.5	0.866							
19	14	-0.5	0.866							

Figure 4.4.1 Approximating the Cube Roots of Unity

Problem 4.4.5 Use (4.4.8) to show that if $y_1 = 0$, then we have $y_n = 0$ for all $n > 1$, and $u(x, 0)$ reduces to the (real) iterator $F(x) = \frac{2x^3+1}{3x^2}$ associated with real solutions of $x^3 - 1 = 0$.

Problem 4.4.6 Use the spreadsheet of Problem 4.4.4 to confirm that for z_1 on the x-axis, the iterates $z_2, z_3, \ldots$ remain on that axis and (with rare exceptions) approach 1.

Problem 4.4.7 Confirm that $-\sqrt[3]{\frac{1}{2}}$ is one of the "rare exceptions" referred to in Problem 4.4.6. Find two other values for z_1 on the x-axis whose iterates fail to approach $z = 1$.

Problem 4.4.8 Find the three complex solutions of $2z^3 + 1 = 0$. Confirm that choosing z_1 as any of these numbers will lead to iterates $z_2, z_3, \ldots$ that fail to approach a cube root of unity.

Problem 4.4.9 Use the spreadsheet of Problem 4.4.4 to confirm that if z_1 lies on the line $y = \sqrt{3}x$, then the iterates $z_2, z_3, \ldots$ lie on the same line and (with rare exceptions) approach $-\frac{1}{2} - i\frac{\sqrt{3}}{2}$. Formulate and confirm an analogous behavior for initial values z_1 lying on the line $y = -\sqrt{3}x$.

Our experience with quadratic equations, combined with the fact that the complex cube roots of unity enjoy a threefold symmetry, may lead us to expect that the three basins of attraction for the iterator $\frac{2z^3+1}{3z^2}$ are separated by the three rays $y = \sqrt{3}x$ with $x > 0$, $y = -\sqrt{3}x$ with $x > 0$, and $y = 0$ with $x < 0$. (See Figure 4.4.2.)

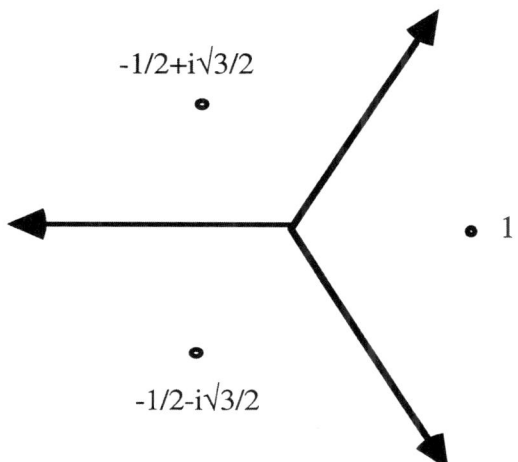

-1/2+i√3/2

• 1

-1/2-i√3/2

Figure 4.4.2 Threefold Symmetry of the Cube Roots of Unity

However, the spreadsheet experiments in the preceding problems show that this is *not* the case. With rare exceptions, points on the ray $y = 0$ with $x < 0$ belong to the basin of attraction of $z = 1$. Similarly, *most* points on the ray $y = \sqrt{3}x$ with $x > 0$ belong to the basin of attraction of $-\frac{1}{2} - i\frac{\sqrt{3}}{2}$, while *most* points on the ray $y = -\sqrt{3}x$ with $x > 0$ belong to the basin of attraction of $-\frac{1}{2} + i\frac{\sqrt{3}}{2}$. Although we have succeeded in locating some exceptional points on these rays that do belong to the boundary of these basins of attraction, the rays themselves cannot constitute such a boundary.

It will turn out that the three basins of attraction do enjoy a form of threefold symmetry corresponding to Figure 4.4.2, but the boundary separating them has a wholly unexpected structure. The spreadsheet of Figure 4.4.3 provides a basis for beginning to explore this structure.

Problem 4.4.10 Use the spreadsheet of Figure 4.4.3 to confirm that initial values of z_1 satisfying $|z_1 - 1| < \frac{1}{2}$ belong to the basin of attraction corresponding to $z = 1$. Find a value of z_1 satisfying $\frac{1}{2} < |z_1 - 1| < \frac{3}{4}$ that lies in the basin of attraction of $-\frac{1}{2} - i\frac{\sqrt{3}}{2}$ and another that lies in the basin of attraction $-\frac{1}{2} + i\frac{\sqrt{3}}{2}$.

Problem 4.4.11 Use Figure 4.4.3 to confirm that iterates of an initial value $z_1 = -0.6 + (0.1)i$ approach 1, while iterates of $z_1 = -0.6 + (0.11)i$ approach $-\frac{1}{2} + i\frac{\sqrt{3}}{2}$. Verify that the line segment $x = -0.6$ with $0.10 < y < 0.11$ contains initial values whose iterates approach $-\frac{1}{2} - i\frac{\sqrt{3}}{2}$.

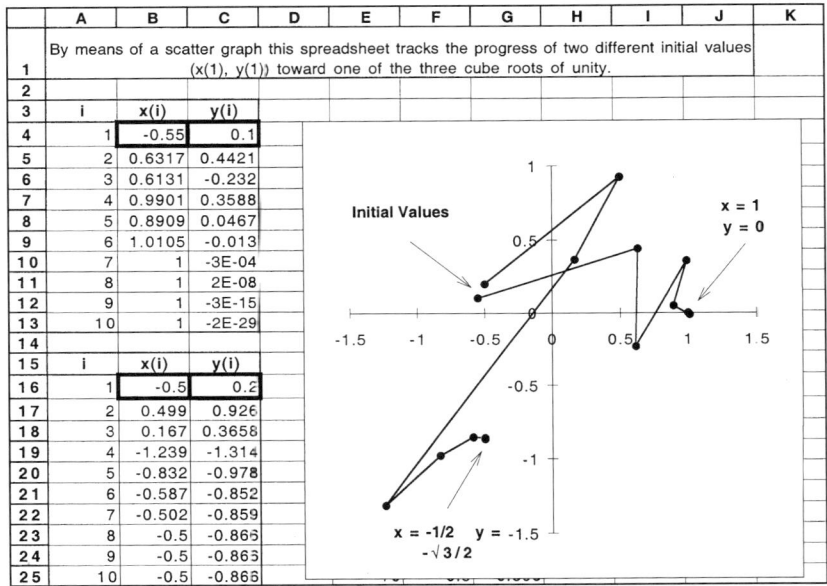

Figure 4.4.3 Iterations Approaching the Cube Roots of Unity

Problem 4.4.12 In Problem 4.4.7 it was shown that the point $x = -\sqrt[3]{\frac{1}{2}}$ and $y = 0$ does not lie in any of the three basins of attraction. Use Figure 4.4.3 to show that it is possible to find initial values z_1 whose distances from this boundary point are less than 0.01 and whose iterates approach *any* of the three cube roots of unity. Repeat when 0.01 is reduced to 0.001.

While the above problems provide some insights into the structure of the basins of attraction associated with $z^3 - 1 = 0$, they do not fully reveal its remarkable complexity. Readers interested in using spreadsheets to probe such matters further will be able to do so via problems at the end of this chapter. These problems will give credence to the claim that there is a single boundary shared by all three basins of attraction. This boundary, called a *Julia set*, has the following property: *Arbitrarily close to each boundary point separating two of the basins of attraction we always find points from the third basin of attraction.*

Visualizing a boundary set with this property is virtually impossible without computer-aided graphics. Yet many properties of Julia sets were derived before computers provided a means for visualizing them. The mathematicians who first studied these remarkable structures (notably Gaston Julia (1893–1978)) never had the opportunity to see the representations that computer technology provides.

In Figure 4.4.5, the basins of attraction for $1, -\frac{1}{2} + i\frac{\sqrt{3}}{2}$ and $-\frac{1}{2} - i\frac{\sqrt{3}}{2}$ are depicted in different shades. The Julia set itself is the *boundary* separating these regions, as depicted in Figure 4.4.4.

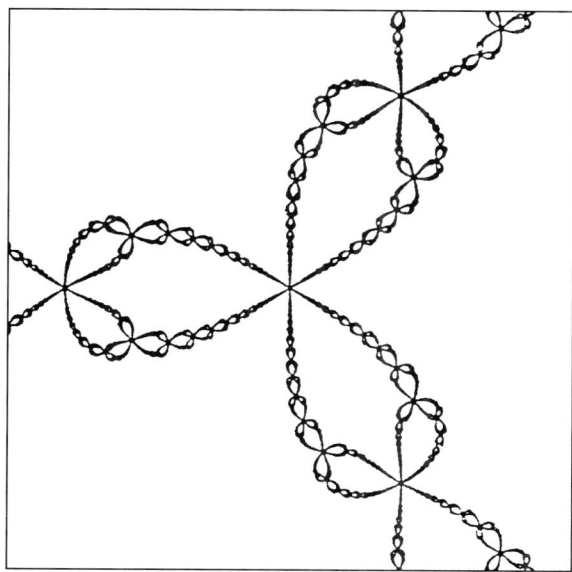

Figure 4.4.4 The Julia Set

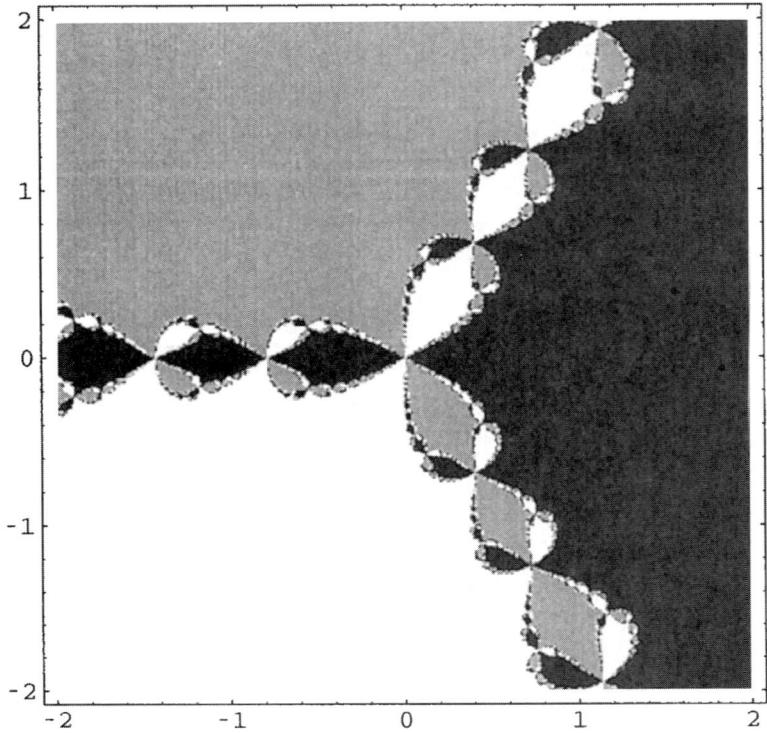

Figure 4.4.5 The Three Basins of Attraction (*Source: "A First Course in Discrete Dynamical Systems," R. Holmgren, Springer-Verlag, NY 1996*)

The fact that every point of the Julia set adjoins all three basins of attraction is made possible only by the endless sequences of tri-petaled rosettes appearing in Figure 4.4.4. Notice that there are infinitely many bulb-shaped regions that get smaller and smaller. At any junction point where three of these bulbs meet, all three basins of attraction are represented.

The structure at such a point is very much like the structure of the entire set near the origin. Thus a magnification of a small part of the Julia set looks very much like the entire set. This property of "self-similarity" or "quasi-self-similarity" is characteristic of a larger family of geometric figures called fractals. A brief overview of fractals is the subject of Section 4.5.

4.5. Iterative Geometry and Fractals   In Section 4.4 we saw how the solution of $z^3 - 1 = 0$ gives rise to a remarkable geometric object, one that is built on the endless repetition of a pattern suggestive of self-similarity. This raises the question of whether there are other geometric shapes, ones not necessarily generated by the solution of algebraic equations, that also exhibit interesting forms of self-similarity.

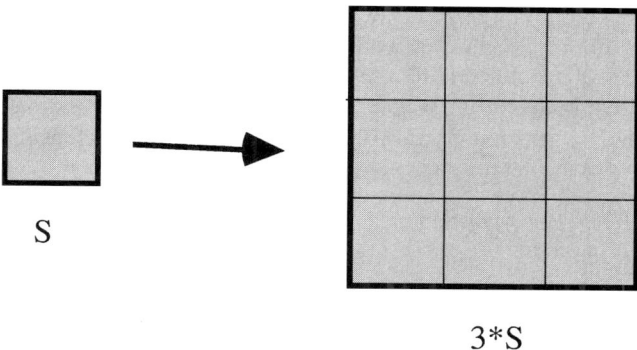

$$3*S$$

Figure 4.5.1 Self-similarity of a Square

In geometry, the concept of similarity is closely related to "magnification." Given a square S, we can consider one that is "three times as large" and denoted by $3 * S$. (See Figure 4.5.1.)

Here S and $3 * S$ are similar figures in the two-dimensional world of plane geometry. This world contains two-dimensional objects (such as the square S), one-dimensional objects (such as a side of S), and even zero-dimensional objects (such as a corner of S). As long as we deal with simple geometric shapes, our intuition is a good guide as to "the dimension of S." However, once we admit objects such as the Julia set into our geometric world, the concept of "dimension" becomes more elusive. Accordingly, it will be useful to inquire as to what it is about a square that leads us to dub it as "two-dimensional," while its sides are "one-dimensional" and a cube is "three-dimensional."

Our answer will be arrived at by comparing a square's threefold magnification $3 * S$ with S itself. In Figure 4.5.1, $3 * S$ is composed of 9 copies of S. However, each side of $3 * S$ is composed of 3 copies of the corresponding side of S. This observation sets S apart from its sides, and it will underlie our characterization of S as "2-dimensional" and its sides as "1-dimensional."

To put this characterization into mathematical form, it is useful to introduce another concept, namely "the measure of S." Based on Figure 4.5.1, we will say that

$$(\text{measure of } 3 * S) = 9 \cdot (\text{measure of } S).$$

It is the observation that the measure of $3 * S$ is 3^2 times the measure of S that will now lead us to $D = 2$. More generally, a geometric object S with the property that

$$(4.5.1) \qquad (\text{measure of } 3 * S) = 3^D \cdot (\text{measure of } S)$$

will be said to have dimension D.

At this point the careful reader may have a sense of unease. In our efforts to give the intuitive notion of "the dimension of S" a mathematical characterization, we have introduced another intuitive concept, namely "the measure of S." A full resolution of this issue would require a mathematical excursion into the realm of "measure theory." As an alternative, we ask the reader to accept the following assertion, one that is central to the concept of measure. It is:

> If A and B are geometric objects and A can be partitioned into N congruent copies of B, then the measure of A equals N times the measure of B.

Returning now to the geometry of Figure 4.5.1, the above property of measure does enable us to conclude that "the measure of $3 * S$ is 9 times the measure of S." This in turn enables us to apply (4.5.1) as a basis for reaching the conclusion that $D = 2$.

Problem 4.5.1 If C is a cube and $3 * C$ is a threefold magnification of C, then (as illustrated by Rubik's cube) $3 * C$ consists of 27 copies of C. Use this fact and (4.5.1) to show that C has dimension $D = 3$.

Problem 4.5.2 If L is a line segment and $3 * L$ is a threefold magnification of L, then $3 * L$ consists of 3 copies of L. Use this fact and (4.5.1) to conclude that L has dimension $D = 1$.

While (4.5.1) enables us to determine the dimension of many geometric objects, we do not want a definition of D to be tied to magnifications by a factor of three. Doubling the size of a square S leads to a square $2 * S$ containing 4 copies of S, and this fact should also lead us to the conclusion that a square has dimension $D = 2$. The following generalization of (4.5.1) makes it possible to use r-fold magnification of a figure S as the basis for determining its dimension:

$$(4.5.2) \qquad (\text{measure of } r * S) = r^D \cdot (\text{measure of } S).$$

According to the property of measure formulated above, if $r * S$ can be partitioned into N congruent copies of S, then

$$N \cdot (\text{measure of } S) = r^D \cdot (\text{measure of } S).$$

If in addition, the measure of S is both *finite* and *nonzero*, then we can cancel the term in parentheses to obtain the alternative characterization

$$N = r^D,$$

or

$$D = \frac{\log N}{\log r}.$$

It is this latter form of (4.5.2) that will be central to our calculations of D.

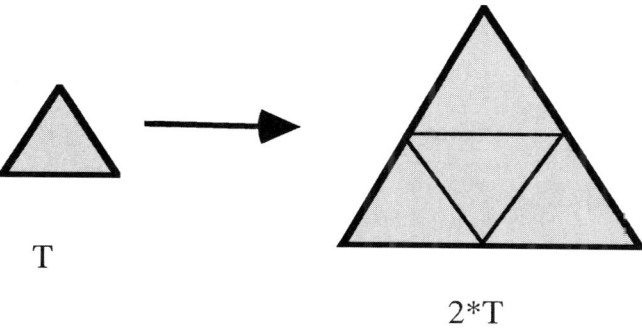

T

2*T

T

Figure 4.5.2 A Doubling of Scale for an Equilateral Triangle

According to (4.5.2) and Figure 4.5.2, a "doubling of scale" for an equilateral triangle T enables us to conclude that T has dimension $D = 2$.

That is, (measure of $2 * T$) = 4. (measure of T), and this leads us to conclude from (4.5.2) that $D = 2$.

Problem 4.5.3 Show in a similar fashion that *any* triangle has dimension $D = 2$.

Problem 4.5.4 For an equilateral triangle T, draw its threefold magnification $3 * T$. Use this sketch and (4.5.2) as a basis for concluding that T has dimension $D = 2$.

So far we have discovered little that is new. Our intuition about the dimension of geometric shapes has been fortified by a characterization of D whose purpose at this point is far from evident. To give meaning to these efforts we shall consider the threefold magnification of a figure S that is "a square with a hole in the middle" (Figure 4.5.3). The "hole" in S is itself in the shape of a square, one whose sides are $\frac{1}{3}$ the length of the sides of S.[1] For brevity, we shall call this "the middle square."

Problem 4.5.5 Suppose that S has outside edge of length 3 so that its shaded area contains 8 copies of a 1×1 shaded square. Show that the shaded portion of $3 * S$ contains 72 copies of such a 1×1 shaded square, and on this basis explain why S has dimension $D = 2$.

---

[1] A subtle question arises in this process: Is the set of points comprising the boundary of the middle square removed? Except when noted otherwise (e.g., in Figure 4.5.7), the boundary will *not* be removed.

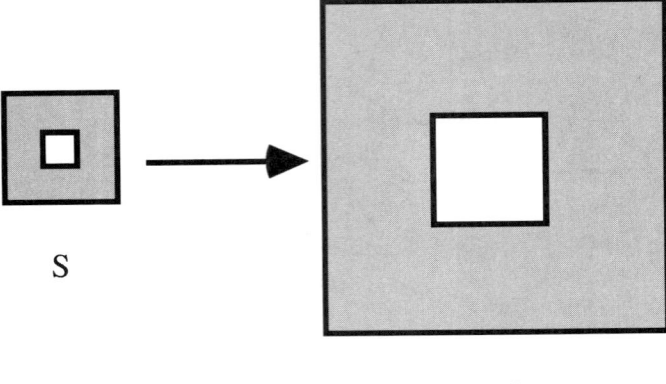

3*S

Figure 4.5.3 A Square Without its 'Middle Square"

Well, another mundane discovery. A square with a hole in it *also* has dimension 2. Indeed, the same is true for the region S_1 of Figure 4.5.4, one in which the 8 shaded 1×1 squares comprising S have their middle squares removed.

Problem 4.5.6 For S_1 as given in Figure 4.5.4, draw $3 * S_1$. By counting copies of small shaded squares and using (4.5.2), conclude that S_1 also has dimension $D = 2$.

The *Sierpiński carpet* is based on the infinite continuation of the process begun in Figures 4.5.4 and 4.5.3. It is the geometric shape obtained by *continuing the process*

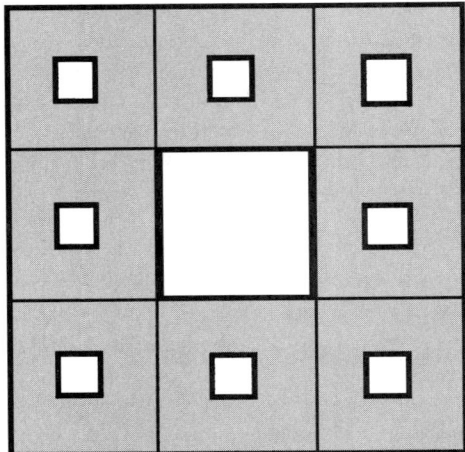

Figure 4.5.4 More Middle Squares Removed yield S_1

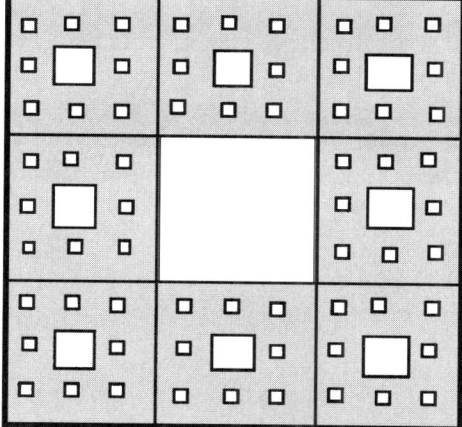

Figure 4.5.5 A Second Approximation of the Sierpiński Carpet yields S_2

of "removing middle squares" indefinitely. The next step consists in subdividing the shaded region in Figure 4.5.4 into 64 congruent squares and removing each of *their* middle squares. This yields Figure 4.5.5.

The Sierpiński carpet, which we shall denote by S_∞, can never be drawn precisely. However, given a starting square that is subject to such an unending "removal of middle squares," it *is* possible to describe precisely which points of that square belong to S_∞ and which do not.[2] While an accurate drawing of S_∞ cannot exist in a finite world, it *is* a well-defined geometric object.

Comparing S_∞ to $3 * S_\infty$ now leads to a crucial observation: $3 * S_\infty$ consists of *eight* copies of S_∞. Whereas all figures obtained from a *finite* number of "removing middle squares" have dimension $D = 2$, the dimension of the Sierpiński carpet satisfies

$$(\text{measure of } 3 * S_\infty) = 8 \cdot (\text{measure of } S_\infty).$$

According to (4.5.2) we now have

$$3^D = 8,$$

or

$$D = \log 8 / \log 3 \approx 1.892789261.$$

As such, we have arrived at a geometric object whose dimension D is not a whole number. Benoit Mandelbrot, the founder of the field now known as fractal geometry, coined the term *fractal* to refer to such geometric objects.

---

[2] It is at this point that the decision not to remove the boundaries of middle squares becomes important.

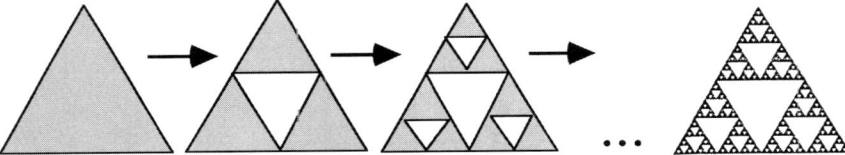

Figure 4.5.6 Moving Toward a Sierpiński Gasket

Problem 4.5.7 Show that the base of the logarithm used above is irrelevant. That is, show that for any base $b > 0$ and numbers $x, y > 0$, we have $\frac{\log_b x}{\log_b y} = \frac{\log_{10} x}{\log_{10} y}$.

Closely related to Sierpiński's carpet is the *Sierpiński gasket* obtained by the endless removal of "middle triangles" in Figure 4.5.2. This process is indicated in Figure 4.5.6.

Problem 4.5.8 Show that the first three geometric objects in Figure 4.5.6 all have dimension $D = 2$.

Problem 4.5.9 Show that an infinite continuation of the process indicated in Figure 4.5.6 leads to a geometric object with dimension $D = \log 3 / \log 2$.

The Sierpiński gasket is arguably the best-known example of a fractal whose dimension D is not an integer. However, closely related to the Sierpiński gasket is another fractal that turns out to have integer dimension! Its construction also begins with an equilateral triangle, but rather than removing middle triangles, we now remove "middle regular hexagons" whose sides are one-third the side of the triangle. The first step of this process produces 3 triangles that are $\frac{1}{3}$ the size of the original (Figure 4.5.7).

Here we have strayed from the convention that the boundary of the figure being removed is to remain part of S. In removing the central hexagon, we also removed the three outer edges of the hexagon (dotted in Figure 4.5.7). This leaves behind

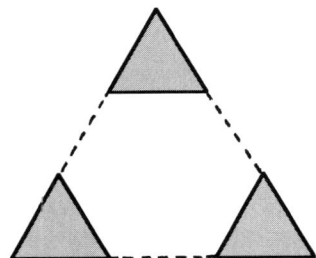

Figure 4.5.7 A Variant of the Sierpiński Gasket (Step 1)

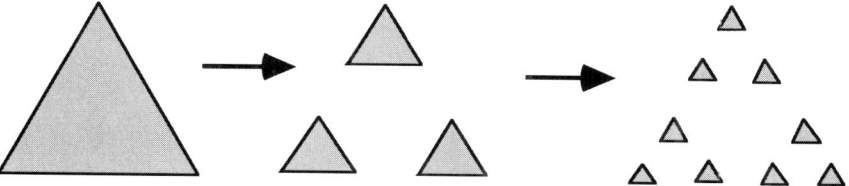

Figure 4.5.8 A Variant of the Sierpiński Gasket (continued)

three triangular regions and their boundaries, but not any edges connecting them. Another step in this process is illustrated in Figure 4.5.8.

As the process of Figure 4.5.8 proceeds indefinitely, we find ourselves left with a geometric object consisting of a very sparse set of points. This is an example of what Mandelbrot dubbed fractal "dust" because of its scattered nature. While the dimension of a single point (or even a set made up of a billion points) is zero, the geometric object defined above will turn out to have measure $D > 0$.

Problem 4.5.10 Find the (surprising) dimension of the geometric object obtained by an indefinite continuation of the process indicated in Figure 4.5.8.

So far, the connection between iterative algebra and fractals has been somewhat tenuous. In approximating solutions of the cubic equation $z^3 - 1 = 0$ in the complex domain, we encountered a geometric object known as a Julia set, one whose structure is suggestive of an unending form of self-similarity. It was on this basis that we began to explore self-similarity among familiar geometric shapes and were led to fractals.

There is, however, a very direct connection between fractals and the concept of chaos as discussed in Chapter 3. This connection can be illustrated in terms of the most famous form of fractal "dust," the so-called Cantor set.

Early in this discussion we observed that line segments have dimension $D = 1$. This is because an r-fold magnification of a line segment produces a new line segment that contains r copies of the original. This means that

(4.5.3)   (measure of $r * S$) $= r \cdot$ (measure of S),

so that $r^D = r$, and $D = 1$.

The construction of the Cantor set proceeds much as our previous examples, except we now begin with a line segment of length 1. Here it will be useful to take this segment to be the closed[3] interval $[0, 1]$. Removing the interior of the "middle third" leaves behind the two closed intervals $[0, \frac{1}{3}]$ and $[\frac{2}{3}, 1]$. If S is such a line segment with its middle third removed, we can compare it with $3 * S$ to conclude that S has dimension $D = 1$ (Figure 4.5.9).

[3] The fact that the interval $[0, 1]$ is closed means that it includes the endpoints 0 and 1.

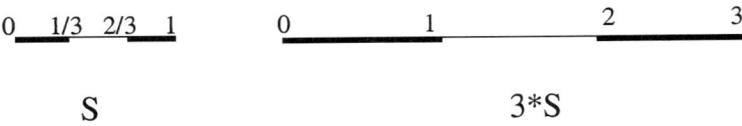

S   **3*S**

Figure 4.5.9 Threefold Magnification of a Segment Sans Middle Third

Problem 4.5.11 In Figure 4.5.9 explain why the measure of $3 * S$ is three times the measure of S and, as a result, S has dimension $D = 1$.

Continuing now to remove middle thirds of the two remaining segments (Figure 4.5.10), we are left with four segments of length $\frac{1}{9}$, namely $[0, \frac{1}{9}]$, $[\frac{2}{9}, \frac{1}{3}]$, $[\frac{2}{3}, \frac{7}{9}]$, and $[\frac{8}{9}, 1]$.

The geometric object obtained by continuing this process indefinitely is called a *Cantor set*, which we shall denote by C. Since $3 * C$ consists of *two* copies of C (separated by a gap of length 1), C has the property that

$$(\text{measure of } 3 * C) = 2 \cdot (\text{measure of } C).$$

Problem 4.5.12 Show that the dimension of the Cantor set is $\log 2 / \log 3$.

The fact that C has dimension $D \approx 0.6309$ does not in itself provide much information about the size of C. From the process used to construct C, it follows that the *endpoints* of the segments removed remain in C. Thus we can be sure that C contains an infinite set of points, including those with coordinates

$$0, \tfrac{1}{3}, \tfrac{2}{3}, \tfrac{1}{9}, \tfrac{2}{9}, \tfrac{7}{9}, \tfrac{8}{9}, \tfrac{1}{27}, \tfrac{2}{27}, \ldots$$

While C is in one sense quite large, this form of "largeness" doesn't add up to much when it comes to measures of length.

Problem 4.5.13 In continuing the process of Figure 4.5.10, show that after n steps there remain 2^n intervals, each of which has length $1/3^n$. Use this fact to show that the cumulative "length" (i.e., the 1-dimensional measure) of the Cantor set is 0.

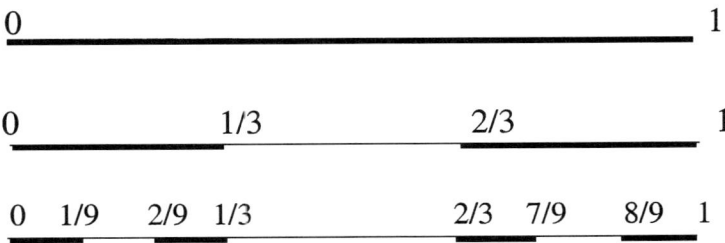

Figure 4.5.10 Removing More Middle Thirds

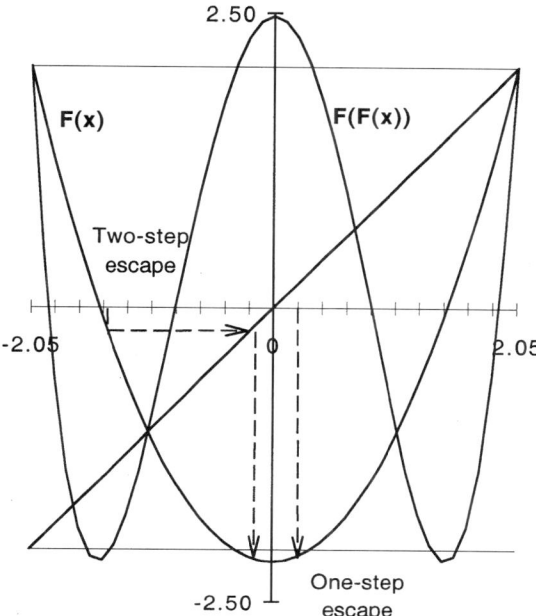

Figure 4.5.11 Escape Windows for $c > 2$

The connection between iterative algebra and the Cantor set goes back to the spreadsheet experiments conducted with the iterator $F(x) = x^2 - c$ in Chapter 3. Here we saw that as c increases from $c < \frac{3}{4}$ to $c > \frac{3}{4}$, the fixed point $x = \frac{1}{2} + \sqrt{c + \frac{1}{4}}$ is transformed from attractor to repeller. As a result, iterates of an initial x_1 satisfying $|x_1| < \frac{1}{2} + \sqrt{c + \frac{1}{4}}$ are trapped between two repellers, and as c continues to increase beyond $\frac{3}{4}$, the iterative scheme $x_{i+1} = F(x_i)$ turns from cycles of length 2 to cycles of length 4 to ..., and eventually to a form of chaos.

At $c = 2$, however, another dramatic change occurs (see Figure 3.5.1 and Problem 3.5.3). For $c > 2$ the graph of $y = F(x)$ dips below the line $y = -\frac{1}{2} - \sqrt{c + \frac{1}{4}}$, and this produces an "escape window" (Figure 4.5.11). That is, if x_1 lies in the interval defined by

$$F(x) < -\tfrac{1}{2} - \sqrt{c + \tfrac{1}{4}},$$

then, after a single iteration, $x_2 = F(x_1)$ lies outside the "critical square" having vertices at $\pm \frac{1}{2} \pm \sqrt{c + \frac{1}{4}}$. Once x_2 finds itself outside this critical square (i.e., if $|x_2| > \frac{1}{2} + \sqrt{c + \frac{1}{4}}$), then its iterates $x_3, x_4, x_5, \ldots$ also lie outside the critical square and rapidly grow toward $+\infty$.

This is, however, not the only escape window that is created. For each $c > 2$ there are *two* intervals where the graph of $F(F(x))$ dips below the line $y =$

$-\frac{1}{2} - \sqrt{c+\frac{1}{4}}$, and these intervals produce two (smaller) escape windows through which an initial x_1 can escape the critical square in 2 steps (see Figure 3.5.2 and Problem 3.5.7). Going on to consider $F(F(F(x)))$, we find that there are *four* intervals in which its graph dips below the critical square, creating four new escape windows through which x_1 can escape the critical square in 3 steps (see Problem 3.5.8), etc.

In order to characterize initial values x_1 whose iterates remain in the the critical square, we must eliminate from $[-\frac{1}{2} - \sqrt{c+\frac{1}{4}}, \frac{1}{2} + \sqrt{c+\frac{1}{4}}]$

> one interval defined by $F(x) < -\frac{1}{2} - \sqrt{c+\frac{1}{4}}$,
>
> two intervals defined by $F(F(x)) < -\frac{1}{2} - \sqrt{c+\frac{1}{4}}$,
>
> (4.5.3)   four intervals defined by $F(F(F(x))) < -\frac{1}{2} - \sqrt{c+\frac{1}{4}}$,
>
> eight intervals defined by $F(F(F(F(x)))) < -\frac{1}{2} - \sqrt{c+\frac{1}{4}}$,
>
> etc.

Such efforts to contain the iterative scheme $x_{i+1} = F(x_i)$ within the critical square resemble the construction of the Cantor set, but there are some important differences. The intervals being removed are not "$\frac{1}{3}$ the length of the intervals produced in the previous step." As a result, the geometric object obtained is *suggestive* of unending self-similarity but does not actually embody this property. Because it is a precise form of self-similarity that underlies the relationship between C and $3*C$, the geometric objects defined by "deleting all escape windows associated with $F(x) = x^2 - c$" are considerably more difficult to analyze.

There is, however, a simple iterator for which "deleting all escape windows" does correspond *exactly* to the procedure used to construct the Cantor set. While this iterator lacks some of the interesting properties associated with $F(x) = x^2 - c$, it does serve to establish a direct connection between fractals and iterative algebra.

The iterator that generates the Cantor set is

(4.5.4) $G(x) = 3|x| - c.$

Unlike the parabola with vertex at $(0, -c)$ that is associated with $F(x) = x^2 - c$, we now have a V-shaped graph that is made up of two rays emanating upward from $(0, -c)$ with slopes ± 3 (Figure 4.5.12).

Problem 4.5.14 Recalling that $|x| = x$ if $x > 0$ and $|x| = -x$ if $x < 0$, show that

$$G(x) = 3x - c \quad \text{if} \quad x > 0 \quad \text{and} \quad G(x) = -3x - c \quad \text{if} \quad x < 0.$$

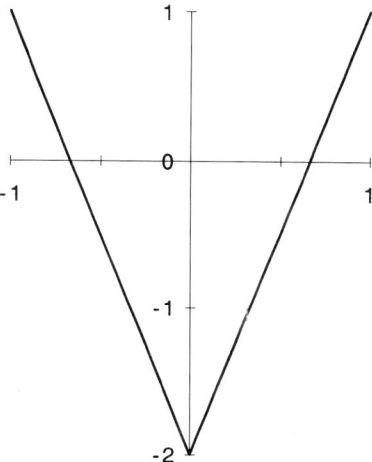

Figure 4.5.12 The Graph of $G(x) = 3|x| - 2$

Use this characterization to show that the fixed points of $G(x) = 3|x| - c$ are $c/2$ and $-c/4$. Explain why both fixed points are repellers.

On the basis of Problem 4.5.14, we associate with $G(x)$ a critical square having vertices at $(\pm c/2, \pm c/2)$. (If ever $|x_i| > c/2$, then its iterates escape $[-c/2, c/2]$ and grow without bound.)

Problem 4.5.15 Show that for any $c > 0$, the values of x for which the graph of $G(x) = 3|x| - c$ lies below the bottom edge of the critical square constitute the "middle third" of the interval $[-c/2, c/2]$.

Problem 4.5.16 Create a spreadsheet in the format of Figure 4.5.13. Use it to confirm that for $-c/6 < x_1 < c/6$, iterates of x_1 escape the critical square in a single step. Is it possible to find an x_1 satisfying $c/6 < |x_1| < c/2$ for which x_2 lies outside the critical square? An x_1 for which x_3 lies outside the critical square?

Problem 4.5.17 Modify the spreadsheet of Problem 4.5.16 to graph both $G(x)$ and $G(G(x))$ relative to the critical square. Show that the graph of $G(G(x))$ is a W-shaped polygon whose graph extends below the line $y = -c/2$ for x in two separate intervals.

Problem 4.5.18 Show that $G(G(x)) = 9x - 4c$ for $x > c/3$ and $-9x + 2c$ for $0 < x < c/3$. On this basis, determine one of the intervals for which $G(G(x)) < -c/2$.

Problem 4.5.19 The graph of a function $G(x)$ is symmetric about the y-axis if $G(x) = G(-x)$. Use this fact to determine the other interval for which

	A	B	C	D	E	F	G	H	I	J	K
1	For G(x) = 3|x| - c, this spreadsheet graphs G(x) and G(G(x)) relative to the critical square with vertices at (±c/2, ±c/2). There is one escape window of length 1/3 and two of length 1/9, corresponding to Figure 4.5.10										
2											
3	c= 2			L=	-1		R=	1			
4											
5	x	x	G(x)	G(G(x))							
6	-1	-1	1	1							
7	-0.95	-0.95	0.85	0.55							
8	-0.9	-0.9	0.7	0.1							
9	-0.85	-0.85	0.55	-0.35							
10	-0.8	-0.8	0.4	-0.8							
11	-0.75	-0.75	0.25	-1.25							
12	-0.7	-0.7	0.1	-1.7							
13	-0.65	-0.65	-0.05	-2							
14	-0.6	-0.6	-0.2	-1.4							
15	-0.55	-0.55	-0.35	-0.95							
16	-0.5	-0.5	-0.5	-0.5							
17	-0.45	-0.45	-0.65	-0.05							
18	-0.4	-0.4	-0.8	0.4							
19	-0.35	-0.35	-0.95	0.85							
20	-0.3	-0.3	-1.1	1.3							
21	-0.25	-0.25	-1.25	1.75							
22	-0.2	-0.2	-1.4	2.2							
23	-0.15	-0.15	-1.55	2.65							
24	-0.1	-0.1	-1.7	3.1							
25	-0.05	-0.05	-1.85	3.55							
26	0	0	-2	4							
27	0.05	0.05	-1.85	3.55							
28	0.1	0.1	-1.7	3.1							
29	0.15	0.15	-1.55	2.65							
30	0.2	0.2	-1.4	2.2	-1	1					
31	0.25	0.25	-1.25	1.75	-1	1					
32	0.3	0.3	-1.1	1.3	-1	1					

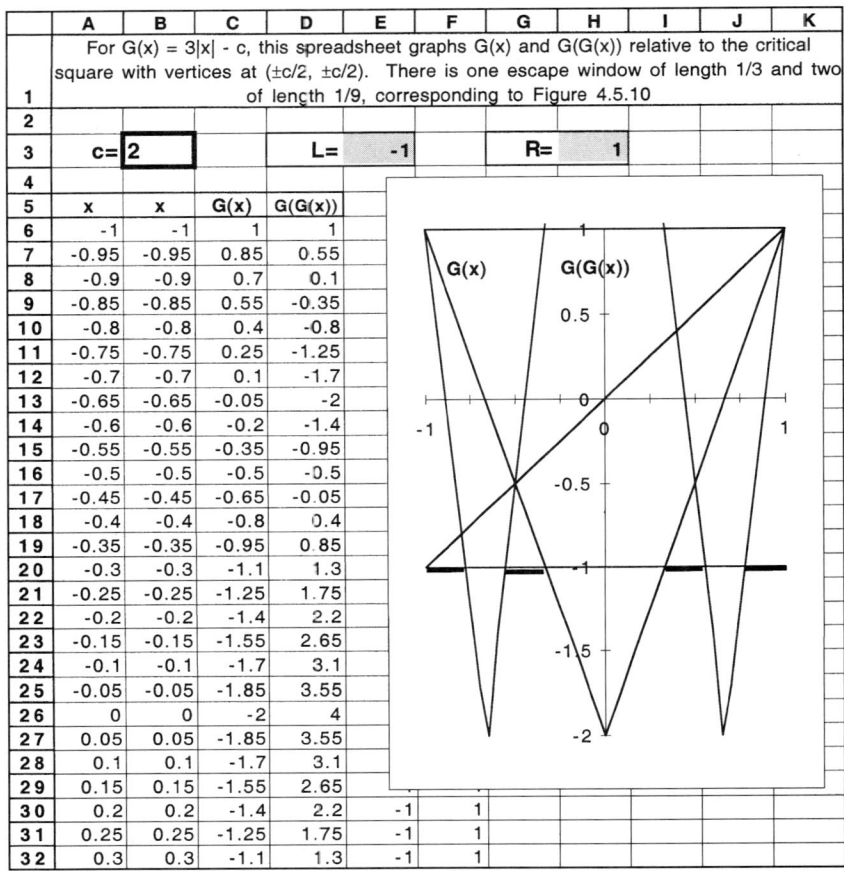

Figure 4.5.13 $G(x)$ and the Critical Square

$G(G(x)) < -c/2$. Confirm that the values of x for which the graph of $G(G(x))$ lies below the bottom edge of the critical square constitute the "middle thirds" of the intervals $[-c/2, -c/6]$ and $[c/6, c/2]$.

Problem 4.5.20 Construct a spreadsheet that graphs $G(x) = a|x| - c$ for arbitrary values of x and c and determine the attracting fixed points of G for $-1 < a < 1$. After generalizing Problems 4.5.14 through 4.5.19 to the case $1 < a < 3$ and $c > 0$, explain why iterates of $G(x)$ fail to exhibit the cyclic behavior associated with $F(x) = x^2 - c$.

4.6. Problems, Exercises, and Projects
A common view of complex numbers is that "they are symbols of the form $a + ib$." Here a and b are taken to be real numbers, while i is an "imaginary unit" with the property that $i^2 = -1$. In

calculating with such complex numbers, "one simply follows the ordinary rules of algebra," but with the proviso that i^2 is replaced by -1 wherever it appears. While such an approach can make computations with complex numbers mechanical and easy, it also creates an air of mystery about what the symbol $a + ib$ really means.

The representation of complex numbers as points in the (x, y)-plane can serve to dispel this sense of mystery. Even more important, there are computations (e.g., powers and roots of complex numbers) that are greatly simplified by a geometric approach to the algebra of complex numbers.

Problem 4.6.1 An identification of real numbers with points on "the real number line" provides us with a geometric representation of the algebra of real numbers. This problem deals with an analogous representation for complex numbers, one in which we identify a complex number $z = a + ib$ with the point (a, b) in the (x, y)-plane. In considering a pair of complex numbers

$$z_1 = (a_1, b_1) \quad \text{and} \quad z_2 = (a_2, b_2),$$

we can formulate the usual rules for addition and multiplication by

(A) $$(a_1, b_1) + (a_2, b_2) = (a_1 + a_2, b_1 + b_2)$$

and

(M) $$(a_1, b_1)(a_2, b_2) = (a_1 a_2 - b_1 b_2, a_1 b_2 + a_2 b_1).$$

(a) Locate the complex numbers $z_1 = 2 - i$, $z_2 = 1 + 3i$, and $z_1 + z_2 = 3 + 2i$ in the complex plane. Confirm that they illustrate the following parallelogram law: $(a_1, b_1) + (a_2, b_2)$ is the fourth vertex of the parallelogram determined by the three vertices $(0, 0)$, (a_1, b_1), and (a_2, b_2).

(b) Show that the rule (M) corresponds to "applying the ordinary rules of algebra to $(a_1 + ib_1)(a_2 + ib_2)$ with the proviso that $i^2 = -1$."

(c) Identifying the points $(x, 0)$ and $(y, 0)$ with the real numbers x and y, show that (A) corresponds to the usual representation of addition on the real number line, and (M) to multiplication.

(d) Noting that $(0, 0)$ acts like a "zero" with respect to addition [i.e., that $(0, 0) + (a, b) = (a, b) + (0, 0) = (a, b)$ for all (a, b)], confirm that $(1, 0)$ plays the role of "multiplicative unit," satisfying

$$(1, 0)(a, b) = (a, b)(1, 0) = (a, b)$$

for all points (a, b).

(e) Show that $(-a, -b)$ is the "additive inverse" of (a, b) in the sense that $(a, b) + (-a, -b)$ yields the "zero" of part (d). [On this basis we will write "$-z$" for $(-a, -b)$].

Problem 4.6.2 Central to our everyday use of arithmetic algorithms is the fact that the real numbers satisfy the commutative, associative, and distributive laws. These "CAD laws of arithmetic" assert that for arbitrary real numbers a, b, and c:

(Commutative)	$a + b = b + a,$
	$ab = ba.$
(Associative)	$(a + b) + c = a + (b + c)$
	$(ab)c = a(bc)$
(Distributive)	$a(b + c) = ab + ac$
	$(a + b)c = ac + bc.$

Making use of these laws for *real* numbers and the definitions (A) and (M) in Problem 4.6.1, confirm the following CAD laws for complex numbers:

(a) $(a_1, b_1) + (a_2, b_2) = (a_2, b_2) + (a_1, b_1)$

(b) $(a_1, b_1)(a_2, b_2) = (a_2, b_2)(a_1, b_1)$

(c) $[(a_1, b_1) + (a_2, b_2)] + (a_3, b_3) = (a_1, b_1) + [(a_2, b_2) + (a_3, b_3)]$

(d) $[(a_1, b_1)(a_2, b_2)](a_3, b_3) = (a_1, b_1)[(a_2, b_2)(a_3, b_3)]$

(e) $(a_1, b_1)[(a_2, b_2) + (a_3, b_3)] = (a_1, b_1)(a_2, b_2) + (a_1, b_1)(a_3, b_3)$

(f) $[(a_1, b_1) + (a_2, b_2)](a_3, b_3) = (a_1, b_1)(a_3, b_3) + (a_2, b_2)(a_3, b_3).$

Multiplication and Division

Problem 4.6.3 Recalling the rule (M) from Problem 4.6.1,

(a) Show that $(0, 1)(0, 1) = (-1, 0)$ and that for all points (a, b) we have

$$(a, b) = (a, 0) + (0, 1)(b, 0).$$

[The above calculation justifies the notation "$i = (0, 1)$" and "$(a, b) = a + ib$."]

(b) Given (a, b), the "multiplicative inverse" of (a, b) is a point (c, d) for which $(a, b)(c, d) = (1, 0)$. Show that if $(a, b) \neq (0, 0)$, then it has multiplicative inverse $(\frac{a}{a^2+b^2}, -\frac{b}{a^2+b^2})$. [The above calculation justifies the notation z^{-1} or $1/z$ for the above multiplicative inverse.]

(c) If $z_1 = (a_1, b_1)$ and $z_2 = (a_2, b_2) \neq (0, 0)$, we define $\frac{z_1}{z_2}$ to be $z_1 z_2^{-1}$. Show that $\frac{z_1}{z_2} z_2 = z_1$. [Remark: While this follows easily from the associative law of multiplication, it can also be checked by the rule (M) from Problem 4.6.1.]

Complex Conjugate and Absolute Value

Problem 4.6.4 Reverting to the traditional notation $z = a + ib$, the *complex conjugate* of z, denoted by $\bar{z}$, is defined as $\bar{z} = a - ib$. The *modulus* of z, or *absolute value* of z, is the distance from (a, b) to $(0, 0)$, or

$$|z| = \sqrt{a^2 + b^2}.$$

Also, if $z = a + ib$ (a and b real), then

$$\text{Re}(z) = a \text{ is called the real part of } z;$$
$$\text{Im}(z) = b \text{ is called the imaginary part of } z.$$

With these definitions, verify the following properties.

(a) $\overline{(z_1 \pm z_2)} = \overline{z_1} \pm \overline{z_2}$, and $\overline{(z_2 z_2)} = (\overline{z_1})(\overline{z_2})$.

(b) $\overline{(\bar{z})} = z$.

(c) $z\bar{z} = |z|^2$.

(d) $|z_1 z_2| = |z_1| \, |z_2|$, and if $z_2 \neq 0$, then $\left|\frac{z_1}{z_2}\right| = \frac{|z_1|}{|z_2|}$.

(e) If $z \neq 0$, then $1/z = \bar{z}/|z|^2$.

(f) $\text{Re}(z) = (z + \bar{z})/2$, $\text{Im}(z) = (z - \bar{z})/(2i)$, $\text{Re}(iz) = -\text{Im}(z)$, and $\text{Im}(iz) = \text{Re}(z)$.

(g) If we represent $z = a + ib$ as the point (a, b) in the (x, y)-plane, then $\bar{z}$ is obtained by *reflecting* this point across the x-axis. Use this to give part (b) a geometric interpretation.

(h) Relying on the geometric interpretation of $\bar{z}$ given in part (g), use the "parallelogram law" of addition to give a geometric interpretation of $\overline{z_1 + z_2} = \overline{z_1} + \overline{z_2}$.

(i) Viewing z_1 and z_2 as points in the (x, y)-plane, explain why $|z_1 - z_2|$ is the distance between z_1 and z_2.

Rotating with Complex Numbers

Problem 4.6.5 Suppose w is a complex number with $|w| = 1$. Then w has the polar representation $w = \cos\theta + i\sin\theta$.

(a) Show that for each z the product wz is obtained geometrically by rotating z through an angle θ about the origin. [Hint: Write z in polar form $z = r\cos\varphi + ir\sin\varphi$.]

(b) Use part (a) to show that if (x, y) is a point in the plane, and (x', y') is the point obtained by rotating (x, y) counterclockwise about the origin through an angle θ, then

$$x' = x\cos\theta - y\sin\theta \quad \text{and} \quad y' = x\sin\theta + y\cos\theta.$$

(c) Based on the concept of rotation, how is iz related to z?

(d) Based on the concept of rotation, how is $-z$ related to z?

(e) If $w = -\frac{1}{2} + i\frac{\sqrt{3}}{2}$, how is wz related to z? How is w^2z related to z? For n a positive integer, how is w^nz related to z?

Some Trigonometric Identities

Problem 4.6.6 De Moivre's theorem (see Problem 4.2.11) can serve as a clever device for establishing certain trigonometric identities. For example, when $n = 2$, we have

$$(*) \qquad \cos 2\theta + i \sin 2\theta = (\cos \theta + i \sin \theta)^2.$$

(a) By expanding the right-hand side and then equating real and imaginary parts of $(*)$, show that

$$\cos 2\theta = \cos^2 \theta - \sin^2 \theta \quad \text{and} \quad \sin 2\theta = 2 \cos \theta \sin \theta.$$

(b) Apply this device with $n = 3$ to obtain

$$\cos 3\theta = \cos^3 \theta - 3 \cos \theta \sin^2 \theta,$$
$$\sin 3\theta = 3 \cos^2 \theta \sin \theta - \sin^3 \theta.$$

(c) Use part (b) to derive

$$\cos 3\theta = 4 \cos^3 \theta - 3 \cos \theta,$$
$$\sin 3\theta = 3 \sin \theta - 4 \sin^3 \theta.$$

(d) Show that $\cos 4\theta$ is expressible as a "polynomial in $\cos \theta$," namely

$$\cos 4\theta = 8 \cos^4 \theta - 8 \cos^2 \theta + 1.$$

Iterative Relations Among Polynomials

In Chapters 1 and 2 we studied iterative schemes of the form $x_{i+1} = F(x_i)$, ones in which the value of each x_i is determined by its immediate predecessor x_{i-1}. Such iterative schemes are special cases of situations in which x_i is determined by *one or more* of its predecessors (these are called recursion relationships). The following problem builds on Problem 4.6.6 to establish an important recursion relationship, not among numbers $x_1, x_2, x_3, \ldots$, but among certain functions $P_1(x), P_2(x), P_3(x), \ldots$.

Problem 4.6.7 This problem will establish that for every positive integer n, there is a polynomial $P_n(x)$ with integer coefficients such that

$$\cos n\theta = P_n(\cos \theta).$$

[These polynomials are called Chebyshev polynomials and play an important role in numerical analysis.]

(a) Recalling the well-known identities $\cos 2\theta = 2\cos^2\theta - 1 = 1 - 2\sin^2\theta$, use Problem 4.6.6 to show that

$$P_1(x) = x, \quad P_2(x) = 2x^2 - 1,$$
$$P_3(x) = 4x^3 - 3x, \quad \text{and} \quad P_4(x) = 8x^4 - 8x^2 + 1.$$

(b) Recalling the formula for the cosine of the sum of two angles, show that

$$\cos(n+1)\theta = \cos n\theta \cos\theta - \sin n\theta \sin\theta,$$
$$\cos(n-1)\theta = \cos n\theta \cos\theta + \sin n\theta \sin\theta,$$

and then use these to establish the identity

$$\cos(n+1)\theta = 2\cos\theta\cos n\theta - \cos(n-1)\theta.$$

(c) Assuming that $P_{n-1}(x)$ and $P_n(x)$ satisfy the definition given above, use part (b) to show that $2\cos\theta\, P_n(\cos\theta) - P_{n-1}(\cos\theta) = \cos(n+1)\theta$. From this deduce that the polynomial $P_{n+1}(x)$, defined by the recursion

$$P_{n+1}(x) = 2x\, P_n(x) - P_{n-1}(x),$$

also satisfies $P_{n+1}(\cos\theta) = \cos(n+1)\theta$.

(d) Recalling from part (a) that $P_1(x) = x$ and $P_2(x) = 2x^2 - 1$, use the recursion relationship $P_{n+1}(x) = 2x\, P_n(x) - P_{n-1}(x)$ to establish that

$$P_3(x) = 4x^3 - 3x, \quad P_4(x) = 8x^4 - 8x^2 + 1,$$

and

$$P_5(x) = 16x^5 - 20x^3 + 5x, \quad P_6(x) = 32x^6 - 48x^4 + 18x^2 - 1.$$

(e) Show that

$$P_m(P_n(x)) = P_n(P_m(x)) = P_{mn}(x).$$

[Hint: Since $P_n(\cos\theta) = \cos n\theta$, we have $P_m(P_n(\cos\theta)) = P_m(\cos n\theta) = \cos(mn\theta)$.]

(f) From part (e) deduce that $P_2(P_2(x)) = P_4(x)$, $P_2(P_4(x)) = P_8(x)$, and in general that

$$P_2(P_{2^{n-1}}(x)) = P_{2^n}(x).$$

Chebyshev Polynomials and the Critical Square

In Chapter 3 the iterator $F(x) = x^2 - 2$ played a central role, separating iterators whose graphs lie above the base of the critical square (i.e., $x^2 - c$ for $c < 2$) and those whose graphs extend below this base ($x^2 - c$ for $c > 2$). As seen in Problem 3.7.14, the iterator $F(x) = x^2 - 2$ and its iterates are all tangent to the base of the critical square for F. They also enjoy some surprising relationships to the Chebyshev polynomials considered in Problem 4.6.7.

Problem 4.6.8 Defining $F^{[1]}(x) = x^2 - 2$ and $F^{[2]}(x) = F(F(x))$, $F^{[3]}(x) = F(F(F(x)))$, and so forth, and recalling that $P_2(x) = 2x^2 - 1$:

(a) Verify that $F^{[1]}(x) = 2P_2(x/2)$. Further, show that

$$F^{[2]}(x) = 2P_2^{[2]}(x/2), \quad F^{[3]}(x) = 2P_2^{[3]}(x/2),$$

and that in general,

$$F^{[n]}(x) = 2P_2^{[n]}(x/2).$$

[Hint: If for some n we have $F^{[n-1]}(x) = 2P_2^{[n-1]}(x/2)$, then

$$F^{[n]}(x) = F(F^{[n-1]}(x)) = F\left(2P_2^{[n-1]}(x/2)\right) = 2P_2(2P_2^{[n-1]}x/2)/2.]$$

(b) Use part (a) and Problem 4.6.7(f) to show that

$$F^{[n]}(x) = 2P_{2^n}(x/2).$$

(c) Check by *direct calculation* that $F^{[2]}(x) = 2P_4(x/2)$.

(d) If $m \geq 2$, show that there are m values of t such that $-1 \leq t \leq 1$ and $P_m(t) = t$.
[Hint: If $0 \leq \theta \leq \pi$ and $\cos m\theta = \cos\theta$, then if we set $t = \cos\theta$, we have

$$P_m(t) = P_m(\cos\theta) = \cos m\theta = \cos\theta = t.$$

Thus if we find m values of θ such that $0 \leq \theta \leq \pi$ and $\cos m\theta = \cos\theta$, we have the desired result.]

(e) Show that there are 2^n values of x with $-2 \leq x \leq 2$ such that $F^{[n]}(x) = x$. That is, the iterator $F(x) = x^2 - 2$ has the property that $F^{[n]}(x)$ has 2^n fixed points in the interval $-2 \leq x \leq 2$. [Hint: From part (b), the condition $F^{[n]}(x) = x$ is equivalent to $P_{2^n}(x/2) = x/2$.]

Another Form for Square Roots

Problem 4.6.9 As seen in Problem 4.3.3, if $k = a + ib$, then it is possible to represent the square root of k in the form $\sqrt[4]{a^2 + b^2}(\cos(\frac{1}{2}\tan^{-1}\frac{b}{a}) + i\sin(\frac{1}{2}\tan^{-1}\frac{b}{a}))$.

(a) Verify directly by *squaring* that the two square roots of $a + ib$ can also be represented in the form

$$\pm\left(\sqrt{\frac{a + \sqrt{a^2 + b^2}}{2}} + i\sqrt{\frac{-a + \sqrt{a^2 + b^2}}{2}}\right), \qquad \text{if} \quad b \geq 0,$$

or

$$\pm \left(\sqrt{\frac{a + \sqrt{a^2 + b^2}}{2}} - i \sqrt{\frac{-a + \sqrt{a^2 + b^2}}{2}} \right), \qquad \text{if } b \leq 0.$$

[Remark: In the above expressions, the symbol $\sqrt{}$ is applied only to nonnegative real numbers and denotes their positive square root.]

(b) Check that part (a) yields $\pm i\sqrt{3}$ as the square roots of -3.

(c) Check that part (a) yields $\pm(5 + 4i)$ as the square roots of $9 + 40i$.

Roots of Polynomials with Real Coefficients

Problem 4.6.10 In the case of a second-degree polynomial with real coefficients, the quadratic formula shows that complex roots occur in "complex conjugate pairs." That is, if $z = A + iB$ is a solution of $az^2 + bz + c = 0$, then so is $\bar{z} = A - iB$. This problem extends these considerations to nth-degree polynomials

$$P(z) = a_n z^n + a_{n-1} z^{n-1} + \cdots + a_1 z + a_0.$$

(a) Extend parts (a) and (b) of Problem 4.6.4 by showing that

$$\overline{(z_1 \pm z_2 \pm \cdots \pm z_n)} = \bar{z}_1 \pm \bar{z}_2 \pm \cdots \pm \bar{z}_n \quad \text{and} \quad \overline{z_1 z_2 \cdots z_n} = \bar{z}_1 \bar{z}_2 \cdots \bar{z}_n.$$

(b) Show that for any complex number z and any positive integer n we have $\overline{(z^n)} = (\bar{z})^n$.

(c) Suppose $a_0, a_1, a_2, \ldots, a_n$ are *real* numbers and z is a complex number. Use parts (a) and (b) to show that

$$\overline{(a_n z^n + a_{n-1} z^{n-1} + \cdots + a_1 z + a_0)}$$
$$= a_n (\bar{z})^n + a_{n-1} (\bar{z})^{n-1} + \cdots + a_1 \bar{z} + a_0.$$

(d) Let $P(z) = a_n z^n + a_{n-1} z^{n-1} + \cdots + a_1 z + a_0$ be a polynomial with *real* coefficients and suppose z_1 is a root of $P(z)$, i.e., that $P(z_1) = 0$. Show that $\overline{z_1}$ is also a root of $P(z)$. [Hint: If $P(z_1) = 0$, then $\overline{P(z_1)} = \bar{0} = 0$. Now apply part (c).]

(e) From part (d) conclude that the complex roots of a polynomial with real coefficients occur in "conjugate pairs," $z_1, \overline{z_1}; z_2, \overline{z_2}; \ldots$.

(f) Deduce from part (e) that a polynomial of *odd* degree with *real* coefficients must have at least one *real* root. [Hint: The fundamental theorem of algebra implies that a polynomial of degree n has n roots (some of which may be multiple roots). Keep in mind that if $z = \bar{z}$, then z is a real number.]

(g) Given that $z = i$ is a root of $P(z) = z^4 - 2z^3 + 3z^2 - 2z + 2$, find the other three roots.

A Root by Any Other Name ...

Problem 4.6.11 For the quadratic equation $z^2 + 3z - 10i = 0$:

(a) By direct substitution into the equation, confirm that $1 + 2i$ and $-4 - 2i$ are solutions.

(b) Show that the solutions of part (a) can be obtained by applying the quadratic formula to the above equation.

Simplifying High Powers with De Moivre

Problem 4.6.12 Recalling De Moivre's theorem from Problem 4.2.11:

(a) Show that for each positive integer n we have

$$(\sqrt{3} + i)^n = 2^n \left(\cos \left(\frac{n\pi}{6} \right) + i \sin \left(\frac{n\pi}{6} \right) \right).$$

(b) Find a simple form for $(\sqrt{3} + i)^{6000}$.

The Equation of the Boundary

Problem 4.6.13 For $k > 0$, the symbol $\sqrt{k}$ denotes the *positive* square root of k. To extend this notation to nonzero complex numbers $k = a + ib$, we make use of Problem 4.6.9(a). In the case $b \geq 0$, we shall write

$$\sqrt{k} = \sqrt{\frac{a + \sqrt{a^2 + b^2}}{2}} + i \sqrt{\frac{-a + \sqrt{a^2 + b^2}}{2}},$$

while if $b \leq 0$,

$$\sqrt{k} = -\sqrt{\frac{a + \sqrt{a^2 + b^2}}{2}} + i \sqrt{\frac{-a + \sqrt{a^2 + b^2}}{2}}.$$

Using these as *definitions* for $\sqrt{k}$:

(a) Show that for $a > 0$ and $b = 0$, the above formulas reduce to the usual definition of $\pm\sqrt{k}$.

(b) Assuming $b \geq 0$, show that the slope of the line through $\sqrt{k}$ and $-\sqrt{k}$ is given by $m = \frac{b}{a + \sqrt{a^2 + b^2}}$. [Observe that $a + \sqrt{a^2 + b^2} > 0$.]

(c) Again referring to Problem 4.6.9(a), show that the slope of the line through $\sqrt{k}$ and $-\sqrt{k}$ has exactly the same form as above, even when $b < 0$.

(d) Show that the perpendicular bisector of the line segment connecting $\sqrt{k}$ and $-\sqrt{k}$ has equation $(a + \sqrt{a^2 + b^2})x + by = 0$, as asserted in (4.3.10).

Basins of Attraction for the General Quadratic

Problem 4.6.14 Here we consider the general quadratic equation $az^2 + bz + c = 0$, where a, b, and c are complex numbers and $a \neq 0$. Denoting the roots by q_1 and q_2, where

$$q_1 = \frac{-b + \sqrt{b^2 - 4ac}}{2a} \quad \text{and} \quad q_2 = \frac{-b - \sqrt{b^2 - 4ac}}{2a},$$

we shall assume that the (possibly complex) term $b^2 - 4ac$ is different from zero and that as a result, $q_1 \neq q_2$. Letting L be the perpendicular bisector of the line segment connecting q_1 and q_2, we shall show that the sequence $z_1, z_2, \ldots$ generated by the iterator $F(z) = \frac{az^2 - c}{2az + b}$ approaches q_1 or q_2, depending on which side of L the initial point z_1 lies. Furthermore, if z_1 lies on L, then the sequence $z_1, z_2, \ldots$ remains on L. Our strategy for showing this will be to reduce the general problem to the case of the simple quadratic $z^2 - k = 0$, which has already been dealt with in Problem 4.3.7.

(a) Show that solving $az^2 + bz + c = 0$ is equivalent to solving

$$\left(z + \frac{b}{2a} \right)^2 - \left(\frac{b^2 - 4ac}{4a^2} \right) = 0.$$

(b) Let $k = \frac{b^2 - 4ac}{4a^2}$ and let L^* be the perpendicular bisector of the line segment joining $\sqrt{k}$ to $-\sqrt{k}$. From Problem 4.3.7 we know that the sequence $w_1, w_2, \ldots$ defined by $w_{n+1} = (w_n + k/w_n)/2, n = 1, 2, \ldots$, approaches $\sqrt{k}$ if w_1 is on the same side of L^* as $\sqrt{k}$ and approaches $-\sqrt{k}$ if w_1 is on the same side of L^* as $-\sqrt{k}$. If we now let $z_n = w_n - b/(2a), n = 1, 2, \ldots$, explain why the sequence $z_1, z_2, \ldots$ approaches q_1 or q_2, depending on the position of w_1 relative to L^*.

(c) With $z_1, z_2, \ldots$ and $w_1, w_2, \ldots$ as in part (b), show that z_1 is on the same side of L as q_1 if and only if w_1 is on the same side of L^* as $\sqrt{k}$, and similarly for q_2 and $-\sqrt{k}$. Conclude that $z_1, z_2, \ldots$ approaches q_1 or q_2, depending on which side of L the initial point z_1 lies (unless z_1 lies on L, in which case all points $z_1, z_2, \ldots$ belong to L).

(d) If $z_n = w_n - b/(2a), n = 1, 2, \ldots$, show that the w_n satisfy the recursion relation

$$w_{n+1} = (w_n + k/w_n)/2, \qquad n = 1, 2, \ldots,$$

if and only if the z_n satisfy

$$z_{n+1} + \frac{b}{2a} = \frac{1}{2} \left(z_n + \frac{b}{2a} + \frac{k}{z_n + \frac{b}{2a}} \right)$$

and that this is equivalent to

$$z_{n+1} = \frac{a z_n^2 - c}{2 a z_n + b}, \qquad n = 1, 2, \ldots.$$

(e) Conclude from the above that the sequence $z_1, z_2, \ldots$ generated by the iterator $F(z) = (az^2 - c)/(2az + b)$ approaches q_1 if z_1 is on the same side of L as q_1 and approaches q_2 if z_2 is on the same side of L as q_2. (If z_1 is on L, then so are all succeeding iterates.)

Problem 4.6.15 As an application of the ideas contained in the previous problem, consider the quadratic equation $z^2 - z = 0$, which has the two solutions $q_1 = 1$ and $q_2 = 0$. In this case the iterator $\frac{az^2 - c}{2az + b}$ reduces to $F(z) = \frac{z^2}{2z - 1}$.

(a) Using this iterator, construct a spreadsheet that generates the sequence of iterates $z_1, z_2, z_3, \ldots$ for arbitrary starting value z_1 and displays these complex numbers graphically, as in Figure 4.3.4. Observe that the line L with $x = \frac{1}{2}$ is the boundary separating the two basins of attraction. [Hint: Verifying that $\frac{z^2}{2z-1} = \frac{1}{4}(2z + \frac{1}{2z-1} + 1)$ can simplify the algebra required to express the iteration rule in terms of its real and imaginary parts.]

(b) Confirm with the spreadsheet that if z_1 belongs to L, then its iterates $z_2, z_3, \ldots$ also belong to L. Then use the iteration rule directly to show that this is always true, i.e., that $\mathrm{Re}(z_{n+1}) = \frac{1}{2}$ whenever $\mathrm{Re}(z_n) = \frac{1}{2}$.

Problem 4.6.16 Problem 4.4.2 dealt with the three complex solutions of $z^3 - 1 = 0$, viz.,

$$1, \quad -\frac{1}{2} + i\frac{\sqrt{3}}{2}, \quad \text{and} \quad -\frac{1}{2} - i\frac{\sqrt{3}}{2}.$$

(a) Denoting $-\frac{1}{2} + i\frac{\sqrt{3}}{2}$ by ω, show that these "cube roots of unity" can be written as 1, ω, and ω^2.

(b) Show that $\omega^3 = 1$, $\omega^2 = 1/\omega = \bar{\omega}$, and $1 + \omega + \omega^2 = 0$.

(c) Representing the three cube roots of unity as points on the unit circle, interpret the equation $1 + \omega + \omega^2 = 0$ geometrically (see Problem 4.4.2).

More on Cardano's Method

In Chapter 1 we used the iterator $F(x) = \frac{2ax^3 + bx^2 - d}{3ax^2 + 2bx + c}$ to approximate real solutions of the general cubic equation $ax^3 + bx^2 + cx + d = 0$, where a, b, c, and d are assumed to be real numbers. While this iterator is attractive as an alternative to the method of Cardano, it fails to provide information about complex solutions, which, as we have seen in Section 4.4, can have quite remarkable properties. The problems that follow revisit Cardano's method and highlight the role that complex numbers play in its application.

Recall that Cardano's method begins by reducing the general cubic to the special form

$$x^3 + px + q = 0,$$

where p and q are real numbers. Setting $x = u + v$, we are led (see Problem 1.11.16) to the equations

(∗) $$u^3 + v^3 = -q \quad \text{and} \quad u^3 v^3 = \frac{-p^3}{27}$$

as key to finding x.

Problem 4.6.17 Let $\omega = -\frac{1}{2} + i\frac{\sqrt{3}}{2}$ be the cube root of unity discussed in the previous problem. Assume that u and v are solutions of (∗), so that $x = u + v$ is a solution of $x^3 + px + q = 0$.

(a) Show that replacing u by ωu and v by $\omega^2 v$ gives another solution of (∗). Then show that replacing u by $\omega^2 u$ and v by ωv gives yet another solution of (∗).

(b) Use part (a) to show that if u and v are solutions of (∗), then the three roots of $x^3 + px + q = 0$ are $u + v$, $\omega u + \omega^2 v$, and $\omega^2 u + \omega v$.

(c) For the special cubic $x^3 + 3x - 1 = 0$, Cardano's method calls for solving

(∗∗) $$u^3 + v^3 = 1 \quad \text{and} \quad u^3 v^3 = -1.$$

Show that these equations lead to $(u^3)^2 - u^3 - 1 = 0$.

(d) Recalling that the positive solution of $t^2 - t - 1 = 0$ is the golden ratio $\tau = (1 + \sqrt{5})/2$, show that $u^3 = \tau$ and $v^3 = -1/\tau$ are solutions of (∗∗) and that

$$x = 0.3221853546$$

is a good approximation for one of the roots of $x^3 + 3x - 1 = 0$.

(e) Use the spreadsheet of Figure 1.8.5 to confirm the result obtained in part (d).

(f) Use part (a) to show that the complex roots of $x^3 + 3x - 1 = 0$ are given by

$$-\frac{1}{2}\left(\sqrt[3]{\tau} - \frac{1}{\sqrt[3]{\tau}}\right) + i\frac{\sqrt{3}}{2}\left(\sqrt[3]{\tau} + \frac{1}{\sqrt[3]{\tau}}\right)$$

and

$$-\frac{1}{2}\left(\sqrt[3]{\tau} - \frac{1}{\sqrt[3]{\tau}}\right) - i\frac{\sqrt{3}}{2}\left(\sqrt[3]{\tau} + \frac{1}{\sqrt[3]{\tau}}\right)$$

[Note that these complex roots are a conjugate pair, as expected, since the equation has real coefficients.]

The Irreducible Case

This problem deals with the case where all three solutions to the cubic $x^3 + px + q = 0$ are real. While Cardano's method does yield all three roots, they tend to emerge as real numbers in disguise, i.e., expressions involving the sum of two *nonreal* complex numbers whose "real nature" may be hard to discern. This phenomenon was a puzzlement to early algebraists and came to be known as *casus irreducibilis*, or the irreducible case in the solution of the cubic.

Problem 4.6.18 For the cubic equation $x^3 - x = 0$, factorization shows that $x = 0$, $x = 1$, and $x = -1$ are its three real solutions. If, however, we apply Cardano's method, the substitution $x = u + v$ leads to the equations

$$u^3 + v^3 = 0 \quad \text{and} \quad u^3 v^3 = 1/27.$$

(a) Noting that $u = i/\sqrt{3}$ and $v = -i/\sqrt{3}$ are solutions of the above equations, show that Cardano's method yields the solution $x = 0$ as the sum of two nonreal complex numbers.

(b) For $\omega = -\frac{1}{2} + i\frac{\sqrt{3}}{2}$, calculate $\omega u + \omega^2 v$ and $\omega^2 u + v$. Confirm that the resulting expressions yield the solution $x = -1$ and $x = 1$ as the sum of two nonreal complex numbers.

Problem 4.6.19 Consider the cubic equation $x^3 - 3x + 1 = 0$. (Note how little this differs from the equation considered in Problem 4.6.17(c)!) Here the substitution $x = u + v$ leads to

(∗) $$u^3 + v^3 = -1 \quad \text{and} \quad u^3 v^3 = 1.$$

(a) For the system above, show that u satisfies $(u^3)^2 + u^3 + 1 = 0$.

(b) Noting that $u^3 = -\frac{1}{2} + i\frac{\sqrt{3}}{2} = \cos(120°) + i\sin(120°)$ is a solution of the above quadratic, show that

$$u = \cos(40°) + i\sin(40°) \quad \text{and} \quad v = \cos(40°) - i\sin(40°)$$

is a solution of (∗) and that $x = 2\cos(40°)$ is a solution of $x^3 - 3x + 1 = 0$. [Hint: Given $u = \cos(40°) + i\sin(40°)$, use $u^3 v^3 = 1$ to find v.]

(c) Recalling Problem 4.6.17(a), show that the other two roots of $x^3 - 3x + 1 = 0$ are

$$x = 2\cos(80°) \quad \text{and} \quad x = 2\cos(160°).$$

(d) Use the spreadsheet format of Figure 1.8.5 to approximate the three real roots of $x^3 - 3x + 1 = 0$:

$$2\cos(40°) \approx 1.532088886,$$
$$2\cos(80°) \approx 0.3472963553,$$
$$2\cos(160°) \approx -1.879385242.$$

Viète's Trigonometric Solution

Problem 4.6.20 Viète (already mentioned in Section 1.2) found a clever way to use trigonometric functions to solve cubic equations in the irreducible case (i.e., when there are three real solutions). As applied to $x^3 - 3x + 1 = 0$, the method is as follows.

(a) Show that a number of the form $x = 2\cos\theta$ is a solution of $x^3 - 3x + 1 = 0$ if

$$4\cos^3\theta - 3\cos\theta = -\tfrac{1}{2}.$$

(b) Using the identity $\cos 3\theta = 4\cos^3\theta - 3\cos\theta$ (see Problem 4.6.6), infer from part (a) that $x = 2\cos\theta$ is a solution of $x^3 - 3x + 1 = 0$ if $\cos 3\theta = -\tfrac{1}{2}$.

(c) From parts (a) and (b) deduce that the solutions of $x^3 - 3x + 1 = 0$ are as given in Problem 4.6.19 above.

A Rough Approximation of the Julia Set

Problem 4.6.21 On a piece of graph paper generate a 21×21 grid of points corresponding to complex numbers of the form

$$\frac{m}{10} + \frac{n}{10}i, \qquad m, n = \pm 1, \pm 2, \dots, \pm 10.$$

(a) Regarding these as initial values for the iterative scheme underlying Figure 4.4.3, color each such z_1 black, red, or blue, depending on whether its iterates approach 1, $-\tfrac{1}{2} + i\tfrac{\sqrt{3}}{2}$, or $-\tfrac{1}{2} - i\tfrac{\sqrt{3}}{2}$, respectively. [Hint: Your work can be cut in half by noting that points colored black are symmetric about the x-axis, while red points below the x-axis are the reflection of blue points above, and vice versa.]

(b) Compare the results of part (a) with Figure 4.4.5.

The Symmetry of a Basin of Attraction

Problem 4.6.22 In Problem 4.6.21 we asserted that the basin of attraction of $z = 1$ is symmetric about the x-axis. In order to confirm this assertion, we shall refer to this set of initial values as A and show that a point z_1 belongs to A if and only if $\overline{z_1}$ belongs to A.

(a) Given an initial value z_1, let $z_2, z_3, \dots$ denote the sequence generated by the iterative scheme $z_{i+1} = F(z_i)$ with $F(z) = \frac{2z^3 + 1}{3z^2}$. Show that if z_1 is replaced by $\overline{z_1}$, the sequence of iterates is replaced by $\overline{z_2}, \overline{z_3}, \dots$. [Hint: Show that $F(\overline{z}) = \overline{F(z)}$.]

(b) Use part (a) to show that if z_1 belongs to A, so does $\overline{z_1}$.

(c) Assuming that $\overline{z_1}$ belongs to A, explain why z_1 must belong to A.

Rotational Symmetry of a Julia Set

Problem 4.6.23 Figure 4.4.4 suggests that in addition to symmetry about the x-axis, the Julia set enjoys a 3-fold *rotational* symmetry. For $F(z) = \frac{2z^3+1}{3z^2}$, let A, B, and C denote the basins of attraction for 1, $-\frac{1}{2} + i\frac{\sqrt{3}}{2}$, and $-\frac{1}{2} - i\frac{\sqrt{3}}{2}$, respectively. To confirm such rotational symmetry, we shall show that B is obtained by rotating A by $120°$ counterclockwise about the origin, C is obtained by rotating B by $120°$ counterclockwise about the origin, and A is obtained by rotating C by $120°$ counterclockwise about the origin.

(a) Recalling the notation $\omega = -\frac{1}{2} + i\frac{\sqrt{3}}{2} = \cos 120° + i \sin 120°$ for one of the cube roots of unity, show that $F(z) = \frac{2z^3+1}{3z^2}$ has the property $F(\omega z) = \omega F(z)$.

(b) Let z_1 belong to A, so that its iterates $z_2, z_3, \ldots$ approach $z = 1$. Explain carefully why the iterates of ωz_1 are $\omega z_2, \omega z_3, \ldots$ and why ωz_1 belongs to B.

(c) Show that a counterclockwise rotation through $120°$ sends B onto C and C onto A.

A Julia Set with Four-fold Rotational Symmetry

Problem 4.6.24 The ambitious reader is now encouraged to adapt the last three problems to the iterator $F(z) = \frac{1}{4}(3z + \frac{1}{z^3})$, whose fixed points are the four solutions of $z^4 - 1 = 0$.

Order Behind Apparent Chaos

Problem 4.6.25 Using the "Babylonian iterator" $F(x) = \frac{x^2+k}{2x}$ with $k = -1$ and x_1 real, one generates sequences of real numbers according to the rule

$$(*) \qquad x_{n+1} = \frac{1}{2}\left(x_n - \frac{1}{x_n}\right), \qquad n = 1, 2, \ldots.$$

While these iterates may appear to jump about chaotically on the real number line (see Problem 4.2.3), there is in fact a geometric pattern that governs their behavior. In order to discover this pattern we shall project the iterates generated by $(*)$ onto a circle of radius 1 centered at $(0, 1)$, using the center of the circle as center of projection.

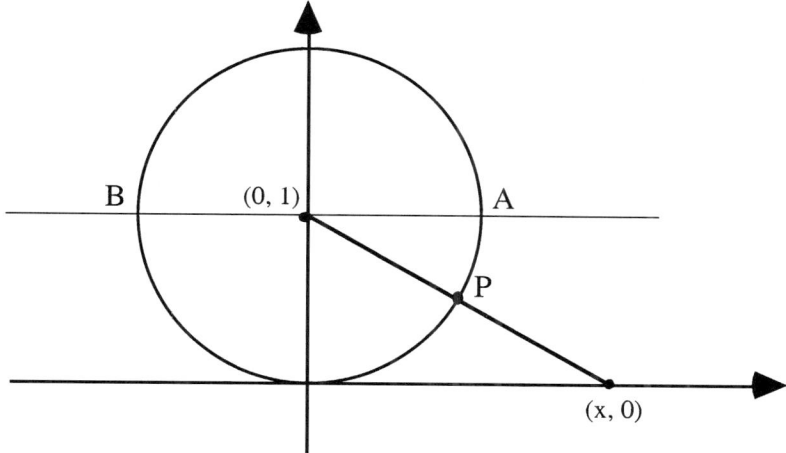

Instead of looking for a pattern among the points $(x, 0)$ on the x-axis, we shall look for a pattern among their projections P.

(a) Consider points $(a, 0)$ and $(b, 0)$ in the (x, y)-plane whose relationship is determined in terms of the line segments connecting them to $(0, 1)$, as in the picture below. Specifically, if the line segment connecting $(a, 0)$ with $(0, 1)$ makes an acute angle θ with the line $y = 1$, then $(b, 0)$ is to make an angle 2θ.

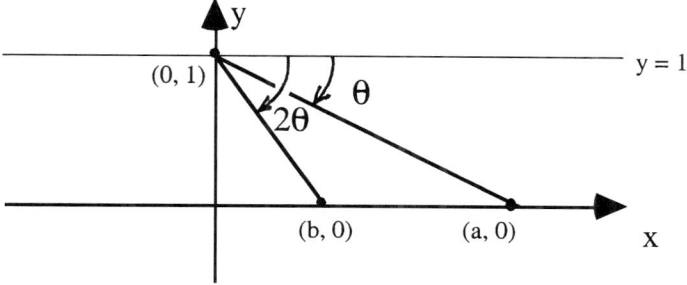

For $a < 0$ we measure θ and choose b as below:

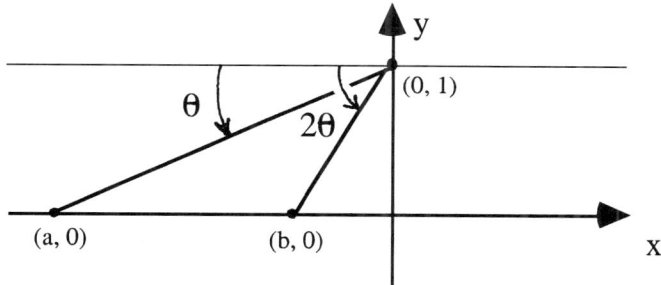

Show that with the above conventions, if $a > 0$, then $\tan\theta = 1/a$ and $\tan 2\theta = 1/b$, while if $a < 0$, then $\tan\theta = -1/a$ and $\tan 2\theta = -1/b$. Deduce that in both cases we have $b = (a - 1/a)/2$. [Hint: Use the identity $\tan 2\theta = \frac{2\tan\theta}{1-\tan^2\theta}$.]

(b) Recalling the first figure above, show that if $x_1, x_2, \ldots$ is a sequence generated by the rule (∗) and $P_1, P_2, \ldots$ is the sequence of projections of $(x_1, 0), (x_2, 0)\ldots$ onto the unit circle centered at $(0, 1)$, then:

If $x_n > 0$, the circular arc from $A = (1, 1)$ to P_{n+1} is twice the arc from A to P_n.

If $x_n < 0$, the circular arc from $B = (-1, 1)$ to P_{n+1} is twice the arc from B to P_n.

(c) Verify that the initial value $x_1 = 1/\sqrt{3}$ leads to the cycle

$$\frac{1}{\sqrt{3}}, \quad -\frac{1}{\sqrt{3}}, \quad \frac{1}{\sqrt{3}}, \quad -\frac{1}{\sqrt{3}}, \ldots$$

Interpret this pattern geometrically in terms of the points $P_1, P_2, \ldots$ on the circle.

More Cantor Dust in the Plane

Problem 4.6.26 This problem deals with another fractal that qualifies as "dust." Much like the set in Problem 4.5.10, it is a planar analogue of the Cantor set.

(a) Starting with a unit square, remove the "middle cross" made up of 5 squares each $\frac{1}{3}$ the size of the initial square:

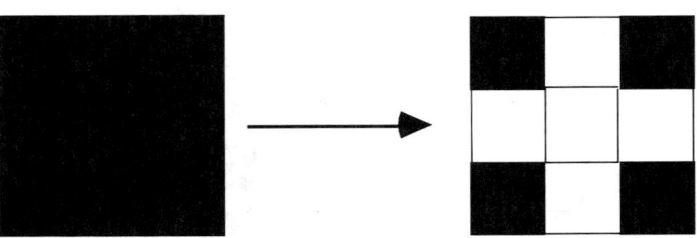

This process leaves 4 little squares, each having area $\frac{1}{9}$ that of the original. Next remove middle crosses from these 4 little squares, leaving behind 16 squares, each with $\frac{1}{81}$ the area of the original. Continuing this process *ad infinitum*, calculate the total area removed.

(b) Denoting by Q the "planar Cantor dust" produced by removing infinitely many such middle crosses, use (4.5.2) to show that Q has dimension

$$D = \frac{\log 4}{\log 3} \approx 1.26186.$$

5

ALGEBRAIC SYSTEMS

5.1. Introduction Until now our "models for change" have been based on difference equations involving a single variable x_i. In fact, however, most real-world phenomena involve the interaction of several variables. For example, in modeling the growth of a population we have made use of the logistic equation $x_{i+1} - x_i = ax_i - bx_i^2$. However, a crucial factor affecting the growth of a population x_i may be its interaction with another population, denoted perhaps by y_j. Similarly, our microeconomic models were based on the price x_i of a single product, one that determines the behavior of producers and consumers in a particular marketplace. However, real markets tend to involve the interaction of several products and their prices.

Such examples suggest that an important next step in dynamic modeling will be to consider several difference equations that are linked to one another. In such a "systems context," we will face the challenge of building into our models meaningful assumptions about how the state of one variable affects the state of others.

Our study of *systems* of difference equations will begin by considering systems of *algebraic* equations. A particular algebraic system has already been encountered in Chapter 1. There we considered the linear system

$$(5.1.1) \qquad \begin{aligned} ax + by &= e, \\ cx + dy &= f, \end{aligned}$$

and noted that every such pair of linear equations in two unknowns can be represented by a pair of lines in the (x, y)-plane. Furthermore, a point of intersection of these lines (if any) corresponds to a solution of (5.1.1) (if any).

Problems reducible to a system of linear equations involving more than one variable can be traced back to ancient times. For instance, the Chinese mathematical treatise *Jiuzhang*, dating to about 200 B.C., poses the following problem:

> The price of 1 acre of good land is 300 pieces of gold; the price of 7 acres of bad land is 500 pieces of gold. If one has purchased 100 acres and the price was 10,000 pieces of gold, how much good land was bought and how much bad land?

The Chinese rule for solution was based on "the method of false position," which we already encountered in Section 1.3. Starting with a guess, "Suppose there are 20 acres of good land and 80 acres of bad" (satisfying the first equation but not the second) and then making another such guess, false position enables one to adjust this initial guess so as to arrive at the solution. This method for solving (5.1.1) is explored in problems at the end of this chapter.

A more contemporary approach would call for letting x denote "acres of good land" and y denote "acres of bad land." This leads to

(5.1.2)
$$x + y = 100,$$
$$300x + \frac{500}{7}y = 10,000,$$

which we now tend to solve by means of substitution or by determinants.

Problem 5.1.1 Represent (5.1.2) as a pair of lines in the first quadrant of the (x, y)-plane.

(a) Find an approximate value for the solution geometrically.

(b) Find a precise value for the solution algebraically.

Another problem from the *Jiuzhang*, one leading to three linear equations in three unknowns, is as follows:

> A man went to market, where he sold 2 oxen and 5 sheep, bought 13 pigs, and came home with a net profit of 1000 pieces of currency. A second man sold 3 oxen and 3 pigs, bought 9 sheep, and broke even. A third man sold 6 sheep and 8 pigs, bought 5 oxen, and returned home having spent 600 pieces of currency. What was the market price of each animal?

Problem 5.1.2 Use modern techniques to solve the above problem from the *Jiuzhang*.

The seventeenth century saw a flourishing of systematic methods for solving such systems of equations. During the eighteenth century these found their way into school curricula, largely through Colin Maclaurin's *A Treatise on Algebra in Three Parts*, which popularized the method for solving (5.1.1) with what we now refer

	A	B	C	D	E	F	G	H	I	J	K
1	This spreadsheet implements Cramer's rule for the system ax + by = e; cx + dy = f, giving values for x and y whenever the determinant ad - bc ≠ 0.										
2											
3	a=1		b=1		e=100			Det=	-228.6		
4											
5	c=300		d=71.4286		f=10000						
6											
7	x=12.50		y=87 5								

Figure 5.1.1 Implementing Cramer's Rule for $n = 2$

to as "Cramer's rule." This method is based on the fact that the solution of (5.1.1) can be written as

$$x = \frac{de - bf}{ad - bc} \quad \text{and} \quad y = \frac{af - ce}{ad - bc}$$

whenever $ad - bc \neq 0$. Happily, for 2 equations in 2 unknowns such rules are easy to implement on a spreadsheet (Figure 5.1.1).

Problem 5.1.3 Create a spreadsheet for solving the system (5.1.1) in the above format.

Even though the concept of a determinant had appeared in the late seventeenth century, Maclaurin did not relate determinants to this method of solution (or to the corresponding method for solving three or four equations as also dealt with in his book). He did note, however, that each term in the numerator of such a "solution by Cramer's rule" involves the right side of the system, whereas the denominator consists only of coefficients of the unknowns. In fact, this denominator can be taken to be the determinant associated with such a system.

It was not until the nineteenth century that Carl Friedrich Gauss began to relate systems of linear equations to the concepts of matrices and determinants. Letting $x = x_1$, $y = x_2$, and renaming the coefficients in (5.1.1), we can write such systems in the form

(5.1.3)
$$a_{11}x_1 + a_{12}x_2 + b_1 = 0,$$
$$a_{21}x_1 + a_{22}x_2 + b_2 = 0.$$

Problem 5.1.4 State the relationships between coefficients of (5.1.1) and (5.1.3). Express the solution of (5.1.3) by using Cramer's rule.

Gauss's notation enables us to use matrix notation to write (5.1.3) as

(5.1.4)
$$\begin{bmatrix} a_{11} & a_{12} \\ a_{21} & a_{22} \end{bmatrix} \begin{bmatrix} x_1 \\ x_2 \end{bmatrix} + \begin{bmatrix} b_1 \\ b_2 \end{bmatrix} = \begin{bmatrix} 0 \\ 0 \end{bmatrix},$$

or as

(5.1.5) $A\mathbf{x} + \mathbf{b} = \mathbf{0},$

with

$$A = \begin{bmatrix} a_{11} & a_{12} \\ a_{21} & a_{22} \end{bmatrix}, \quad \mathbf{x} = \begin{bmatrix} x_1 \\ x_2 \end{bmatrix}, \quad \mathbf{b} = \begin{bmatrix} b_1 \\ b_2 \end{bmatrix}, \quad \text{and} \quad \mathbf{0} = \begin{bmatrix} 0 \\ 0 \end{bmatrix}.$$

Here (5.1.5) is reminiscent of the first degree polynomial equation

(5.1.6) $ax + b = 0.$

Observing that (5.1.6) has the solution $x = -b/a$, we can use this familiar example to motivate a matrix analogue. Assuming that the solution of (5.1.5) can be written as

(5.1.7) $\mathbf{x} = -A^{-1}\mathbf{b},$

Problems 5.1.4 and 5.1.5 lead us to a definition of A^{-1}.

Problem 5.1.5 Recalling the solution of $A\mathbf{x} + \mathbf{b} = \mathbf{0}$ derived in Problem 5.1.4, show that Cramer's rule is embodied in (5.1.7) and the definition

(5.1.8) $$A^{-1} = \begin{bmatrix} \dfrac{a_{22}}{D} & \dfrac{-a_{12}}{D} \\ \dfrac{-a_{21}}{D} & \dfrac{a_{11}}{D} \end{bmatrix},$$

where $D = a_{11}a_{22} - a_{12}a_{21}$ is the determinant of A.

Problem 5.1.6 With $I = \begin{bmatrix} 1 & 0 \\ 0 & 1 \end{bmatrix}$ and with A and A^{-1} as given above, confirm that $AA^{-1} = A^{-1}A = I$.

Problem 5.1.7 Construct a spreadsheet that calculates A^{-1} for an arbitrary 2×2 matrix A whose determinant is different from zero. Use this spreadsheet to solve (5.1.2) by finding the inverse of $\begin{bmatrix} 1 & \frac{1}{500} \\ 300 & \frac{500}{7} \end{bmatrix}$.

Having established the close connection between A^{-1} and Cramer's rule, there arises the question of whether, in the age of computers, it is also possible to solve $A\mathbf{x} + \mathbf{b} = \mathbf{0}$ by iterative techniques such as those considered in Chapter 2. To address this question we will begin by using iteration to solve the disarmingly simple polynomial equation $ax + b = 0$. As in Chapter 1, this calls for finding an iterator $F(x)$ for which $x = -b/a$ is an attracting fixed point.

With such an iterator in hand, efforts to extend this method to the matrix equation $A\mathbf{x} + \mathbf{b} = \mathbf{0}$ will be hard to resist. The fact that spreadsheets can be used

	A	B	C	D	E	F	G	H	I	J	K	L	M	N	O
1		This spreadsheet iterates a column vector x(1) by an arbitrary 2 x 2 matrix.													
2															
3		Enter A below			x(1)	x(2)	x(3)	x(4)	x(5)	x(6)	x(7)	x(8)	x(9)	x(10)	x(11)
4		2	-1		2	5	11	23	47	95	191	383	767	1535	3071
5		0	1		-1	-1	-1	-1	-1	-1	-1	-1	-1	-1	-1

Figure 5.1.2 A Matrix Iterator

to implement matrix iterations adds to the lure of such an approach. The following problems prepare us for such efforts by developing a spreadsheet format for "matrix iteration."

Problem 5.1.8 For $A = \begin{bmatrix} 2 & -1 \\ 0 & 1 \end{bmatrix}$ and $\mathbf{x} = \begin{bmatrix} 2 \\ -1 \end{bmatrix}$, find $A\mathbf{x}$, $A(A\mathbf{x})$, and $A(A(A\mathbf{x}))$.

Problem 5.1.9 In the format of Figure 5.1.2, create a spreadsheet that calculates $A\mathbf{x}$, $A(A\mathbf{x}) = A^2\mathbf{x}$, $A(A(A\mathbf{x})) = A^3\mathbf{x}, \ldots, A^{10}\mathbf{x}$ for an arbitrary 2×2 matrix A and 2×1 column vector $\mathbf{x}$.

5.2. Linear Systems
Given the ease with which we can solve

(5.2.1)
$$ax + b = 0$$

by *al-jabr*, the idea of solving this equation by iteration may seem like a dubious enterprise. However, if such techniques should enable us to bypass Cramer's rule with an iteration-based method for solving

(5.2.2)
$$A\mathbf{x} + \mathbf{b} = 0,$$

then such an approach may turn out to have merit after all. With this in mind, we interrupt our consideration of systems to examine the simple scalar equation (5.2.1).

In order to find a function $F(x)$ with a fixed point that corresponds to the solution of (5.2.1), we simply add x to both sides of this equation. This yields

(5.2.3)
$$(a + 1)x + b = x$$

and leads us to observe that the solution $x = -b/a$ of (5.2.1) is the unique fixed point of

(5.2.4)
$$F(x) = (a + 1)x + b.$$

There now arises the important question of whether $x_0 = -b/a$ is an attractor for $F(x)$. As shown in Chapter 1, this depends on the slope of the graph of $F(x)$

at the fixed point $(-b/a, -b/a)$. Since the graph of $F(x)$ is a straight line with slope $a + 1$ for all x, the condition

(5.2.5) $-1 < a + 1 < 1,$

or equivalently,

(5.2.6) $-2 < a < 0,$

is both necessary and sufficient for x_0 to be an attractor (see also Problem 2.3.10).

Another way of arriving at (5.2.5) is to implement $x_{i+1} = (a + 1)x_i + b$, for $i = 1, 2, \ldots$. Here we note that for any initial value x_1, the iterations

$$x_2 = (a + 1)x_1 + b,$$
$$x_3 = (a + 1)[(a + 1)x_1 + b] + b$$
$$= (a + 1)^2 x_1 + (a + 1)b + b,$$

. . .

give rise to

$$x_n = (a + 1)^{n-1} x_1 + (a - 1)^{n-2} b + (a + 1)^{n-3} b + \cdots + (a + 1)b + b$$
$$= (a + 1)^{n-1} x_1 + b[(a + 1)^{n-2} + (a + 1)^{n-3} + \cdots + (a + 1) + 1].$$

Problem 5.2.1

(a) Verify that if $r \neq 1$, then

$$1 + r + r^2 + \cdots + r^{n-1} = \frac{1 - r^n}{1 - r}.$$

(b) Sum the geometric series in last expression for x_n above to show that

$$x_n = (a + 1)^{n-1} x_1 + \frac{b[1 - (a + 1)^{n-1}]}{-a}.$$

(c) Explain why x_n approaches $-b/a$ whenever $-2 < a < 0$.

But what about cases in which (5.2.6) is *not* satisfied? Here the equation $ax + b = 0$ still has the solution $x = -b/a$! Can we use iteration to find this solution in such circumstances?

By way of arriving at a positive answer, we consider two cases:

Case 1: If $a \leq -2$, we rewrite (5.2.1) in the form $(a/N)x + b/N = 0$ and choose N sufficiently large so that $-2 < a/N < 0$. Then, applying the iterator $F_N(x) = ((a/N) + 1)x + b/N$ to an arbitrary initial x_1, we obtain a sequence $x_2, x_3, \ldots$ that approaches $-(b/N)/(a/N) = -b/a$.

Case 2: If $a > 0$, we define $\tilde{a} = -a$ such that $\tilde{a} < 0$. Then (5.2.1) becomes $\tilde{a}x - b = 0$. If $\tilde{a} \leq -2$, we apply the transformation of Case 1. However, for any $\tilde{a} < 0$ we are assured that iterations of either $F(x) = (\tilde{a} + 1)x - b$ or of some $F_N(x) = (\tilde{a}/N)x - b/N$ will approach the solution $x = -\tilde{a}/(-b) = -a/b$.

Problem 5.2.2 By subtracting x_i from both sides of $x_{i+1} = (a + 1)x_i + b$, find a first-order difference equation with equilibrium value $x_0 = -b/a$ that is equivalent to this iterative scheme. Show that Case 1 above corresponds to a refinement $\Delta t \mapsto \Delta t/N$ of this difference equation.

Having established an iterative scheme for solving any equation of the form $ax + b = 0$, we turn to the matrix case. Adapting Problem 5.2.1 to systems of the form $A\mathbf{x} + \mathbf{b} = 0$ turns out to be quite straightforward. Rewriting (5.2.2) as

(5.2.7)
$$(A + I)\mathbf{x} + \mathbf{b} = \mathbf{x}$$

and choosing an initial column vector $\mathbf{x}_1$, we define $\mathbf{F}(\mathbf{x}) = (A + I)\mathbf{x} + \mathbf{b}$ to obtain

(5.2.8)
$$\mathbf{x}_2 = (A + I)\mathbf{x}_1 + \mathbf{b},$$
$$\mathbf{x}_3 = (A + I)[(A + I)\mathbf{x}_1 + \mathbf{b}] + \mathbf{b},$$
$$\mathbf{x}_4 = (A + I)[(A + I)[(A + I)\mathbf{x}_1 + \mathbf{b}] + \mathbf{b}] + \mathbf{b},$$

. . .

The question now before us is whether the column vectors $\mathbf{x}_2, \mathbf{x}_3, \mathbf{x}_4, \ldots$ will approach the solution of (5.2.2) that is given by Cramer's rule. This is answered in the following generalization of Problem 5.2.1.

Problem 5.2.3 Given a matrix $A \neq -I$, let $B = A + I$.

(a) Verify that $(I - B)(I + B + B^2 + \cdots + B^{n-1}) = I - B^n$.

(b) The condition $\det A \neq 0$ assures that A has an *inverse matrix* A^{-1} such that $AA^{-1} = A^{-1}A = I$. Then $(-A)^{-1} = -A^{-1} = (I - B)^{-1}$. Show that part (a) gives

$$I + B + B^2 + \cdots + B^{n-1} = (I - B)^{-1}(I - B^n).$$

(c) Equation (5.2.8) of this section gives

$$\mathbf{x}_n = \mathbf{b} + B\mathbf{b} + B^2\mathbf{b} + \cdots + B^{n-2}\mathbf{b} + B^{n-1}\mathbf{x}_1.$$

Show that this can be written

$$\mathbf{x}_n = B^{n-1}\mathbf{x}_1 + (I - B)^{-1}(I - B^{n-1})\mathbf{b}$$
$$= (A + I)^{n-1}\mathbf{x}_1 + (-A)^{-1}(I - (A + I)^{n-1})\mathbf{b}.$$

	A	B	C	D	E	F	G	H	I	J	K	L	M	N	O
1	This spreadsheet implements the matrix iterations x(i+1) = (A+l)x(i) + b for arbitrary A, b, and x(1). It also calculates the solution of Ax + b = 0 by Cramer's rule.														
2															
3	b	Enter A below			x(1)	x(2)	x(3)	x(4)	x(5)	x(6)	x(7)	x(8)	x(9)	x(10)	x(11)
4	1	-0.5	0.3		1	2.1	2.14	2.12	2.1	2.09	2.07	2.06	2.05	2.0428	2.0359
5	-1	0.5	-0.6		2	0.3	0.17	0.14	0.12	0.1	0.08	0.07	0.06	0.0485	0.0408
6														x(20)	x(50)
7			D=	0.15				Solution of Ax+b=0	2					2.008	2
8								by Cramer's rule	0					0.009	5E-05

Figure 5.2.1 Solving $Ax + b = 0$ by Iteration

(d) Assuming that $(A + I)^n$ approaches the zero matrix as n grows without bound, deduce from (c) that $\mathbf{x}_n$ approaches $-(A^{-1})\mathbf{b}$, i.e., the solution given by Cramer's rule (see Problem 5.1.5).

In Figure 5.2.1, the iterative scheme $\mathbf{x}_{i+1} = (A + I)\mathbf{x}_i + \mathbf{b}$ has been put into spreadsheet format. It enables you to specify the entries of the 2×2 matrix A, the column vector $\mathbf{b}$, and to choose an arbitrary initial column vector $\mathbf{x}_1$. For purposes of comparison, this spreadsheet also calculates the solution of $A\mathbf{x} + \mathbf{b} = 0$ by Cramer's rule.

Experimentation with Figure 5.2.1 suggests that bypassing Cramer's rule with an iterator of the form (5.2.7) is indeed a possibility. For the entries

$$A = \begin{bmatrix} -0.5 & 0.3 \\ 0.5 & -0.6 \end{bmatrix}$$

and arbitrary choices of $\mathbf{b}$ and $\mathbf{x}_1$, the 36th iteration of x_1 by

$$(5.2.9) \qquad\qquad F(\mathbf{x}) = (A + I)\mathbf{x} + \mathbf{b}$$

tends to be remarkably close to the solution calculated by Cramer's rule. It is only when the above entries for A are substantially altered that solution by iteration begins to fail.

But then, some such difficulties were to be expected. Even in the scalar case, the condition $|a + 1| < 1$ is essential to solving $ax + b = 0$ by iteration. Accordingly, something akin to $|a + 1| < 1$ will be essential to solving (5.2.2) by matrix iteration.

In Section 5.3 below, we give an appropriate definition to "$\|A + I\| < 1$," one that assures that iterations of an initial $\mathbf{x}_1$ by (5.2.9) will in fact approximate solutions of $A\mathbf{x} + \mathbf{b} = 0$.

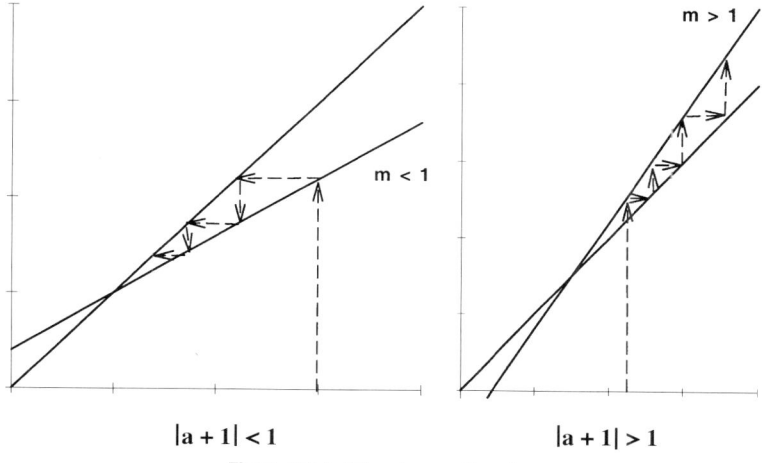

$$|a + 1| < 1 \qquad\qquad |a + 1| > 1$$

Figure 5.3.1 Attractor vs. Repeller

5.3. Contraction Matrices   Our hopes for solving $A\mathbf{x} + \mathbf{b} = \mathbf{0}$ seem to depend on an appropriate definition of

$$(5.3.1) \qquad\qquad \|A + I\| < 1.$$

Here it may be useful to return to the scalar case where the condition $|a + 1| < 1$ can be given a useful geometric interpretation.

The graph of the iterator $F(x) = (a + 1)x + b$ is a straight line through $(0, b)$ with slope $m = a + 1$. In applying the staircase method to this iterator at its fixed point $x_0 = -b/a$, we have seen that $(-b/a, -b/a)$ is an attractor for F if and only if $|a + 1| < 1$.

Since staircase diagrams are hard to visualize for matrix equations, we seek another geometric representation of the dichotomy in Figure 5.3.1. It will be based on the fact that $|a + 1| < 1$ assures that $F(x) = (a + 1)x + b$ "moves points closer together." That is, if $|a + 1| < 1$ and $x \neq y$, then

$$(5.3.2) \qquad\qquad |F(x) - F(y)| < |x - y|.$$

By contrast, if $|a + 1| > 1$, then $F(x) = (a + 1)x + b$ "moves points farther apart." That is, if $|a + 1| > 1$ and $x \neq y$, then

$$(5.3.3) \qquad\qquad |F(x) - F(y)| > |x - y|.$$

The dichotomy illustrated in Figure 5.3.2 can also be established as follows.

Problem 5.3.1 For $F(x) = (a + 1)x + b$:

(a) Show that $|F(x) - F(y)| = |(a + 1)(x - y)|$.

(b) Use (a) to show that $|a + 1| < 1$ guarantees (5.3.2) whenever $x \neq y$.

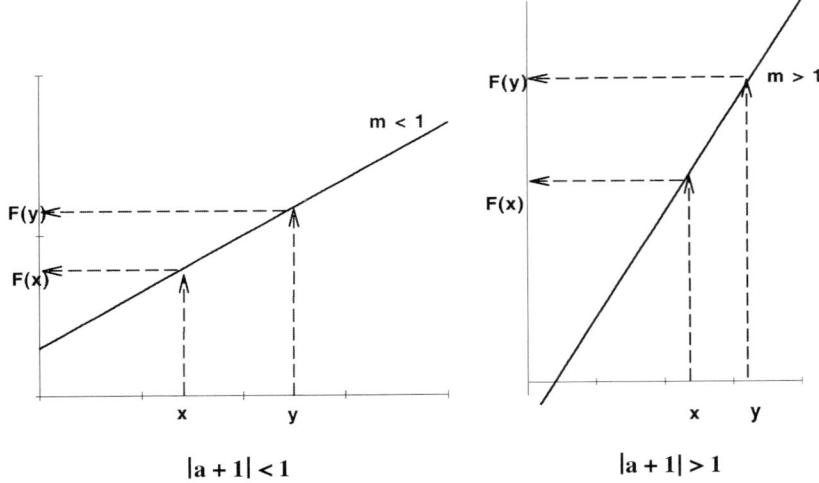

$$|a + 1| < 1 \qquad\qquad |a + 1| > 1$$

Figure 5.3.2 Contraction vs. Dilation

Functions satisfying (5.3.2) are called *contraction mappings*, and they play an important role in various areas of mathematics. In the case considered above, the fact that $F(x)$ is linear enabled us to establish conditions under which it is a contraction mapping on the entire x-axis. However, for more general F it may happen that (5.3.2) is satisfied for some x and y, but not for others.

Problem 5.3.2 For $k > 0$, show that the "Babylonian iterator" $F(x) = (x + k/x)/2$ is a contraction mapping whenever $x > \sqrt{k/3}$ but not when $0 < x < \sqrt{k/3}$.

To see why contraction mappings turn fixed points into attractors, we need only replace y by a fixed point in (5.3.2). If $y = x_0$ and $F(x_0) = x_0$, then (5.3.2) becomes

$$|F(x) - x_0| < |x - x_0|.$$

This inequality corresponds, in functional notation, to the staircase diagram in Figure 5.3.1: If $|a + 1| < 1$, then $F(x_i)$ is closer to x_0 than was x_i.

On this basis, it becomes natural to interpret (5.3.1) in the sense of (5.3.2), i.e., that the "matrix $A + I$ moves points closer together." What is new is that "points" are now

  located in a space of dimension 2;

  represented by column vectors $\mathbf{x}$ and $\mathbf{y}$.

Setting $B = A + I$, we seek to define (5.3.1) in such a way that it becomes equivalent to the condition that "the distance between $B\mathbf{x}$ and $B\mathbf{y}$ is less than the distance between $\mathbf{x}$ and $\mathbf{y}$." Recalling the definition $\|\mathbf{x}\| = \sqrt{x_1^2 + x_2^2}$ and using the notation $\|\mathbf{x} - \mathbf{y}\|$ for the distance between vectors $\mathbf{x}$ and $\mathbf{y}$, this becomes

(5.3.4) $\|B\mathbf{x} - B\mathbf{y}\| < \|\mathbf{x} - \mathbf{y}\|$

for all column vectors $\mathbf{x} = \text{col}[x_1, x_2]$ and $\mathbf{y} = \text{col}[y_1, y_2]$, $\mathbf{x} \neq \mathbf{y}$. It follows from Problem 5.8.19 at the end of this chapter that if $\mathbf{x}_0$ is a fixed point for $\mathbf{F}(\mathbf{x}) = B\mathbf{x} + \mathbf{b}$, then (5.3.4) does indeed assure that iterations of an initial $\mathbf{x}_1$ will approach $\mathbf{x}_0$.

These observations raise the following important question:

What conditions on $a_{11}, a_{12}, a_{21}, a_{22}$ will assure that $B\mathbf{x} = (A + I)\mathbf{x}$ is a contraction mapping in the sense of (5.3.4)?

Problem 5.3.3 Show that (5.3.4) is satisfied if and only if $\|B\mathbf{z}\| < \|\mathbf{z}\|$, for all $\mathbf{z} \neq 0$. [Hint: Setting $\mathbf{z} = \mathbf{x} - \mathbf{y}$, use the fact that $B\mathbf{z} = B\mathbf{x} - B\mathbf{y}$.]

In light of Problem 5.3.3, we are now faced with the simpler task of determining conditions under which

$$B = \begin{bmatrix} b_{11} & b_{12} \\ b_{21} & b_{22} \end{bmatrix}$$

is a *contraction matrix* satisfying

(5.3.5) $$\|B\mathbf{x}\| < \|\mathbf{x}\|, \quad \text{for} \quad \mathbf{x} \neq 0.$$

An important tool in this regard is the Cauchy inequality

(5.3.6) $$(a_1 b_1 + a_2 b_2)^2 \leq (a_1^2 + a_2^2)(b_1^2 + b_2^2)$$

for all real a_1, a_2, b_1, b_2. This inequality follows geometrically from the fact that the expression $a_1 b_1 + a_2 b_2$ appearing on the left side of (5.3.6) is the "dot product"

(5.3.7) $$\mathbf{a} \cdot \mathbf{b} = \|\mathbf{a}\| \|\mathbf{b}\| \cos \theta,$$

where θ is the angle between the vectors $\mathbf{a} = (a_1, a_2)$ and $\mathbf{b} = (b_1, b_2)$.

Problem 5.3.4 Use (5.3.7) to establish the inequality (5.3.6).

Problem 5.3.5 Starting with the fact that $(a_1 b_2 - a_2 b_1)^2 \geq 0$, give an algebraic proof of (5.3.6).

Returning now to the contraction matrix criterion (5.3.5), we note that for $\mathbf{x} = \text{col}[x_1, x_2]$, $\|B\mathbf{x}\| < \|\mathbf{x}\|$ for $\mathbf{x} \neq 0$ is equivalent to $\|B\mathbf{x}\|^2 < \|\mathbf{x}\|^2$, which becomes

$$(b_{11}x_1 + b_{12}x_2)^2 + (b_{21}x_1 + b_{22}x_2)^2 < x_1^2 + x_2^2$$

if $(x_1, x_2) \neq (0, 0)$. By Cauchy's inequality,

$$(b_{11}x_1 + b_{12}x_2)^2 \leq (b_{11}^2 + b_{12}^2)(x_1^2 + x_2^2)$$

and

$$(b_{21}x_1 + b_{22}x_2)^2 \leq (b_{21}^2 + b_{22}^2)(x_1^2 + x_2^2).$$

Adding the last two equations yields

$$(b_{11}x_1 + b_{12}x_2)^2 + (b_{21}x_1 + b_{22}x_2)^2$$

(5.3.8)
$$\leq (b_{11}^2 + b_{12}^2 + b_{21}^2 + b_{22}^2)(x_1^2 + x_2^2).$$

The above inequality motivates us to define a rather unorthodox measure of "the size of B" by

(5.3.9)
$$|||B||| = (b_{11}^2 + b_{12}^2 + b_{21}^2 + b_{22}^2)^{1/2}.$$

Taking the square root of both sides of (5.3.8), we now have

(5.3.10)
$$\|B\mathbf{x}\| \leq |||B||| \cdot \|\mathbf{x}\|.$$

Problem 5.3.6 Show that the condition

$$|||B||| = (b_{11}^2 + b_{12}^2 + b_{21}^2 + b_{22}^2)^{1/2} < 1$$

is sufficient to ensure that B is a contraction matrix.

Applying Problem 5.3.6 in the case $B = A + I$, the corresponding condition for A can be formulated as

(5.3.11)
$$(1 + a_{11})^2 + a_{12}^2 + a_{21}^2 + (1 + a_{22})^2 < 1$$

The spreadsheet in Figure 5.3.3 calculates $-A^{-1}\mathbf{b}$, indicates whether or not (5.3.11) is satisfied, and implements iterations of $\mathbf{x}_1$ by $\mathbf{F}(\mathbf{x}) = (A + I)\mathbf{x} + \mathbf{b}$.

Experimentation with this spreadsheet should confirm the fact that whenever (5.3.11) holds, $A\mathbf{x} + \mathbf{b} = 0$ can be solved by iteration. Such experimentation may also indicate, however, that while (5.3.11) is a *sufficient* condition for $\mathbf{F}(\mathbf{x}) = (A + I)\mathbf{x} + \mathbf{b}$ to be a contraction mapping, it is not a *necessary* one.

	A	B	C	D	E	F	G	H	I	J	K	L	M	N	O
1	This spreadsheet implements the matrix iterations x(i+1) = (A+I)x(i) + b for arbitrary A, b, and x(1). It also calculates |||A+I||| and the solution of Ax + b = 0 by Cramer's rule.														
2															
3	b	Enter A below			x(1)	x(2)	x(3)	x(4)	x(5)	x(6)	x(7)	x(8)	x(9)	x(10)	x(11)
4	1	-0.5	0.3		1	2.1	2.1	2.1	2.1	2.1	2.1	2.1	2.1	2.04	2.04
5	-1	0.5	-0.6		2	0.3	0.2	0.1	0.1	0.1	0.1	0.1	0.1	0.05	0.04
6														x(20)	x(50)
7		|||A+I|||=	0.87					Solution of Ax+b=0	2					2.008	2
8		D=	0.15					by Cramer's rule	0					0.009	5E-05

Figure 5.3.3 Iterations with a Contraction Matrix

Problem 5.3.7 Construct a spreadsheet in the format of Figure 5.3.3. Find a matrix A that does *not* satisfy (5.3.11) but for which iterations of an initial $\mathbf{x}_1$ by $\mathbf{F}(\mathbf{x}) = (A + I)\mathbf{x} + \mathbf{b}$ do nevertheless approach $-A^{-1}\mathbf{b}$.

As shown in problems at the end of this chapter, there does exist a necessary *and* sufficient condition for B to be a contraction matrix, namely that

$$(5.3.12) \qquad \|||B|||^2 + \sqrt{\|||B|||^4 - 4D^2} < 2,$$

where D denotes the determinant of B. We invite you to test $B = A + I$ against (5.3.12) as well as (5.3.11), thereby giving credence to this last assertion.

Problem 5.3.8 Modify the spreadsheet constructed in Problem 5.3.7 to include a test of A against (5.3.12). Confirm that your answer to Problem 5.3.7 satisfies (5.3.12).

But what if a given matrix A fails to satisfy both (5.3.11) and (5.3.12)? Can we still solve $A\mathbf{x} + \mathbf{b} = \mathbf{0}$ by iteration?

The answer, as indicated by Problems 5.8.16 and 5.8.17 at the end of this chapter, is a qualified "yes." The linear system $A\mathbf{x} + \mathbf{b} = \mathbf{0}$ is equivalent to $(1/N)A\mathbf{x} + (1/N)\mathbf{b} = \mathbf{0}$, and this makes it possible to replace (5.3.11) by

$$(5.3.11)' \quad (N + a_{11})^2 + a_{12}^2 + a_{21}^2 + (N + a_{22})^2 < N^2 \text{ for some } N > 0.$$

Similarly, $A\mathbf{x} + \mathbf{b} = \mathbf{0}$ is equivalent to $(-A)\mathbf{x} - \mathbf{b} = \mathbf{0}$, making it possible to replace (5.3.11) by

$$(5.3.11)'' \qquad (1 - a_{11})^2 + a_{12}^2 + a_{21}^2 + (1 - a_{22})^2 < 1.$$

Unfortunately, these strategies are not as effective for systems as they were in the case of a single equation $ax + b = 0$. As elaborated upon in problems at the end of this chapter, there are cases where all such efforts to approximate solutions of $A\mathbf{x} + \mathbf{b} = \mathbf{0}$ by iterations of $\mathbf{F}(\mathbf{x}) = (A + I)\mathbf{x} + \mathbf{b}$ will fail.

The fact that this form of iteration does not work in all cases creates a need for other approaches as well. In the next section we again return to the scalar equation $ax + b = 0$, emulating the solution $x = -b/a$ by giving direct meaning to the matrix expression $\mathbf{x} = -A^{-1}\mathbf{b}$ as the solution of $A\mathbf{x} + \mathbf{b} = \mathbf{0}$.

Central to such a development is the definition and computation of the inverse matrix A^{-1}, and here we shall see that iteration again plays an important role.

5.4. The Matrix Inverse   In order to give meaning to

$$(5.4.1) \qquad \mathbf{x} = -A^{-1}\mathbf{b}$$

as the solution of $A\mathbf{x} + \mathbf{b} = 0$, it is essential that (5.4.1) embody the solution given by Cramer's rule. For $n = 2$, this means that (5.4.1) must be equivalent to

(5.4.2)
$$x_1 = \frac{a_{12}b_2 - b_1 a_{22}}{a_{11}a_{22} - a_{12}a_{21}}$$
$$x_2 = \frac{b_1 a_{21} - a_{11}b_2}{a_{11}a_{22} - a_{12}a_{21}}.$$

Letting $D = a_{11}a_{22} - a_{12}a_{21}$ denote the determinant of A, Problems 5.1.5 and 5.1.6 showed that Cramer's rule is embodied in the definition

(5.4.3)
$$A^{-1} = \begin{bmatrix} \dfrac{a_{22}}{D} & \dfrac{-a_{12}}{D} \\ \dfrac{-a_{21}}{D} & \dfrac{a_{11}}{D} \end{bmatrix}$$

if $D \neq 0$, and that $A A^{-1} = A^{-1} A = I$ when the inverse is so defined.

Because it will also be important to establish a correspondence between Cramer's rule and A^{-1} for larger systems (i.e., for n linear equations in n unknowns), we will develop an iterative procedure for calculating (5.4.3) from the system of equations corresponding to $A\mathbf{x} + \mathbf{b} = 0$. Then, assuming that this procedure generalizes to $n > 2$, we will be able to harness technology to calculate A^{-1} for larger $n \times n$ matrices.

With this goal in mind, we formulate a sequence of "elementary row operations" (EROs) by which the 2×2 system

(5.4.4)
$$a_{11}x_1 + a_{12}x_2 = -b_1,$$
$$a_{21}x_1 + a_{22}x_2 = -b_2$$

can be transformed into (5.4.2) whenever $D \neq 0$. The simplest such ERO is the interchange of the two equations, e.g., one that replaces

$$\begin{matrix} 3x_1 + 2x_2 = 5, \\ 2x_1 - 4x_2 = 7 \end{matrix} \quad \text{with} \quad \begin{matrix} 2x_1 - 4x_2 = 7, \\ 3x_1 + 2x_2 = 5. \end{matrix}$$

The following problem formalizes a direct consequence of this ERO.

Problem 5.4.1 Assuming $D \neq 0$ and given the right to interchange equations, explain why we may assume that $a_{11} \neq 0$.

Given that $a_{11} \neq 0$, we can now execute a series of EROs whose end result is a display of values of x_1 and x_2 that constitute a solution of (5.4.4). This process is illustrated by the following example.

To solve

(5.4.5)
$$\begin{matrix} x_1 - x_2 = 4, \\ 2x_1 + x_2 = 5, \end{matrix}$$

we begin by subtracting twice the first equation ($2x_1 - 2x_2 = 8$) from the second.

This eliminates x_1 from the second equation to yield

$$x_1 - x_2 = 4,$$
$$3x_2 = -3.$$

Now adding one-third of the second equation ($x_2 = -1$) to the first eliminates x_2 from the first equation and yields

$$x_1 = 3,$$
$$3x_2 = -3.$$

Finally, dividing through the second equation by 3 displays the solutions as

$$x_1 = 3,$$
$$x_2 = -1.$$

To show that this process can be applied whenever

$$D = a_{11}a_{22} - a_{12}a_{21} \neq 0,$$

it remains to formulate these EROs for the general system

$$a_{11}x_1 + a_{12}x_2 = -b_1$$
$$a_{21}x_1 + a_{22}x_2 = -b_2.$$

Step 1. Eliminate the x_1 term in the second equation: Multiply the first equation by $-\frac{a_{21}}{a_{11}}$ and add the resulting equation to the second to obtain the new system

$$a_{11}x_1 + a_{12}x_2 = -b_1,$$
$$\frac{D}{a_{11}}x_2 = \frac{a_{21}}{a_{11}}b_1 - b_2.$$

Step 2. Eliminate the x_2 term in the first equation: Multiply the (modified) second equation above by $-\frac{a_{11}a_{12}}{D}$ and add the resulting equation to the (original) first equation to obtain the new system[1]

$$a_{11}x_1 = -\left(1 + \frac{a_{12}a_{21}}{D}\right)b_1 + \frac{a_{11}a_{12}}{D}b_2,$$
$$\frac{D}{a_{11}}x_2 = \frac{a_{21}}{a_{11}}b_1 - b_2.$$

Step 3. Solve for x_1 and x_2 separately: Divide the first equation by a_{11} and the second by D/a_{11} to obtain

$$x_1 = -\frac{a_{22}}{D}b_1 + \frac{a_{12}}{D}b_2,$$
$$x_2 = \frac{a_{21}}{D}b_1 - \frac{a_{11}}{D}b_2.$$

[1] If $a_{12} = 0$, this step can be skipped.

Problem 5.4.2 Fill in the algebraic details underlying Steps 1, 2, and 3.

Problem 5.4.3 Confirm that the above steps result in solution by Cramer's rule as given by (5.4.2).

Problem 5.4.4 Confirm that the above steps correspond to $\mathbf{x} = -A^{-1}\mathbf{b}$, where

$$\mathbf{x} = \begin{bmatrix} x_1 \\ x_2 \end{bmatrix}, \quad A^{-1} = \begin{bmatrix} \dfrac{a_{22}}{D} & \dfrac{-a_{12}}{D} \\ \dfrac{-a_{21}}{D} & \dfrac{a_{11}}{D} \end{bmatrix}, \quad \text{and} \quad \mathbf{b} = \begin{bmatrix} b_1 \\ b_2 \end{bmatrix}.$$

While such EROs were rather complicated to formulate for the general system (5.4.4), their execution tends to be straightforward in specific cases.

Problem 5.4.5 Apply the above three steps to the system

$$3x_1 + 2x_2 = -1,$$
$$x_1 + x_2 = 1.$$

Recalling that EROs enabled us to solve

(5.4.5)
$$x_1 - x_2 = 4,$$
$$2x_1 + x_2 = 5,$$

for $\begin{bmatrix} x_1 \\ x_2 \end{bmatrix} = \begin{bmatrix} 3 \\ -1 \end{bmatrix}$ and that this solution is closely related to the fact that

(5.4.6)
$$\begin{bmatrix} 1 & -1 \\ 2 & 1 \end{bmatrix}^{-1} = \begin{bmatrix} \frac{1}{3} & \frac{1}{3} \\ -\frac{2}{3} & \frac{1}{3} \end{bmatrix},$$

there now arises the question of whether EROs enable us to arrive at (5.4.6) as well. In other words, can EROs be used to compute the inverse of a matrix? To show that this is indeed possible, we write (5.4.4) in the form $A\mathbf{x} = I(-\mathbf{b})$, or

(5.4.7)
$$\begin{bmatrix} a_{11} & a_{12} \\ a_{21} & a_{22} \end{bmatrix} \begin{bmatrix} x_1 \\ x_2 \end{bmatrix} = \begin{bmatrix} 1 & 0 \\ 0 & 1 \end{bmatrix} \begin{bmatrix} -b_1 \\ -b_2 \end{bmatrix}.$$

Problem 5.4.6 Confirm that the EROs corresponding to Steps 1, 2, and 3 above can be accomplished by multiplying (from the left) both sides of (5.4.7) by

$$M_1 = \begin{bmatrix} 1 & 0 \\ -\dfrac{a_{21}}{a_{11}} & 1 \end{bmatrix}, \quad M_2 = \begin{bmatrix} 1 & -\dfrac{a_{11}a_{12}}{D} \\ 0 & 1 \end{bmatrix}, \quad M_3 = \begin{bmatrix} \dfrac{1}{a_{11}} & 0 \\ 0 & \dfrac{a_{11}}{D} \end{bmatrix}.$$

Since the sequence of elementary row operations used to solve (5.4.4) is equivalent to

Step 1. M_1 operates on both $A\mathbf{x}$ and $I(-\mathbf{b})$;

Step 2. M_2 operates on both $M_1 Ax$ and $M_1(-b)$
Step 3. M_3 operates on both $M_2 M_1 Ax$ and $M_2 M_1(-b)$,

these operations transform $Ax = -b$ into $Ix = A^{-1}(-b)$. On this basis, we conclude that

$$A^{-1} = M_3 M_2 M_1.$$

Recalling that this method requires $a_{11} \neq 0$, we note that such an "interchange of two equations" can also be accomplished by matrix multiplication.

Problem 5.4.7 Find a 2×2 matrix that interchanges the two rows of A, i.e., find a matrix M_0 for which $M_0 \begin{bmatrix} a_{11} & a_{12} \\ a_{21} & a_{22} \end{bmatrix} = \begin{bmatrix} a_{21} & a_{22} \\ a_{11} & a_{12} \end{bmatrix}$.

Problem 5.4.8 For the system (5.4.5) calculate $M_3 M_2 M_1$ and confirm that $M_3 M_2 M_1$ equals the inverse matrix given in (5.4.6).

Problem 5.4.9 Confirm that for a general 2×2 matrix A with nonzero determinant, $M_3 M_2 M_1$ equals the inverse matrix A^{-1}.

These observations lead to a simple spreadsheet format (Figure 5.4.1) for computing A^{-1}, one that does not make explicit use of the matrices M_1, M_2, M_3.

After entering both A and I in horizontal alignment, subject both A and I to elementary row operations that transform A into I. At the conclusion, the matrix I will have been transformed into A^{-1}.

	A	B	C	D	E	F	G	H	I	J
	This spreadsheet provides a format for inverting a 2 x 2 matrix A (with D ≠ 0) by means of elementary row operations on A and I. These EROs convert I to the inverse of A at the same time as they convert A to I.									
1										
2										
3		Enter A below			Enter I below			D=	6	
4										
5		3	0		1	0				
6		1	2		0	1				
7										
8		Apply M(1) to both matrices								
9										
10		3	0		1	0				
11		0	2		-0.333	1				
12										
13		Apply M(2) to both matrices								
14										
15		3	0		1	0				
16		0	2		-0.333	1				
17										
18		Apply M(3) to both matrices								
19										
20		1	0		0.3333	0				
21		0	1		-0.167	0.5				

Figure 5.4.1 Using EROs to Compute A^{-1}

What is important about Figure 5.4.1 is that its format and underlying ideas extend directly to finding the inverse of any $n \times n$ matrix A for which the corresponding system of n linear equations $Ax + b = 0$ can be solved for arbitrary **b**. In courses on "linear algebra" it is shown that the solvability of systems such as

$$\begin{bmatrix} a_{11} & a_{12} & a_{13} \\ a_{21} & a_{22} & a_{23} \\ a_{31} & a_{32} & a_{33} \end{bmatrix} \begin{bmatrix} x_1 \\ x_2 \\ x_3 \end{bmatrix} + \begin{bmatrix} b_1 \\ b_2 \\ b_3 \end{bmatrix} = \begin{bmatrix} 0 \\ 0 \\ 0 \end{bmatrix}$$

is directly linked to the *determinant* of the above matrix A. This determinant serves as a common denominator for x_1, x_2, and x_3 in Cramer's rule when $n = 3$.

The spreadsheet of Figure 5.4.2 illustrates the use of EROs to calculate the inverse of a 3×3 matrix.

Problem 5.4.9 Recalling the problem from the *Jiuzhang* solved in Problem 5.1.2:

(a) Formulate this problem in the form $Ax + b = 0$.
(b) Use the format of Figure 5.4.2 to compute A^{-1}.
(c) Use this inverse matrix to find the solution.

	A	B	C	D	E	F	G	H	I	J
1		This spreadsheet provides a format for inverting a 3 x 3 matrix A (with D ≠ 0) by means of elementary row operations on A and I. These EROs convert I to the inverse of A at the same time as they convert A to I. Please choose a(1,1) ≠ 0.								
2										
3			Enter A below				Enter I below		D=	1 2
4										
5		1	3	-2		1	0	0		
6		0	1	1		0	1	0		
7		2	0	2		0	0	1		
8										
9		1	3	-2		1	0	0		
10		0	1	1		0	1	0		
11		0	-6	6		-2	0	1		
12										
13		1	0	-5		1	-3	0		
14		0	1	1		0	1	0		
15		0	0	12		-2	6	1		
16										
17		1	0	0		0.1667	-0.5	0.4167		
18		0	1	0		0.1667	0.5	-0.083		
19		0	0	12		-2	6	1		
20										
21		1	0	0		0.1667	-0.5	0.4167		
22		0	1	0		0.1667	0.5	-0.083		
23		0	0	1		-0.167	0.5	0.0833		

Figure 5.4.2 Using EROs to Invert a 3 × 3 Matrix

	Agricultural Requirements	Energy Requirements	Household Requirements	TOTALS
Agricultural Outputs	$x_{11} = 10$ bushels	$x_{12} = 20$ bushels	$x_{13} = 70$ bushels	$x_1 = 100$ bushels
Energy Outputs	$x_{21} = 20$ barrels	$x_{22} = 10$ barrels	$x_{23} = 20$ barrels	$x_2 = 50$ barrels
Household Outputs	$x_{31} = 220$ person hours	$x_{32} = 50$ person hours	$x_{33} = 30$ person hours	$x_3 = 300$ person hours

Figure 5.5.1 An Input–Output Table

While it is satisfying to harness technology in the solution of such historic examples, Section 5.5 will show that modern applications of the matrix inverse go well beyond the kinds of problems posed in the *Jiuzhang*.

5.5. Input–Output Economics

As a modern application of the ideas discussed in this chapter, we return to the economic arena and "input–output economics" as developed by Wassily Leontief during the 1930s. Here we represent an economy as a complex system, one that converts resources (inputs) such as iron, labor, water, and petroleum into goods (outputs) such as food, cars, and clothing. By studying this process in terms of components called "sectors," Leontief was able to analyze, in quantitative terms, the interdependence (and self-dependence) of the various sectors of an economy.

By way of describing the underlying ideas, we begin with an artificially simple economy, one that consists of three sectors[1] labeled Agriculture, Energy, and Households. Assuming a daily output of 100 bushels of wheat, 50 barrels of fuel, and 300 person-hours of labor, we represent the outputs of these three sectors by $x_1 = 100$, $x_2 = 50$, and $x_3 = 300$.

What input–output analysis now calls for is a breakdown as to how these outputs are distributed among the various sectors. Such information can be conveyed by means of a 3×4 table such as that shown in Figure 5.5.1.

Relating Agriculture, Energy, and Household to the indices 1, 2, and 3, respectively, the table in Figure 5.5.1 assigns a value x_{ij} to *the number of units of commodity i required to sustain the above output of commodity j*. Thus, for example, the values $x_{11} = 10$, $x_{21} = 20$, and $x_{31} = 220$ reflect the fact that it requires 10 bushels of wheat, 20 barrels of fuel, and 220 man-years of labor to sustain the production of 100 bushels of wheat.

At this point it becomes important to acknowledge an essential difference between the first two sectors (Agriculture and Energy) and the third (Households). Sectors 1 and 2 represent commodities (food and fossil fuels) that are required to sustain the community's well-being. According to Figure 5.5.1, this particular community requires 70 bushels of wheat and 20 barrels of fossil fuel. In return,

---

[1] Leontief's applications of input–output modeling involved hundreds of sectors.

it is obligated to provide a total of 300 person-hours of labor to sustain this level of consumption. Regarding $x_{13} = 70$ and $x_{23} = 20$ as *requirements* that we seek to impose on the economy, we refer to these as *exogenous* (externally imposed) variables.

This distinction between Households and the other sectors leads to a change of notation. In what follows, we shall introduce the variables y_1 and y_2, setting

$$x_{13} = y_1 = 70,$$
$$x_{23} = y_2 = 20.$$

As for $x_{31} = 220$, $x_{32} = 50$, $x_{33} = 30$, and $x_3 = 300$, let us assume that the community is somewhat flexible in the amount of labor it is able to commit to sustaining its economy. On this basis, our analysis will, for the moment, suppress these variables. These changes enable us to summarize the first two rows in Figure 5.5.1 as

$$x_{11} + x_{12} + y_1 = x_1,$$
$$x_{21} + x_{22} + y_2 = x_2,$$

which can also be written

(5.5.1)
$$(x_1 - x_{11}) - x_{12} = y_1,$$
$$-x_{21} + (x_2 - x_{22}) = y_2.$$

Note that in (5.5.1) the exogenous variables $y_1 = 70$ and $y_2 = 20$ have been isolated on the right side of the equations.

To simplify this last system of equations further, we make an additional definition. Recalling that x_{ij} denotes the amount of commodity i required to sustain the production of x_j units of commodity j, we define the economy's "structural coefficients" by

(5.5.2)
$$a_{ij} = \frac{x_{ij}}{x_j} \quad \text{for} \quad i = 1, 2 \quad \text{and} \quad j = 1, 2.$$

It now follows that a_{ij} denotes the amount of commodity i required to sustain the production of *a single unit* of commodity j. That is, since it takes 20 barrels of fuel to produce 100 bushels of wheat ($x_{21} = 20$ and $x_1 = 100$), it takes 0.2 barrels of fuel to produce a single bushel of wheat.[2] Such is the rationale for introducing the structural coefficient $a_{21} = x_{21}/x_1$.

This last change in notation enables us to write the system (5.5.1) as

(5.5.3)
$$(1 - a_{11})x_1 - a_{12}x_2 = y_1,$$
$$-a_{21}x_1 + (1 - a_{22})x_2 = y_2.$$

---

[2] In terms of real-world economics, American agriculture now uses about 100 calories of fossil fuels to produce 10 calories of food. This ratio might provide the basis for one of the structural coefficients underlying an input–output model for the American economy.

Recalling the values of x_{ij} and x_i from Figure 5.5.1, this system becomes

(5.5.4)
$$0.9x_1 - 0.4x_2 = y_1,$$
$$-0.2x_1 + 0.8x_2 = y_2.$$

Problem 5.5.1 Use (5.5.4) to confirm that an imposition of the exogenous values $y_1 = 70$ and $y_2 = 20$ corresponds to $x_1 = 100$ and $x_2 = 50$.

What makes the system (5.5.3) so important is that it embodies the *structure* of this particular economy. Should the population of this community grow, the exogenous variables y_1 and y_2 can be expected to grow as well. The question, "What effect will population growth have on the economy?" can then be addressed in terms of the system (5.5.3) (assuming, of course, that the structure of the economy remains unchanged).

By way of a specific example, suppose that the population of this community were to double. A doubling of the exogenous variables to $y_1 = 140$ and $y_2 = 40$ would require that the entire economy double its output, from $x_1 = 100$ to 200 and from $x_2 = 50$ to 100.

But what if this community were to accept severe energy conservation measures, ones that maintain the current level of fuel consumption by Households at $y_2 = 20$ barrels/year even as the population doubles? If the community's food requirements were still to increase to 140 bushels of wheat, such conservation measures would not maintain fuel production at its pregrowth level of $x_2 = 50$ barrels. Rather, Agriculture and Energy would both experience increased fuel demands (even though the Households do not), and we would now determine x_1 and x_2 by solving the system

(5.5.5)
$$0.9x_1 - 0.4x_2 = 140,$$
$$-0.2x_1 + 0.8x_2 = 20.$$

Problem 5.5.2 Solve (5.5.5) for x_1 and x_2. Explain why x_1 fails to double, even though Households require twice as much food. Explain the increase in x_2.

In order to extend these ideas to economies with more than three sectors, matrix notation becomes indispensable. Such notation would have us write (5.5.3) as

(5.5.6) $(I - A)\mathbf{x} = \mathbf{y},$

where $\mathbf{x}$ and $\mathbf{y}$ are 3×1 column vectors, I is the identity matrix, and A is the 3×3 matrix whose entries are the structural coefficients a_{ij}. Fortunately, in the beginning of this chapter we developed two tools for solving such systems: iteration and finding $(I - A)^{-1}$.

Problem 5.5.3 Rewriting (5.5.5) as

$$-0.9x_1 + 0.4x_2 + 140 = 0,$$
$$0.2x_1 - 0.8x_2 + 20 = 0,$$

use iteration to approximate a solution of this system.

	Agricultural Requirements	Manufacturing Requirements	Energy Requirements	Household Requirements	Totals
Agricultural Outputs	10 bushels	10 bushels	20 bushels	70 bushels	110 bushels
Manufacturing Outputs	5 yards	10 yards	15 yards	20 yards	50 yards
Energy Outputs	20 barrels	20 barrels	10 barrels	20 barrels	70 barrels
Household Outputs	220 person hours	80 person hours	50 person hours	30 person hours	380 person hours

Figure 5.5.2 A Four-Sector Economy

With matrix tools at hand, let us introduce a Manufacturing Sector into the above economy based on Agriculture, Energy, and Households. Such a four-sector economy will confront us with a system of three equations in three unknowns, one whose solution can be used to illustrate the matrix methods that carry over to larger systems as well.

We begin by representing a four-sector economy by means of a 4×5 table such as that of Figure 5.5.2.

Problem 5.5.4 Noting that $x_{11} = 10$, $x_{12} = 10$, $x_{13} = 20$, etc., while $x_1 = 110$, $x_2 = 50$, etc., determine the matrix A whose entries are the structural coefficients $a_{ij} = x_{ij}/x_j$.

Making use of Problem 5.5.4 and generalizing the case $n = 2$, we are led to the system of equations

$$0.9091x_1 - 0.2x_2 - 0.286x_3 = 70,$$
(5.5.7) $$-0.0451x_1 + 0.8x_2 - 0.214x_3 = 20,$$
$$-0.182x_1 - 0.4x_2 + 0.8571x_3 = 20.$$

Problem 5.5.5 Solve (5.5.7) by finding $(I - A)^{-1}$ in the spreadsheet format of Figure 5.5.3.

Problem 5.5.6 Solve (5.5.7) by creating a spreadsheet that solves an equivalent system by matrix iteration.

Problem 5.5.7 Suppose that population growth obligates the economy represented in Figure 5.5.2 to provide households with twice as much food. Assuming that household consumption of fuel and manufactured goods stays constant, by what percentage would the fuel and manufacturing sectors have to increase their total outputs?

5.6. Matrix Iteration

In Section 5.2 we considered the possibility of solving linear systems of the form $A\mathbf{x} + \mathbf{b} = 0$ by matrix iteration. Here the concept of contraction matrix was central to bypassing Cramer's rule by means of matrix

	A	B	C	D	E	F	G	H	I	J
1	This spreadsheet uses EROs to invert the matrix of of coefficients appearing in equation (5.5.7). It then applies this inverse matrix to the column vector COL[70, 20, 20].									
2										
3			Enter A below				Enter I below			D= 0.4832
4										
5		0.9091	-0.2	-0.286		1	0	0		
6		-0.045	0.8	-0.214		0	1	0		
7		-0.182	-0.4	0.8571		0	0	1		
8										
9		0.9091	-0.2	-0.286		1	0	0		
10		0	0.7901	-0.228		0.0496	1	0		
11		0	-0.44	0.7998		0.2002	0	1		
12										
13		0.9091	-3E-17	-0.344		1.0126	0.2531	0		
14		0	0.7901	-0.228		0.0496	1	0		
15		0	0	0.6728		0.2278	0.557	1		
16										
17		0.9091	-3E-17	6E-17		1.129	0.5377	0.511		
18		0	0.7901	0		0.1269	1.1889	0.3392		
19		0	0	0.6728		0.2278	0.557	1		
20										
21		1	-3E-17	6E-17		1.2419	0.5915	0.5621		
22		0	1	0		0.1606	1.5048	0.4293		
23		0	0	1		0.3387	0.8279	1.4864		
24										
25		1.2419	0.5915	0.5621	70		110			
26		0.1606	1.5048	0.4293	20	=	50			
27		0.3387	0.8279	1.4864	20		70			

Figure 5.5.3 Using EROs to Compute $(I - A)^{-1}$

iteration. There are, however, many situations where matrix iteration arises in its own right, without reference to the concept of "contraction."

By way of an important illustration, we return to the subject of population dynamics. Here we developed models for exponential growth (based on the difference equation $x_{i+1} - x_i = ax_i$) and considered various generalizations (such as the Verhulst equation $x_{i+1} - x_i = ax_i - bx_i^2$). These models took into account only the size of the population as a whole, disregarding effects that a population's age structure may have on its dynamics. However, especially in the study of human populations, the question of whether a population is primarily youthful, mature, or aged can be of central importance in efforts to predict demographic change. Such age structure is often conveyed in terms of "population pyramids" such as those depicted in Figure 5.6.1.

A country experiencing rapid population growth is likely to have a population distribution similar to that of the pyramid at the left. Here the population is largely youthful and even if individuals of childbearing age average exactly one child, the population will continue to grow. By contrast, in a country where much of the population is beyond childbearing age, such a "replacement" fertility rate will be accompanied by a decline in total population.

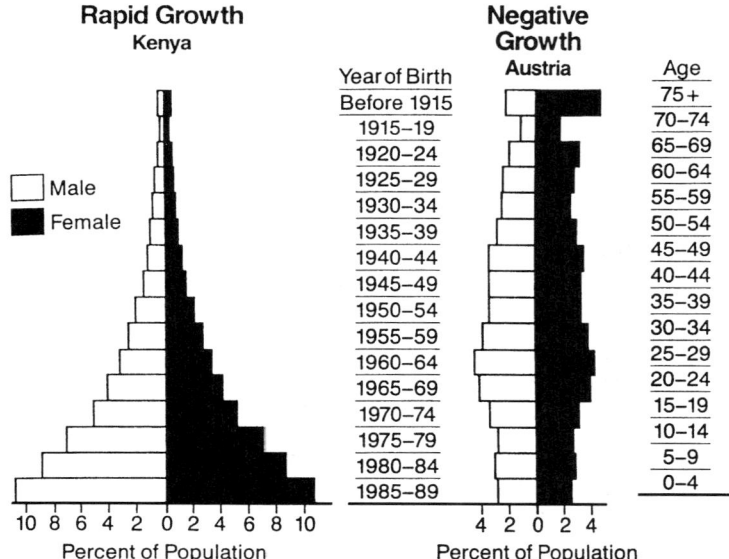

Figure 5.6.1 Two Population Pyramids (*Source: R. Schwartz, "Mathematics and Global Survival," 4th ed. Ginn Press, 1997.*)

When age distribution is not considered, a population with fixed fertility would be modeled by a difference equation

(5.6.1) $x_{i+1} - x_i = a x_i$

with solution

(5.6.2) $x_i = x_1 (1 + a)^{(i-1)}.$

However, taking age distribution into account can lead to more complicated solutions. Rather than satisfying (5.6.2) from the outset, populations with varying age distributions will approach this solution gradually. These phenomena correspond to allowing for "delays" in equations of the form (5.6.1).[1]

In order to develop a mathematical understanding of these phenomena, it is useful to go back to the Middle Ages and the works of a remarkable mathematician named Leonardo of Pisa (ca. 1180–1250). Given the name "Fibonacci" by the nineteenth-century editor of his works, Leonardo's most famous book was the *Liber Abaci,* or the "book of the abacus." Its most famous problem is the following:

> How many pairs of rabbits can be bred in one year from one pair? A certain person places one pair of rabbits in a certain place surrounded on all sides by a wall. We want to know how many pairs can be bred from that pair in one year, assuming it is their nature that each month they give birth to another pair and, *in the second month after birth,* each new pair can also breed [italics added].

[1] Recall also the rather dramatic impact of delays on the solution of the logistic equation, as depicted in the spreadsheet of Figure 2.4.5.

Months	Baby Pairs	Adult Pairs	Total Pairs
1	1	0	1
2	0	1	1
3	1	1	2
4	1	2	3
5	2	3	5
6	3	5	8
7	5	8	13

Figure 5.6.2 A Table of Fibonacci Numbers

Had Fibonacci omitted the phrase "and in the second month after birth," the problem would have been a less interesting one. For then each rabbit pair would account for a new rabbit pair each month, giving rise to

$$(5.6.3) \qquad x_{i+1} - x_i = x_i.$$

The resulting number of pairs would be given by the sequence $1, 2, 4, 8, 16, \ldots$ with $x_{n+1}/x_n = 2$. However, in Fibonacci's version we encounter a population consisting of "babies" and "adults," one in which only adult pairs are able to breed. Starting with one pair of babies, the outcome is described in Figure 5.6.2:

Appearing in the shaded column of Figure 5.6.2 are the famous Fibonacci numbers

$$1, 1, 2, 3, 5, 8, 13, 21, 34, 55, 89, \ldots$$

defined by $x_1 = 1$, $x_2 = 1$, and

$$(5.6.4) \qquad x_{i+1} = x_i + x_{i-1} \qquad \text{for} \quad i = 2, 3, 4, \ldots.$$

Writing (5.6.4) as a difference equation

$$(5.6.5) \qquad x_{i+1} - x_i = x_{i-1},$$

we see that (5.6.5) is closely related to (5.6.3). That is, (5.6.5) is a *delay* difference equation that is closely related to the usual model for exponential growth.

Problem 5.6.1 Recalling the definition of $\Delta\Delta x_i = x_{i+2} - 2x_{i+1} + x_i$, write (5.6.5) as a second-order difference equation of the form $\Delta\Delta x_i = f(x_i, \Delta x_i)$.

While (5.6.4) leads to difference equations of the type considered in Chapter 2, it will be more fruitful to study it in the context of matrix notation. Here the fact that there are two categories of rabbits (babies and adults) makes it natural to represent the population by a column vector $\mathbf{x}_i = \begin{bmatrix} b_i \\ a_i \end{bmatrix}$ whose components denote the number of babies and the number of adults, respectively, at the beginning of

	A	B	C	D	E	F	G	H	I	J	K	L	M	N	O
1	This spreadsheet models the growth of Fibonacci's population of rabbits and thereby generates the famous Fibonacci numbers 1, 1, 2, 3, 5, 8, 13, 21, etc. It also allows one to vary the entries in A and the initial population.														
2															
3				x(1)	x(2)	x(3)	x(4)	x(5)	x(6)	x(7)	x(8)	x(9)	x(10)	x(11)	x(12)
4	0	1		1	0	1	1	2	3	5	8	13	21	34	55
5	1	1		0	1	1	2	3	5	8	13	21	34	55	89
6	Total=x(i)=			1	1	2	3	5	8	13	21	34	55	89	144
7	x(i+1)/x(i)=			1	2	1.5	1.67	1.6	1.63	1.62	1.62	1.62	1.62	1.618	1.618

Figure 5.6.3 Age-Dependent Population Growth for Rabbits

the ith month. Here $x_i = b_i + a_i$ denotes the total population, and Fibonacci's problem takes the form

$$(5.6.6) \qquad \begin{bmatrix} b_{i+1} \\ a_{i+1} \end{bmatrix} = \begin{bmatrix} 0 & 1 \\ 1 & 1 \end{bmatrix} \begin{bmatrix} b_i \\ a_i \end{bmatrix},$$

or

$$(5.6.7) \qquad \mathbf{x}_{i+1} = A\mathbf{x}_i \quad \text{with} \quad A = \begin{bmatrix} 0 & 1 \\ 1 & 1 \end{bmatrix}.$$

Such equations can be represented in spreadsheet format as shown in Figure 5.6.3.

Figure 5.6.3 suggests that Fibonacci's rabbits will also grow exponentially, but not as fast as if the entire population doubled each month. That is, the ratio x_{i+1}/x_i now appears to approach a value ≈ 1.62 rather than 2.

Could this new behavior of x_{i+1}/x_i have been predicted from (5.6.4)? To answer this question we use (5.6.4) to write

$$(5.6.8) \qquad x_{i+2}/x_{i+1} = 1 + x_i/x_{i+1}.$$

Assuming that $x_{i+2}/x_{i+1} \approx r$ for large values of n, (5.6.8) suggests that r will satisfy

$$(5.6.9) \qquad r = 1 + \frac{1}{r}.$$

Problem 5.6.2 Derive (5.6.9) from (5.6.8) and then solve for r.

The mechanism we have developed to study Fibonacci's rabbits allows for a variety of generalizations and applications. Some of these go back to the work of Aldo Leopold, an American conservationist seeking to save endangered species by providing them with refuges where they could regenerate without being hunted or otherwise preyed upon. In his book *Game Management*, Leopold noted that different animals require different amounts of time to reach maturity. Once mature,

CLASSIFICATION OF SPECIES FOR BREEDING POTENTIAL

Breeding Age	Average number of young per year									
	1	1.5[12]	2	3	4	5	6-8	9-11	12-14	15-20
1 year	BANDTAIL PIGEON PASSENGER PIGEON[1]		Horned Owl Goshawk[c]	WOODCOCK[c]	MOURNING DOVE WILSON SNIPE[c] SNOWSHOE[3] Gray Fox Cooper's Hawk Sharpshin	Crow Mink Weasel	CANVASBACK BLUE GROUSE SAGE HEN FOX SQUIRREL[c] GRAY SQUIRREL[c] Red Fox	RIVER DUCKS SCAUP PHEASANT RUFFED GROUSE PINNATED GR. Skunk	BOBWHITE GAMBEL QUAIL SCALED QUAIL SHARPTAIL GR. REDHEAD COTTONTAIL[5] Housecat[2]	HUNGARIAN P. CALIF. QUAIL
2 years	MT. GOAT	DEER[6] MOOSE[6]	ANTELOPE[7]		Bobcat Lynx		CANADA GOOSE Wolf[8] Coyote[8]		WILD TURKEY	
3 years	GRIZZLY[9] ELK BUFFALO[10]	MT. SHEEP[6]	Cougar			SWANS[11]				
4 years	BLACK BEAR[9]									

Figure 5.6.4 Breeding Potentials of Endangered Species (*Source: 'Game Management,"Aldo Leopold, Scribner & Sons, 1933.*)

they have different numbers of young each year. On this basis, he sought to estimate the *unimpeded growth rates* of various species. He wrote,

This rate, which we may call for short the breeding potential, depends *theoretically* on four properties:

1. The minimum breeding age.
2. The maximum breeding age.
3. The number of young per year.
4. The longevity beyond maximum breeding age.

This assumes a population perfectly balanced as to sex and age classes. If not so balanced, the breeding rate is further affected by:

5. The sex and age composition of the population.
6. The mating habits as related to that composition.

Noting the complexity of efforts to incorporate all six properties, he began with
a simpler model very similar to that of Fibonacci. Letting

$$T = \text{minimum breeding age},$$
$$q = \text{number of young per adult per year},$$

Leopold proceeded to develop a series of *breeding potential charts* that "classify
the species according to the principal characteristics determining their breeding
potentials." (See Figure 5.6.4.)

For $T = 1$, Leopold was in fact studying the matrix equation

(5.6.10)
$$\begin{bmatrix} b_{i+1} \\ a_{i+1} \end{bmatrix} = \begin{bmatrix} 0 & q \\ 1 & 1 \end{bmatrix} \begin{bmatrix} b_i \\ a_i \end{bmatrix}$$

and the corresponding population $x_i = b_i + a_i$.

Problem 5.6.3 Modify the spreadsheet in Figure 5.6.3 to model (5.6.10) for varying values of q. Use this spreadsheet to estimate x_{i+1}/x_i for large values of i.

Problem 5.6.4 Show that (5.6.10) gives rise to $x_{i+2} = x_{i+1} + q x_i$ for $i = 1, 2, 3, \ldots$. Use this fact to estimate x_{i+1}/x_i for large values of i. [Hint: If $r \approx x_{i+1}/x_i$ for large values of i, then for large values of i, $r \approx x_{i+2}/x_{i+1}$ and $1/r \approx x_i/x_{i+1}$. Use these observations to relate the quadratic equation $y^2 = y + q$ to the problem at hand.]

But what if $T = 2$? Here it may be useful to replace Fibonacci's rather fecund rabbits by another species, say buffaloes, that require *two* units of time (years) to reach maturity and then bear an average of q young per adult per year. Since such a population can be thought of as having "babies, teens, and adults," it becomes natural to represent it in terms of a 3×1 column vector

$$\mathbf{x}_i = \begin{bmatrix} b_i \\ t_i \\ a_i \end{bmatrix}$$

whose components sum to the total population $x_i = b_i + t_i + a_i$. Corresponding to (5.6.10) we now have

$$(5.6.11) \qquad \begin{bmatrix} b_{i+1} \\ t_{i+1} \\ a_{i+1} \end{bmatrix} = \begin{bmatrix} 0 & 0 & q \\ 1 & 0 & 0 \\ 0 & 1 & 1 \end{bmatrix} \begin{bmatrix} b_i \\ t_i \\ a_i \end{bmatrix}$$

Problem 5.6.5 Create a spreadsheet that models (5.6.11) for varying initial populations and values of q. Use this spreadsheet to estimate x_{i+1}/x_i for large values of i.

Problem 5.6.6 Generalize Problem 5.6.4 to estimate x_{i+1}/x_i for large values of i. [Hint: Relate the ratio $r = x_{i+1}/x_i$ to a cubic equation.]

Building on these ideas, problems at the end of this chapter ask you to generalize from $T = 2$ (buffaloes) to $T = 3$ (elephants) and $T = 4$ (dinosaurs), etc. However, (5.6.11) also lends itself to other forms of generalization, ones that are central to efforts to relate matrix iteration to human demography.

Neither Fibonacci's model for rabbits nor Leopold's "unimpeded growth rates" contended with issues of mortality. However, to be valid for extended periods of time and allow populations to reach a "steady state," such models will have to allow for mortality among adults. Realism may also call for mortality and varying fertility rates among other age groups.

These ideas are readily illustrated in terms of the buffalo model (5.6.11). Letting α denote the fraction of babies that survive to adolescence, β the fraction of teens that survive to adulthood, and γ the annual survival rate of adults, we modify the matrix of (5.6.11) to obtain

$$(5.6.12) \qquad \begin{bmatrix} b_{i+1} \\ t_{i+1} \\ a_{i+1} \end{bmatrix} = \begin{bmatrix} 0 & \tau & q \\ \alpha & 0 & 0 \\ 0 & \beta & \gamma \end{bmatrix} \begin{bmatrix} b_i \\ t_i \\ a_i \end{bmatrix}.$$

Note that (5.6.12) even allows for "teenage pregnancies" by including a term τ that denotes the average number of baby pairs generated by each teen pair.

Problem 5.6.7 Generalize the spreadsheet of Problem 5.6.5 to allow for variables α, β, γ, and τ as in (5.6.12).

With such background, it is straightforward to generalize to a human population broken down into, say, 12 different age groups,

0 to 6 years,   7 to 12 years,   ...,   73 to 78 years,   and   79+ years.

Here we would introduce constants $\alpha_1, \ldots, \alpha_{12}$ denoting the fraction of each cohort that survives for the full 6 years and $\tau_1, \ldots, \tau_{12}$ denoting the average number of babies generated by each individual in that cohort.

Problem 5.6.8 Letting x_i denote the 12×1 column vector whose components describe the size of the 12 cohorts at time i, describe the 12×12 matrix A for which the rule $x_{i+1} = Ax_i$ incorporates the changes that this population undergoes in a six-year period.

We have now seen that for rabbits and buffaloes, the quotient x_{i+1}/x_i approaches a constant value. For rabbits, this constant was found by solving a quadratic equation (Problem 5.6.2), while for buffaloes it was the solution of a cubic (Problem 5.6.6). On this basis, it is reasonable to expect that a population being modeled by Problem 5.6.8 also enjoys this property, but that the corresponding constant will be the solution of a 12th-degree polynomial equation.

In order to characterize such an equation it is useful to return to Fibonacci's rabbits as described by

$$(5.6.13) \qquad \begin{bmatrix} b_{i+1} \\ a_{i+1} \end{bmatrix} = \begin{bmatrix} 0 & 1 \\ 1 & 1 \end{bmatrix} \begin{bmatrix} b_i \\ a_i \end{bmatrix}$$

and make the following observation: If $x_i = b_i + a_i$ and we start with $b_1 = 1$ and $a_1 = 0$, then

$$\frac{x_2}{x_1} = 1 \quad \text{and} \quad \frac{x_3}{x_2} = 2 > \frac{x_2}{x_1}.$$

If, however, we start with $b_1 = 0$ and $a_1 = 1$, then

$$\frac{x_2}{x_1} = 2 \quad \text{and} \quad \frac{x_3}{x_2} = \frac{3}{2} < \frac{x_2}{x_1}.$$

If we were to start with a large number of rabbits, it should be possible to find an initial mix $b_1 : a_1$ for which $x_3/x_2 = x_2/x_1$. Indeed, we may even hope to find an initial mix $b_1 : a_1$ for which $b_3/b_2 = b_2/b_1 = a_3/a_2 = a_2/a_1$, which in turn implies $x_3/x_2 = x_2/x_1$.

The key to doing this turns out to be finding a value λ for which

$$(5.6.14) \qquad\qquad \frac{b_2}{b_1} = \frac{a_2}{a_1} = \lambda.$$

Problem 5.6.9 Show that if (5.6.14) is satisfied, then we also have $b_2 : a_2 = b_1 : a_1$ and as a result,

$$(5.6.15) \qquad\qquad \begin{aligned} \mathbf{x}_2 &= \lambda\mathbf{x}_1, \\ \mathbf{x}_3 &= \lambda\mathbf{x}_2, \\ \mathbf{x}_4 &= \lambda\mathbf{x}_3, \\ &\cdots \end{aligned}$$

Problem 5.6.10 Show that if (5.6.14) is satisfied, then $x_{i+1}/x_i = \lambda$ for $i = 1, 2, 3, \ldots$.

In order to satisfy (5.6.14), it is necessary to find an initial vector $\mathbf{x} \neq \mathbf{0}$ for which $A\mathbf{x} = \lambda\mathbf{x}$, where $A = \begin{bmatrix} 0 & 1 \\ 1 & 1 \end{bmatrix}$. This can also be written $(A - \lambda I)\mathbf{x} = \mathbf{0}$, or

$$(5.6.16) \qquad \begin{bmatrix} -\lambda & 1 \\ 1 & 1-\lambda \end{bmatrix}\begin{bmatrix} b \\ a \end{bmatrix} = \begin{bmatrix} 0 \\ 0 \end{bmatrix}, \qquad \text{where} \quad \mathbf{x} = \begin{bmatrix} b \\ a \end{bmatrix}.$$

Problem 5.6.11 Show that (5.6.16) has solution $\mathbf{x} \neq \mathbf{0}$ if and only if

$$\det\begin{bmatrix} -\lambda & 1 \\ 1 & 1-\lambda \end{bmatrix} = \lambda^2 - \lambda - 1 = 0,$$

and that in this case, the solution $\mathbf{x}$ is a multiple of $\begin{bmatrix} 1 \\ \lambda \end{bmatrix}$.

The vector $\mathbf{p} = \begin{bmatrix} 1 \\ \lambda \end{bmatrix}$ arising in Problem 5.6.11 is called an *eigenvector* of the matrix $A = \begin{bmatrix} 0 & 1 \\ 1 & 1 \end{bmatrix}$. In the context of age-dependent population dynamics, eigenvectors correspond to stable population pyramids, ones in which the relative sizes of the various age cohorts remain constant. The fact that these ideas extend to buffaloes is shown in problems at the end of this chapter.

5.7. Synthesis In Chapter 2 we saw that first-order difference equations

$$(5.7.1) \qquad\qquad x_{i+1} - x_i = f(x_i),$$

with $f(x)$ of the form $ax^2 + bx + c$, arise in both microeconomics (the Walrasian assumption when producers respond quadratically) and population dynamics (the Verhulst equation). In this chapter we have seen how matrix theory underlies both Leontief's input–output economics and the dynamics of age-dependent population growth. In other words, it seems that identical mathematical tools play different but important roles in both economics and ecology.

This observation can simply be regarded as yet another example of the importance and universality of mathematics. However, it lends itself to another interpretation as well: The extent to which the disciplines of economics and ecology tend to exist in isolation, despite such a sharing of common tools, undermines the effectiveness of both disciplines. For while the Walrasian assumption may provide an elegant and, in some contexts, useful basis for economic modeling, the demand for a particular commodity is surely linked to the size of the population being served as well as the price of the commodity. Conversely, efforts to address environmental issues such as biodiversity and global warming in purely biological and physical terms have severe limitations. It is, after all, the interaction of human economic activity with the earth and its biosphere that is at the root of such issues.

In the "empty world" of the eighteenth century, one in which human civilization had rather limited impact on the earth and its resources, economics and ecology could well be conceived and studied as separate disciplines. But as our industrial economy becomes a dominant factor in determining the direction of environmental change, such separation becomes increasingly artificial.

In this context, the fact that economics and ecology employ common mathematical tools can be regarded as a hopeful sign, one that provides a basis for studying their interaction. This is a theme we return to in Chapter 7, but it is also one that we use matrix theory to illustrate at this time.

Central to input–output economics are the variables $y_1, y_2, \ldots$ that represent demands for food, energy, manufactured goods, etc., that households place on the various sectors of an economy. Rather than insisting that these variables be determined by price, input–output economics acknowledges that demand for these commodities may be determined by factors *outside* the economic model under consideration. That is, it allows for *exogenous* variables such as the size and consumption patterns of a particular population. Given that populations are subject to change, this leads to a dynamic form of input–output modeling, one in which the economy is linked to a particular form of ecological change.

By way of illustrating such a synthesis, let us return to the four-sector economy of Section 5.5, where y_1, y_2, and y_3 represent a particular population's demand for food, manufactured goods, and energy, respectively. Since these variables are now subject to change with time (i.e., with change in an index i), it will be useful to replace the symbols y_1, y_2, and y_3 by f, m, and e, respectively. In this notation, the demand placed on an economy during the ith period will be represented by means of a column vector

$$\mathbf{d}_i = \begin{bmatrix} f_i \\ m_i \\ e_i \end{bmatrix}.$$

To keep the model simple, let us begin by considering such an economy's interaction with a population that, like Fibonacci's rabbits, consists of two age groups (babies and adults) and grows according to the rule

(5.7.2)
$$\begin{bmatrix} b_{i+1} \\ a_{i+1} \end{bmatrix} = \begin{bmatrix} 0 & q \\ \beta & \gamma \end{bmatrix} \begin{bmatrix} b_i \\ a_i \end{bmatrix},$$

or

$$\begin{bmatrix} b_{i+1} \\ a_{i+1} \end{bmatrix} = \begin{bmatrix} 0 & q \\ \beta & \gamma \end{bmatrix}^i \begin{bmatrix} b_1 \\ a_1 \end{bmatrix}.$$

Let us also suppose that babies require 1, 1, and 3 units of food, manufactured goods, and energy each year, whereas adults require 2, 4, and 5 units of the same commodities. The demand that such a population places on the economy at time i is then given by

(5.7.3)
$$\mathbf{d}_i = \begin{bmatrix} f_i \\ m_i \\ e_i \end{bmatrix} = \begin{bmatrix} 1 & 2 \\ 1 & 4 \\ 3 & 5 \end{bmatrix} \begin{bmatrix} b_i \\ a_i \end{bmatrix}.$$

Recalling the hypothetical four-sector economy defined by Figure 5.5.2, we would find the inputs that such exogenous variables require by solving

(5.7.4)
$$\begin{aligned} 0.9091x_1 - 0.2x_2 - 0.286x_3 &= f_i, \\ -0.0451x_1 + 0.8x_2 - 0.214x_3 &= m_i, \\ -0.182x_1 - 0.4x_2 + 0.8571x_3 &= e_i. \end{aligned}$$

Here we would again do well to replace the variables x_1, x_2, and x_3 by new variables F, M, and E. This makes it easier to represent the fact that these required inputs change with time by appending the index i as a subscript. Then, recalling the inverse of the matrix of coefficients in (5.7.4) given in Figure 5.5.3, we obtain

(5.7.5)
$$\begin{bmatrix} F_i \\ M_i \\ E_i \end{bmatrix} = \begin{bmatrix} 1.242 & 0.591 & 0.562 \\ 0.161 & 1.505 & 0.430 \\ 0.339 & 0.828 & 1.487 \end{bmatrix} \begin{bmatrix} f_i \\ m_i \\ e_i \end{bmatrix}.$$

In summary, we obtain the dynamic model

(5.7.6)
$$\begin{bmatrix} F_i \\ M_i \\ E_i \end{bmatrix} = \begin{bmatrix} 1.242 & 0.591 & 0.562 \\ 0.161 & 1.505 & 0.430 \\ 0.339 & 0.828 & 1.487 \end{bmatrix} \begin{bmatrix} 1 & 2 \\ 1 & 4 \\ 3 & 5 \end{bmatrix} \begin{bmatrix} 0 & q \\ \beta & \gamma \end{bmatrix}^{i-1} \begin{bmatrix} b_1 \\ a_1 \end{bmatrix}$$

for the inputs that this particular economy will require during the ith period in order to meet the needs of this particular population.

Problem 5.7.1 Build a spreadsheet corresponding to (5.7.6) that calculates $\mathbf{d}_i$ for $1 \le i \le 20$ and allows one to change the values of q, β, and γ.

These ideas generalize readily to a society with more age cohorts (e.g., 0 to 6 year olds, 7 to 12 year olds, 13 to 18 year olds, etc.) and a more complex economy with a larger number of sectors. Given n different age cohorts and m different economic sectors, an $m \times n$ matrix D can link the size of the various age cohorts with demands on the economy according to the rule $\mathbf{y} = D\mathbf{x}$. Leontief's structural matrix then transforms these demands into the "inputs" that this economy will require.

In trying to adapt these ideas beyond a single country, one would also have to contend with the fact that the populations of different countries (e.g., "industrialized" vs. "third world") place different demands on the global economy. Industrialized countries tend to have smaller rates of population growth, but they place a higher per capita demand on the global economy. The following problem suggests a context for addressing such issues in a model similar to that of (5.7.6).

Problem 5.7.2 Letting $b_1 = 4$ billion denote the current "third world" population and $a_1 = 2$ billion denote the current "industrialized" population, suppose b_i grows at 2%/year while a_i grows at 1%. Suppose also that third world individuals receive, respectively, 3, 1, and 2 units of food, manufactured goods, and energy per year, while individuals in the industrialized world receive 5, 4, and 7 such units. Finally, suppose that each year 3% of individuals in third world countries make the transition to industrialized levels of reproduction and consumption. Create a spreadsheet that determines the inputs required for the economy of Problem 5.7.1.

5.8. Problems, Exercises, and Projects   This chapter began with the observation that the system of linear equations

$$ax + by = e,$$
$$cx + dy = f$$

can be written in the form $A\mathbf{x} + \mathbf{b} = \mathbf{0}$ and (sometimes) solved in terms of the iterative scheme $\mathbf{x}_{i+1} = (A + I)\mathbf{x}_i + \mathbf{b}$. As shown in Problem 5.2.3, the key to using iteration to circumvent Cramer's rule is to ensure that for each vector $\mathbf{x}$, $(A+I)^n\mathbf{x}$ approaches the zero vector as n grows without bound. A sufficient condition for $A+I$ to have this "contraction property" was established in Problem 5.3.6, while a necessary and sufficient condition was stated (without proof) as (5.3.12). The somewhat challenging task of establishing (5.3.12) is the subject of Problem 5.8.9 below.

These problems begin by establishing some alternative ways of ensuring that $(A + I)^n\mathbf{x}$ approaches the zero vector. As such, these problems also provide a means of ensuring that iterates generated by the scheme $\mathbf{x}_{i+1} = (A + I)\mathbf{x}_i + \mathbf{b}$ approach the solution of $A\mathbf{x} + \mathbf{b} = \mathbf{0}$. To distinguish these conditions from the "contraction matrices" considered in Section 5.3, we will refer to them with different terminology.

Reduction Matrices

Problem 5.8.1 Consider a population of rabbits in which each adult pair generates f pairs of babies each month, the monthly survival rate of babies is s, and the monthly mortality rate among adults is m. Letting b_i and a_i denote the number of babies and adults at time i, and setting $p = 1 - m$:

(a) Show that (5.6.13) should be replaced by $\begin{bmatrix} b_{i+1} \\ a_{i+1} \end{bmatrix} = \begin{bmatrix} 0 & f \\ s & p \end{bmatrix} \begin{bmatrix} b_i \\ a_i \end{bmatrix}$.

(b) Extend Problem 5.6.9 to this more general situation.

(c) Extend Problem 5.6.10 to this more general situation.

(d) Show that $b_{i+1}/b_i = a_{i+1}/a_i = \lambda$ if and only if $\begin{bmatrix} 0 & f \\ s & p \end{bmatrix} \begin{bmatrix} b_1 \\ a_1 \end{bmatrix} = \lambda \begin{bmatrix} b_1 \\ a_1 \end{bmatrix}$.

(e) Recalling Problem 5.6.11, show that $b_{i+1}/b_i = a_{i+1}/a_i = \lambda$ if and only if

$$\det \begin{bmatrix} -\lambda & f \\ s & p - \lambda \end{bmatrix} = 0.$$

(f) Noting that $\det \begin{bmatrix} -\lambda & f \\ s & p-\lambda \end{bmatrix} = \lambda^2 - p\lambda - fs$ and recalling that b_i and a_i are necessarily nonnegative, find conditions on f, s, and p that determine whether $\lambda < 1$ or $\lambda > 1$.

Problem 5.8.2 In Problem 5.8.1, the total number of rabbits at time i is given by $r_i = b_i + a_i$. As such, the inequality $r_{i+1} < r_i$ corresponds to a reduction in the total number of rabbits (adults plus babies) during the ith period of time. On this basis, we will say that a 2×2 matrix B is a "reduction matrix" if

$$\begin{bmatrix} y_1 \\ y_2 \end{bmatrix} = \begin{bmatrix} b_{11} & b_{12} \\ b_{21} & b_{22} \end{bmatrix} \begin{bmatrix} x_1 \\ x_2 \end{bmatrix} \text{ implies that } |y_1| + |y_2| < |x_1| + |x_2|$$

$$\text{for every vector } \begin{bmatrix} x_1 \\ x_2 \end{bmatrix}.$$

(a) Show that a matrix B all of whose elements are nonnegative is a reduction matrix whenever $b_{11} + b_{21} < 1$ and $b_{12} + b_{22} < 1$.

(b) Show that a matrix B all of whose elements are real numbers is a reduction matrix whenever $|b_{11}| + |b_{21}| < 1$ and $|b_{12}| + |b_{22}| < 1$.

(c) Given a matrix B whose elements satisfy $|b_{11}| + |b_{21}| < 1$ and $|b_{12}| + |b_{22}| \geq 1$, find a vector $\begin{bmatrix} x_1 \\ x_2 \end{bmatrix}$ for which $\begin{bmatrix} y_1 \\ y_2 \end{bmatrix} = \begin{bmatrix} b_{11} & b_{12} \\ b_{21} & b_{22} \end{bmatrix} \begin{bmatrix} x_1 \\ x_2 \end{bmatrix}$ satisfies $|y_1| + |y_2| \geq |x_1| + |x_2|$.

(d) Show that the structural matrix corresponding to the 3-sector economy of Figure 5.5.1 is a reduction matrix.

(e) Give an example of a 3-sector economy whose structural matrix is *not* a reduction matrix.

(f) Give an example of a matrix B that satisfies the requirements for a reduction matrix but not (5.3.12).

Problem 5.8.3 Recalling that a population of buffaloes can be represented by a column vector $\text{col}[b_i, t_i, a_i]$ and that the number of buffaloes in such a population is given by $b_i + t_i + a_i$, generalize parts (a)–(c) of Problem 5.8.2 to 3×3 matrices.

Problem 5.8.4 Suppose $B = \begin{bmatrix} b_{11} & b_{12} \\ b_{21} & b_{22} \end{bmatrix}$ is a reduction matrix and that ρ is a number, $0 < \rho < 1$, for which $|b_{11}| + |b_{21}| \le \rho$ and $|b_{12}| + |b_{22}| \le \rho$.

(a) Show that if $C = B^2$ and $C = \begin{bmatrix} c_{11} & c_{12} \\ c_{21} & c_{22} \end{bmatrix}$, then

$$|c_{11}| + |c_{21}| \le \rho^2 \quad \text{and} \quad |c_{12}| + |c_{22}| \le \rho^2.$$

(b) Show that if $C = B^n$ and $C = \begin{bmatrix} c_{11} & c_{12} \\ c_{21} & c_{22} \end{bmatrix}$, then

$$|c_{11}| + |c_{21}| \le \rho^n \quad \text{and} \quad |c_{12}| + |c_{22}| \le \rho^n.$$

(c) Show that if $B = A + I$ is a reduction matrix, then the elements of B^n all approach 0 an n increases without bound.

Compression Matrices

Problem 5.8.5 Referring again to Problem 5.8.2, consider a population of rabbits for which a baby and an adult can be housed in a single hutch, but no hutch is allowed to contain more than one baby or more than one adult. The number of hutches required to house such a population at time i is given by $h_i = \max[b_i, a_i]$. As such, the inequality $h_{i+1} < h_i$ corresponds to a reduction in the number of hutches required to house such a population during the ith period of time. On this basis, we will say that a 2×2 matrix B is a "compression matrix" if

$$\begin{bmatrix} y_1 \\ y_2 \end{bmatrix} = \begin{bmatrix} b_{11} & b_{12} \\ b_{21} & b_{22} \end{bmatrix} \begin{bmatrix} x_1 \\ x_2 \end{bmatrix} \text{ implies that } \max[|y_1|, |y_2|] < \max[|x_1|, |x_2|]$$

for every vector $\begin{bmatrix} x_1 \\ x_2 \end{bmatrix}$.

(a) Show that a matrix B all of whose elements are nonnegative is a compression matrix whenever $b_{11} + b_{12} < 1$ and $b_{21} + b_{22} < 1$.

(b) Show that a matrix B all of whose elements are real numbers is a compression matrix whenever $|b_{11}| + |b_{12}| < 1$ and $|b_{21}| + |b_{22}| < 1$.

(c) Given a matrix B whose elements satisfy $|b_{11}| + |b_{12}| < 1$ and $|b_{21}| + |b_{22}| \ge 1$, find a vector $\begin{bmatrix} x_1 \\ x_2 \end{bmatrix}$ for which $\begin{bmatrix} y_1 \\ y_2 \end{bmatrix} = \begin{bmatrix} b_{11} & b_{12} \\ b_{21} & b_{22} \end{bmatrix} \begin{bmatrix} x_1 \\ x_2 \end{bmatrix}$ satisfies $\max[|y_1|, |y_2|] \ge \max[|x_1|, |x_2|]$.

(d) Give an example of a matrix B that satisfies the requirements for a compression matrix but not (5.3.12).

(e) Recalling that a population of buffaloes can be represented by a column vector $\mathrm{col}[b_i, t_i, a_i]$, suppose a buffalo shelter can house any combination of a baby and/or a teen and/or an adult, but no more than a single buffalo in any one of these categories. Use this analogy to generalize parts (a)–(c) above to 3×3 matrices.

Problem 5.8.6 Use the spreadsheet format of Figure 5.3.3 to solve

$$-9x + 8y = 76,$$
$$5x - 7y = -43$$

by iteration. [Hint: Divide both equations by 10 before forming $A + I$.]

Problem 5.8.7 Use the spreadsheet format of Figure 5.3.3 to solve

$$-9x + 8y = 76,$$
$$5x + 7y = -43$$

by iteration. [Hint: Multiply through the second equation by -1.]

Criteria for Contraction

Problem 5.3.6 showed that the condition $b_{11}^2 + b_{12}^2 + b_{21}^2 + b_{22}^2 < 1$ is sufficient to ensure that a 2×2 matrix $B = \begin{bmatrix} b_{11} & b_{12} \\ b_{12} & b_{22} \end{bmatrix}$ will be a contraction matrix. It was on this basis that we defined a rather unusual measure by which to measure the "size" of a matrix, namely $\|\|B\|\| = \sqrt{b_{11}^2 + b_{12}^2 + b_{21}^2 + b_{22}^2}$. This definition enables us to formulate the result of Problem 5.3.6 as

"$\|\|B\|\|^2 < 1$ is a *sufficient* condition to ensure

that B is a contraction matrix."

The next two problems establish the stronger result (see 5.3.12) that

$$\|\|B\|\|^2 + \sqrt{\|\|B\|\|^4 - 4D^2} < 2 \text{ is a } necessary \text{ and } sufficient \text{ condition}$$

for B to be a contraction matrix.

While this condition and the problems required to establish it are rather complicated, we shall also see that they allow considerable simplification in some special cases.

A Consequence of Cauchy's Inequality

Problem 5.8.8 Recalling Cauchy's inequality (5.3.6):

(a) Show that if a_1 and a_2 are real numbers and β is any angle, then

$$-\sqrt{a_1^2 + a_2^2} \le a_1 \cos \beta + a_2 \sin \beta \le \sqrt{a_1^2 + a_2^2}.$$

(b) Given numbers a_1 and a_2, show that there always exist angles β_1 and β_2 such that

$$a_1 \cos \beta_1 + a_2 \sin \beta_1 = \sqrt{a_1^2 + a_2^2} \quad \text{and}$$
$$a_1 \cos \beta_2 + a_2 \sin \beta_2 = -\sqrt{a_1^2 + a_2^2}.$$

[Hint: If $(a_1, a_2) \neq (0, 0)$, then

$$\left(\frac{a_1}{\sqrt{a_1^2 + a_2^2}}, \frac{a_2}{\sqrt{a_1^2 + a_2^2}} \right)$$

is a point on the unit circle $x^2 + y^2 = 1$.]

A Necessary and Sufficient Condition for Contraction

Problem 5.8.9 Recall that $B = \begin{bmatrix} b_{11} & b_{12} \\ b_{21} & b_{22} \end{bmatrix}$ is a contraction matrix if and only if

$$(b_{11}x_1 + b_{12}x_2)^2 + (b_{21}x_1 + b_{22}x_2)^2 < x_1^2 + x_2^2$$

for all $(x_1, x_2) \neq (0, 0)$. This can be rewritten as

$$(b_{11}^2 + b_{21}^2)x_1^2 + 2(b_{11}b_{12} + b_{21}b_{22})x_1x_2 + (b_{12}^2 + b_{22}^2)x_2^2 < x_1^2 + x_2^2.$$

(a) Setting $a = b_{11}^2 + b_{21}^2$, $b = b_{11}b_{12} + b_{21}b_{22}$, and $c = b_{12}^2 + b_{22}^2$, show that the preceding condition takes the form

$$a \left(\frac{x_1}{\sqrt{x_1^2 + x_2^2}} \right)^2$$

(*) $$+ 2b \left(\frac{x_1}{\sqrt{x_1^2 + x_2^2}} \right) \left(\frac{x_2}{\sqrt{x_1^2 + x_2^2}} \right) + c \left(\frac{x_2}{\sqrt{x_1^2 + x_2^2}} \right)^2 < 1.$$

(b) Defining an angle θ by

$$\cos \theta = \frac{x_1}{\sqrt{x_1^2 + x_2^2}} \quad \text{and} \quad \sin \theta = \frac{x_2}{\sqrt{x_1^2 + x_2^2}},$$

show that (*) is the same as $a \cos^2 \theta + 2b \cos \theta \sin \theta + c \sin^2 \theta < 1$, for all θ.

(c) Use trigonometric double angle formulas and part (b) to show that (*) is equivalent to

$$\frac{a + c}{2} + \frac{a - c}{2} \cos 2\theta + b \sin 2\theta < 1.$$

for all θ.

(d) Deduce from part (c) that (*) holds for all $(x_1, x_2) \neq (0, 0)$ if and only if

$$\frac{a + c}{2} + \sqrt{\left(\frac{a - c}{2} \right)^2 + b^2} < 1.$$

[Hint: Recall Problem 5.8.8.]

(e) From part (d) deduce that B is a contraction matrix if and only if

$$\frac{1}{2}\left[b_{11}^2 + b_{21}^2 + b_{12}^2 + b_{22}^2\right.$$

$$\left. + \sqrt{\left(b_{11}^2 + b_{21}^2 - b_{12}^2 - b_{22}^2\right)^2 + 4\left(b_{11}b_{12} + b_{21}b_{22}\right)^2}\right] < 1$$

and that this is equivalent to

$$\| B \|\|^2 + \sqrt{\|\| B \|\|^4 - 4D^2} < 2,$$

where D is the determinant of B. [The last condition is established by showing $\|\| B \|\|^4 - 4D^2 = (b_{11}^2 + b_{21}^2 - b_{12}^2 - b_{22}^2)^2 + 4(b_{11}b_{12} + b_{21}b_{22})^2$, an expression that is clearly nonnegative.]

Problem 5.8.10 In the process of establishing the necessary and sufficient condition (5.3.12), it was seen that $\|\| B \|\|^4 - 4D^2$ cannot be negative. As an alternative approach, show that $|\det B| \le \frac{\|\| B \|\|^2}{2}$ by:

(a) Using Cauchy's inequality to show that

$$D^2 = (b_{11}b_{22} - b_{12}b_{21})^2 \le \left(b_{11}^2 + b_{12}^2\right)\left(b_{12}^2 + b_{22}^2\right).$$

(b) Using the A-G inequality to show that

$$\left(b_{11}^2 + b_{12}^2\right)\left(b_{12}^2 + b_{22}^2\right)$$

$$\le \left(\frac{(b_{11}^2 + b_{12}^2) + (b_{21}^2 + b_{22}^2)}{2}\right)^2 = \left(\frac{\|\| B \|\|^2}{2}\right)^2.$$

Problem 5.8.11 Suppose that B is a 2×2 diagonal matrix, i.e., $B = \begin{bmatrix} b_{11} & 0 \\ 0 & b_{22} \end{bmatrix}$. Show that condition (5.3.12) reduces to

$$b_{11}^2 + b_{22}^2 + \left|b_{11}^2 - b_{22}^2\right| < 2,$$

and that this is equivalent to the condition that both $|b_{11}| < 1$ and $|b_{22}| < 1$.

Problem 5.8.12 Without appealing to (5.3.12), show directly from (5.3.5) that $B = \begin{bmatrix} b_{11} & 0 \\ 0 & b_{22} \end{bmatrix}$ is a contraction matrix if and only if $|b_{11}| < 1$ and $|b_{22}| < 1$.

Problem 5.8.13 According to Problem 5.3.6, $B = \begin{bmatrix} b_{11} & b_{12} \\ b_{21} & b_{22} \end{bmatrix}$ is a contraction matrix if $\|\| B \|\|^2 < 1$. Now use Problems 5.8.9 and 5.8.10 to show that:

(a) If $\|\| B \|\|^2 \ge 2$, then B is not a contraction matrix.
(b) If $1 \le \|\| B \|\|^2 < 2$, then B is a contraction matrix if and only if

$$\|\| B \|\|^2 < 1 + D^2,$$

where $D = \det B$.

Problem 5.8.14 Establish part (a) of Problem 5.8.13 without reference to the condition (5.3.12). [Hint: If $b_{11}b_{12} + b_{21}b_{22} \geq 0$, consider $\|Bx\|^2$, where $x = \begin{bmatrix} 1 \\ 1 \end{bmatrix}$. If $b_{11}b_{12} + b_{21}b_{22} \leq 0$, consider $\|Bx\|^2$, where $x = \begin{bmatrix} 1 \\ -1 \end{bmatrix}$.]

A Special Matrix

Problem 5.8.15 Let $B = \begin{bmatrix} a & b \\ -b & a \end{bmatrix}$, where a and b are real numbers. Show that B is a contraction matrix if and only if $a^2 + b^2 < 1$.

Contraction After Refinement

Problem 5.8.16 Let $A = \begin{bmatrix} -p & q \\ r & -s \end{bmatrix}$, where $p > 0$ and $s > 0$. Let $B = I + \frac{1}{N}A$, where N is a positive integer to be fixed later.

(a) Show that $\||B\||^2 < 2$ if $N > \frac{\|A\|^2}{2(p+s)}$.

(b) Suppose A also satisfies $(q + r)^2 < 4ps$. Show that B is a contraction matrix for all sufficiently large values of N.

(c) Let $A = \begin{bmatrix} -4 & 5 \\ 2 & -4 \end{bmatrix}$. Show that for $N = 6$, B is neither a contraction matrix nor a reduction matrix nor a compression matrix.

(d) For A as in part (c) and $N > 6$, show that B is a contraction matrix.

(e) For A as in part (c), show that B is neither a reduction matrix nor a compression matrix for any positive integer N.

Problem 5.8.17 Solve the linear system

$$4x - 5y = 11,$$
$$2x - 4y = 8$$

(a) by EROs.

(b) by Cramer's Rule

(c) by iteration, making use of part (d) of Problem 5.8.16.

A Recalcitrant Matrix

Problem 5.8.18 As mentioned at the end of Section 5.3, there exist matrices A for which neither $Ax + b = 0$ nor $(-A)x + (-b) = 0$ allows a solution by iteration, regardless of the degree of refinement. Show that $A = \begin{bmatrix} 2 & 0 \\ 0 & -2 \end{bmatrix}$ is an example of such a "recalcitrant matrix."

Powers of A Contraction Matrix

Throughout this chapter we have made use of the fact that if B is a contraction matrix and x is any vector, then $B^n x$ approaches 0 as n grows without bound (see part (d) of Problem 5.2.3). The following two problems confirm this property of contraction matrices.

Problem 5.8.19 Let B be a contraction matrix and suppose that there exists a number ρ, $0 < \rho < 1$, for which $\|B\mathbf{x}\| \le \rho\|\mathbf{x}\|$ for all $\mathbf{x}$.

(a) Show that for each vector $\mathbf{x}$, $\|B^n\mathbf{x}\| \le \rho^n\|\mathbf{x}\|$, $n = 1, 2, 3, \ldots$, and that as a result, $B^n\mathbf{x}$ approaches $\mathbf{0}$ as n grows without bound.

(b) Let $A = \begin{bmatrix} a_{11} & a_{21} \\ a_{12} & a_{22} \end{bmatrix}$ and suppose $k > 0$ is such that $\|A\mathbf{x}\| \le k\|\mathbf{x}\|$ for all $\mathbf{x}$. Show that all entries of A must have absolute value $\le k$. [Hint: Consider $\|A\mathbf{x}\|^2$, where $\mathbf{x} = \begin{bmatrix} 1 \\ 0 \end{bmatrix}$ and $\mathbf{x} = \begin{bmatrix} 0 \\ 1 \end{bmatrix}$.]

(c) Deduce from parts (a) and (b) that if B is a contraction matrix, then the entries of B^n approach 0 as n grows without bound, i.e., that B^n approaches the zero matrix.

Given a contraction matrix B and a particular vector $\mathbf{x}$, there surely exists a number ρ, $0 < \rho < 1$, for which $\|B\mathbf{x}\| \le \rho\|x\|$. But is there a single such ρ, one for which $\|B\mathbf{x}\| \le \rho\|x\|$ for all $\mathbf{x}$ (as was assumed in Problem 5.8.19 above)? Establishing this subtle but important elaboration of what it means to be a contraction matrix can call for some rather advanced mathematical ideas. However, in the case of 2×2 matrices an answer can also be based on Problem 5.8.9.

Problem 5.8.20 Confirm that parts (a)–(c) of Problem 5.8.9 correspond to maximizing $\|B\mathbf{x}\|$ as $\mathbf{x}$ ranges over all vectors with $\|\mathbf{x}\| = 1$, and that parts (d)–(f) correspond to showing that this maximum is $\frac{1}{2}(\|\|B\|\|^2 + \sqrt{\|\|B\|\|^4 - 4D^2})$.

(a) Show that setting $\rho = \frac{1}{2}(\|\|B\|\|^2 + \sqrt{\|\|B\|\|^4 - 4D^2})$ leads to $\|B\mathbf{x}\| \le \rho\|\mathbf{x}\|$ for *all* $\mathbf{x}$.

(b) For this value of ρ, show that (5.3.12) is equivalent to $\rho < 1$.

The Method of False Position

Problem 5.8.21 The method of false position, or *regula falsi*, was a technique for solving certain algebraic equations found in many nineteenth-century textbooks. Variants of this method form the basis for certain iterative schemes used in numerical analysis. Confronted with the problem of solving $f(x) = c$, where f is a given function, one tries a guess x_1. Unless $f(x_1) = c$, this guess generates an "error" $e_1 = f(x_1) - c$, which is noted. Next one tries another guess x_2 to obtain $f(x_2)$ and another error $e_2 = f(x_2) - c$. Assuming $e_1 \ne e_2$, the method now asserts that the solution of the original problem $f(x) = c$ is given by the formula $x = \frac{x_1 e_2 - x_2 e_1}{e_2 - e_1}$.

(a) Suppose $c = 0$ and we are trying to find a zero of a function f. Show that the method of false position gives $x = \frac{x_1 f(x_2) - x_2 f(x_1)}{f(x_2) - f(x_1)}$.

(b) Show that for any *linear* function $f(x) = ax + b$, the method of false position yields the correct solution of $f(x) = c$.

(c) Give an example showing that if f is *not* linear, the method of false position cannot be counted on to provide a solution of $f(x) = c$.

(d) In the figure below, show that the line through (x_1, y_1) and (x_2, y_2) intersects the x-axis at $\left(\frac{x_1 y_2 - x_2 y_1}{y_2 - y_1}, 0\right)$. Explain why this is the basis for finding the zero of a linear function by the method of false position.

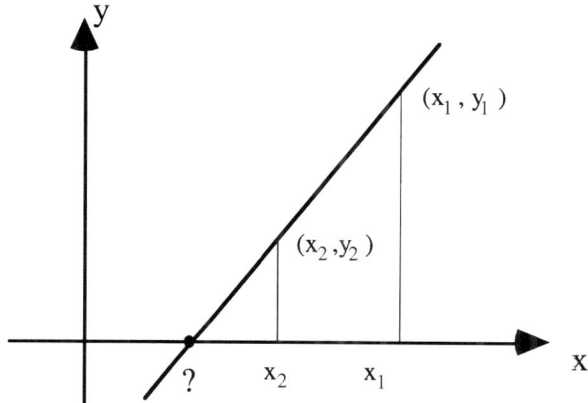

[Note: As an alternative to finding the equation of the line through (x_1, y_1) and (x_2, y_2), it is possible to obtain the result by using similar right triangles.]

Problem 5.8.22 A variant of the method of false position can also be used to solve a pair of linear equations

$$(*) \qquad \begin{aligned} ax + by + c &= 0, \\ Ax + By + C &= 0. \end{aligned}$$

Suppose we choose a pair of points (x_1, y_1) and (x_2, y_2) that are solutions of the first equation, $ax + by + c = 0$. Substituting these into $Ax + By + C = 0$ generates a pair of "errors"

$$e_1 = Ax_1 + By_1 + C \quad \text{and} \quad e_2 = Ax_2 + By_2 + C.$$

Assuming $e_1 \neq e_2$, show that

$$x = \frac{x_2 e_1 - x_1 e_2}{e_1 - e_2} \quad \text{and} \quad y = \frac{y_2 e_1 - y_1 e_2}{e_1 - e_2}$$

(a) is also a solution of $ax + by + c = 0$.
(b) is a solution of $(*)$.

Problem 5.8.23 Apply the above method to solve the land problem from *Jiuzhang* that is stated in Section 5.1.

*Problem 5.8.24 Readers familiar with analytic geometry will be able to develop a geometric interpretation of the method described in Problem 5.8.22.

(a) Show that if (x_1, y_1, z_1) and (x_2, y_2, z_2) are two points in (x, y, z)-space, with $z_1 \neq z_2$, then the line through these two points intersects the plane $z = 0$ in a point $(x, y, 0)$ such that

(**) $$x = \frac{x_2 z_1 - x_1 z_2}{z_1 - z_2} \quad \text{and} \quad y = \frac{y_2 z_1 - y_1 z_2}{z_1 - z_2}.$$

(b) Show that if the pair of points $(x_1, y_1, 0)$ and $(x_2, y_2, 0)$ both belong to a line $ax + by + c = 0$ in the plane $z = 0$, then $(x, y, 0)$ also belongs to this line.

(c) For (x_1, y_1) and (x_2, y_2) as in part (b), let

$$z_1 = Ax_1 + By_1 + C \quad \text{and} \quad z_2 = Ax_2 + By_2 + C.$$

Then (x_1, y_1, z_1) and (x_2, y_2, z_2) are points on the plane $z = Ax + By + C$, and the line through these two points intersects the plane $z = 0$ in the point $(x, y, 0)$ satisfying (**). Explain why the first two coordinates of this point satisfy $Ax + By + C = 0$.

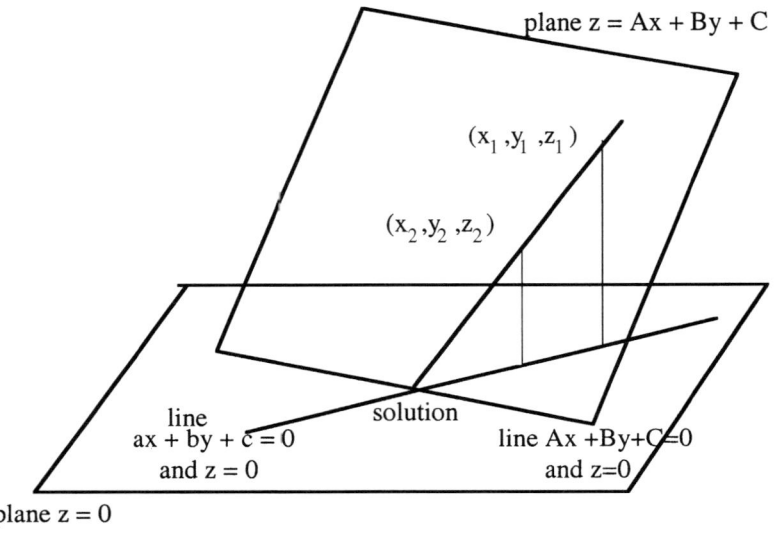

Economies Whose Structural Matrices Contract, Reduce, and Compress

Problem 5.8.25 The system 5.5.3 is of the form $(I - A)\mathbf{x} = \mathbf{y}$, where A is the structural matrix corresponding to the economy of Figure 5.5.1.

(a) Verify the matrix identity $I - (I - A) = A$.

(b) Show that (5.5.3) can be solved by iteration whenever the structural matrix A is a contraction matrix.

(c) Confirm that the structural matrix corresponding to Figure 5.5.1 is a contraction matrix (as well as a reduction matrix and a compression matrix) and then solve (5.5.3) by iteration.

(d) Explain why we might expect the structural matrix corresponding to a consumer-oriented economy to be a contraction matrix.

Problem 5.8.26 Confirm that the structural matrix corresponding to the four-sector economy of Figure 5.5.2 is a reduction matrix in the sense of Problem 5.8.3. Solve (5.5.7) with a spreadsheet that implements matrix iteration by a 3 × 3 matrix.

Determinants and the Area of a Parallelogram

Problem 5.8.27 Consider a parallelogram $OPQR$ in the (x, y)-plane with vertices $O = (0, 0)$, $P = (a, c)$, $R = (b, d)$, and $Q = (a + b, c + d)$, as in the figure below.

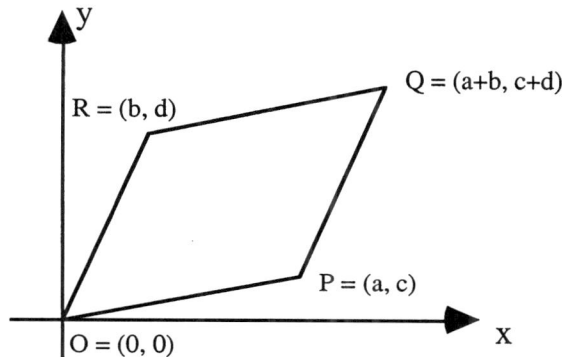

The purpose of this problem is to show that the area A of this parallelogram is $|ad - bc|$, which is also the absolute value of the determinant of the matrix $\begin{bmatrix} a & b \\ c & d \end{bmatrix}$. Note that the columns of this determinant are the coordinates of the points P and R that define the parallelogram.

(a) Assuming $d \neq 0$, show that the line through P and Q intersects the x-axis at the point S with coordinates $(\frac{ad - bc}{d}, 0)$. [Hint: Apply Problem 5.8.21(d).]

(b) Using the figure below, show that the parallelogram $OSR'R$ has the same area as the parallelogram $OPQR$, where R' is the point at which the

horizontal line through R intersects the line segment PQ. [Hint: Prove
that $\triangle OSP$ is congruent to $\triangle RR'Q$.]

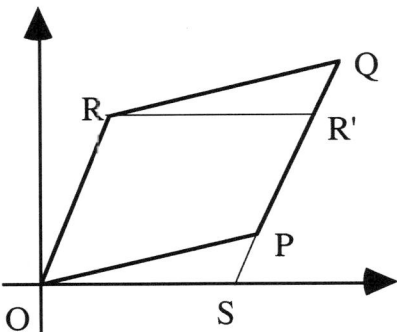

(c) Recalling that $P = (a, c)$, $R = (b, d)$, and $Q = (a + b, c + d)$, show
that the distance OS is $|\frac{ad - bc}{d}|$ and that the distance from R' to the
x-axis is $|d|$. From this deduce that the area of the parallelogram is
$|ad - bc|$, as we set out to prove.

A Geometrical Interpretation of the Determinant

A 2×2 matrix $B = \begin{bmatrix} a & b \\ c & d \end{bmatrix}$ gives rise to a mapping of the (x_1, x_2)-plane into itself
by associating to each point $x = \begin{bmatrix} x_1 \\ x_2 \end{bmatrix}$ the image point $Bx = \begin{bmatrix} a & b \\ c & d \end{bmatrix}\begin{bmatrix} x_1 \\ x_2 \end{bmatrix} = \begin{bmatrix} ax_1 + bx_2 \\ cx_1 + dx_2 \end{bmatrix}$.
Such a mapping of the (x_1, x_2)-plane to the (x_1, x_2)-plane is called a *linear mapping* (or a *linear transformation*). An important property of such linear mappings
is that if $\det B \neq 0$, the image of a straight line is again a straight line and the
image of a pair of parallel lines is again a pair of parallel lines. The following problem provides a heuristic explanation of the relation between the determinant of a
matrix and the magnification that areas undergo under the corresponding linear
mapping.

Problem 5.8.28 Making use of the above assertions about linear mappings,

(a) Show that $Bx = \begin{bmatrix} a & b \\ c & d \end{bmatrix}\begin{bmatrix} x_1 \\ x_2 \end{bmatrix}$ maps the unit square with vertices
$(0, 0)$, $(1, 0)$, $(1, 1)$, $(0, 1)$ onto the parallelogram with vertices
$(0, 0)$, (a, c), $(a + b, c + d)$, (b, d).

(b) Show that the area of the image of the unit square in part (a) is $\pm \det B$.

(c) Show that the $k \times k$ square S with vertices $(0, 0)$, $(k, 0)$, (k, k), $(0, k)$ is
mapped onto a parallelogram T such that the area$(T) = |\det B| \cdot$ area(S).

(d) Show that any square S with sides parallel to the axes is mapped onto a
parallelogram T such that area$(T) = |\det B| \cdot$ area(S). [Hint: Translate
the lower left vertex of the square to the origin and apply part (c).]

(e) Let S be a region that can be partitioned into congruent squares with sides parallel to the axes. If T is the image of S, explain why
$$\text{area}(T) = |\det B| \cdot \text{area}(S).$$

Since the answer to part (e) applies regardless of the size or number of congruent squares, it is reasonable to ask whether the result
$$\text{area}(T) = |\det B| \cdot \text{area}(S)$$

can be extended to more general regions S. That is, if a region S can be (arbitrarily) closely approximated by tiny squares and its image T is also (arbitrarily) closely approximated by the corresponding parallelograms, then the above relation should remain valid. The following problem is based on the assumption that this line of reasoning can be applied when S is a circle.

Problem 5.8.29 Show that the linear mapping given by $B = \begin{bmatrix} a & 0 \\ 0 & b \end{bmatrix}$ maps the region bounded by the unit circle $x_1^2 + x_2^2 = 1$ onto the region bounded by the ellipse $\frac{x_1^2}{a^2} + \frac{x_2^2}{b^2} = 1$. Deduce from this the formula for the area of an ellipse.

Some Further Properties of the Determinant

Problem 5.8.30 Given 2×2 matrices A and B, let S be a region in the (x_1, x_2)-plane, S^* the image of S under B, and let S^{**} be the image of S^* under A.

(a) Show that S^{**} is the image of S under the single linear mapping AB.

(b) Show that $\text{area}(S^{**}) = |\det(AB)| \cdot \text{area}(S)$.

(c) Use parts (a) and (b) to deduce that $|\det(AB)| = |(\det A) \cdot (\det B)|$.

(d) For $A = \begin{bmatrix} a_{11} & a_{12} \\ a_{21} & a_{22} \end{bmatrix}$ and $B = \begin{bmatrix} b_{11} & b_{12} \\ b_{21} & b_{22} \end{bmatrix}$, use matrix multiplication to establish the stronger result that $\det(AB) = (\det A) \cdot (\det B)$.

(e) Assuming $A = \begin{bmatrix} a_{11} & a_{12} \\ a_{21} & a_{22} \end{bmatrix}$ has an inverse A^{-1} satisfying $AA^{-1} = I$, show that $\det(A^{-1}) = 1/(\det A)$.

(f) From part (d) deduce that for every positive integer n, $\det(A^n) = (\det A)^n$.

Fibonacci and Determinants

Problem 5.8.31 Recall the Fibonacci matrix $A = \begin{bmatrix} 0 & 1 \\ 1 & 1 \end{bmatrix}$ arising in (5.6.7), and the sequence of Fibonacci numbers $x_1, x_2, x_3, \ldots$ defined by
$$x_1 = x_2 = 1 \quad \text{and} \quad x_{i+2} = x_{i+1} + x_i, \qquad i = 1, 2, \ldots.$$

(a) Show that for each integer $n > 1$, $A^n = \begin{bmatrix} x_{n-1} & x_n \\ x_n & x_{n+1} \end{bmatrix}$. [Hint: If this relation holds for some n, then $A^{n+1} = \begin{bmatrix} x_{n-1} & x_n \\ x_n & x_{n+1} \end{bmatrix} \begin{bmatrix} 0 & 1 \\ 1 & 1 \end{bmatrix}$.]

(b) Recalling that $\det(A^n) = (\det A)^n$, show that the Fibonacci numbers satisfy

$$x_{n-1}x_{n+1} - x_n^2 = (-1)^n, \qquad n = 2, 3, \ldots.$$

Eigenvalues and Eigenvectors

Problem 5.8.32 The eigenvalues of the matrix $A = \begin{bmatrix} 0 & q \\ 1 & 1 \end{bmatrix}$ arising in (5.6.10) are the roots of the equation $\det(A - \lambda I) = \det \begin{bmatrix} -\lambda & q \\ 1 & 1-\lambda \end{bmatrix} = 0$.

(a) Confirm that $\lambda_1 = \frac{1+\sqrt{1+4q}}{2}$ and $\lambda_2 = \frac{1-\sqrt{1+4q}}{2}$ are the eigenvalues of A and that

$$\mathbf{v}_1 = \begin{bmatrix} q \\ \lambda_1 \end{bmatrix}, \qquad \mathbf{v}_2 = \begin{bmatrix} q \\ \lambda_2 \end{bmatrix}$$

are the corresponding *eigenvectors* (i.e., that $A\mathbf{v}_i = \lambda_i \mathbf{v}_i$ for $i = 1, 2$).

(b) Suppose we generate a sequence of "population vectors" $\mathbf{x}_i = \begin{bmatrix} b_i \\ a_i \end{bmatrix}$ by the iterative scheme $\begin{bmatrix} b_{i+1} \\ a_{i+1} \end{bmatrix} = \begin{bmatrix} 0 & q \\ 1 & 1 \end{bmatrix}\begin{bmatrix} b_i \\ a_i \end{bmatrix}$. Show that the initial vector $\mathbf{x}_1 = \begin{bmatrix} q \\ \lambda_1 \end{bmatrix}$ results in a "stable population pyramid" in the sense that

$$\frac{b_{i+1}}{a_{i+1}} = \frac{b_i}{a_i} \quad \text{and} \quad \frac{a_{i+2} + b_{i+2}}{a_{i+1} + b_{i+1}} = \frac{a_{i+1} + b_{i+1}}{a_i + b_i} \qquad \text{for} \quad i = 1, 2, \ldots.$$

Problem 5.8.33 The "buffalo matrix"

$$A = \begin{bmatrix} 0 & 0 & q \\ 1 & 0 & 0 \\ 0 & 1 & 1 \end{bmatrix}$$

arose in (5.6.11).

(a) Show that the eigenvalues of the buffalo matrix are solutions of the cubic equation $\lambda^3 - \lambda^2 - q = 0$.

(b) Recall that a cubic with real coefficients always has at least one real root. Show that if λ is a real root of the cubic in part (a) then

$$\mathbf{v} = \begin{bmatrix} q\lambda \\ q \\ \lambda^2 \end{bmatrix}$$

is an eigenvector of A.

(c) Suppose population vectors of the form

$$\mathbf{x}_i = \begin{bmatrix} b_i \\ t_i \\ a_i \end{bmatrix}$$

are generated by the iterative scheme $\mathbf{x}_{i+1} = A\mathbf{x}_i$, $i = 1, 2, \ldots$. Show that a choice of

$$\mathbf{v} = \begin{bmatrix} q\lambda \\ q \\ \lambda^2 \end{bmatrix}$$

as initial vector $\mathbf{x}_1$ results in a stable population pyramid, i.e., that

$$\frac{b_{i+1}}{b_i} = \frac{t_{i+1}}{t_i} = \frac{a_{i+1}}{a_i} \quad \text{and} \quad \frac{b_{i+2} + t_{i+2} + a_{i+2}}{b_{i+1} + t_{i+1} + a_{i+1}} = \frac{b_{i+1} + t_{i+1} + a_{i+1}}{b_i + t_i + a_i}$$

for $i = 1, 2, \ldots$.

Elephant Matrices

Problem 5.8.34 Consider creatures called "elephants" with a minimum breeding age $T = 3$. This population consists of babies, children (1-year-olds), teens (2-year-olds), and adults who are 3+ years old and, on the average, give rise to q babies each year per adult.

(a) Representing such a population at time i by a population vector

$$\mathbf{x}_i = \begin{bmatrix} b_i \\ c_i \\ t_i \\ a_i \end{bmatrix},$$

show that the growth of such a population satisfies $\mathbf{x}_{i+1} = A\mathbf{x}_i$ with

$$A = \begin{bmatrix} 0 & 0 & 0 & q \\ 1 & 0 & 0 & 0 \\ 0 & 1 & 0 & 0 \\ 0 & 0 & 1 & 1 \end{bmatrix}.$$

(b) Defining $e_i = b_i + c_i + t_i + a_i$, show that $e_{i+4} = e_{i+3} + qe_i$, $i = 1, 2, \ldots$.

(c) Assuming that the population ratio e_{i+1}/e_i approaches a value r as i grows without bound, show that r satisfies the quartic equation $r^4 - r^3 - q = 0$. [Hint: Divide the equation of part (b) through by e_i. Note that for large i, $e_{i+4}/e_i = (e_{i+4}/e_{i+3})(e_{i+3}/e_{i+2})(e_{i+2}/e_{i+1}) \cdot (e_{i+1}/e_i) \approx r^4$.]

(d) With A as in part (a), show that the eigenvalues of A are the roots of the quartic equation $\lambda^4 - \lambda^3 - q = 0$. [Remark: This problem requires an ability to evaluate the determinant of a 4×4 matrix.]

(e) Assuming that λ is an eigenvalue of A, show that

$$\mathbf{v} = \begin{bmatrix} q\lambda^2 \\ q\lambda \\ q \\ \lambda^3 \end{bmatrix}$$

is an eigenvector, satisfying $A\mathbf{v} = \lambda\mathbf{v}$.

(f) Generalize Problem 5.8.33(c) to obtain conditions for a stable population matrix for elephants.

(g) Setting $q = 1$, we are led to a variant of the famous Fibonacci numbers. Here Fibonacci's fecund rabbits are replaced by rather slow-to-breed elephants. Starting with one pair of babies, we obtain the "elephant numbers"

$$1, 1, 1, 1, 2, 3, 4, 5, 7, 10, 14, 19, 26, 36, 50, 69, 95, 131, \ldots,$$

where the generating rule is

$$e_1 = e_2 = e_3 = e_4 = 1 \quad \text{and} \quad e_{i+4} = e_{i+3} + e_i$$
$$\text{for} \quad i = 1, 2, \ldots.$$

Using a spreadsheet similar to that of Figure 5.6.3 to calculate such x_i, check the relation $x_{i+1}/x_i \approx r$, where $r = 1.380277569\ldots$ is the real positive root of $r^4 - r^3 - 1 = 0$.

Dinosaur Matrices

Problem 5.8.35 The ambitious reader is now invited to extend prior results concerning rabbits, buffaloes, and elephants to a population of dinosaurs whose minimum breeding age is $T = 4$.

6

SYSTEMS OF DIFFERENCE EQUATIONS

6.1. Introduction   In Chapter 2 we made a transition from iterations of the form $x_{i+1} = F(x_i)$ to difference equations of the form $x_{i+1} - x_i = f(x_i)$. This was done by rewriting the iterative scheme $x_{i+1} = F(x_i)$ as

$$x_{i+1} - x_i = F(x_i) - x_i$$

and then defining $f(x) = F(x) - x$. In this chapter we apply the same strategy to vector-valued functions in order to study *systems* of difference equations of the form

(6.1.1) $$\mathbf{x}_{i+1} - \mathbf{x}_i = \mathbf{f}(\mathbf{x}_i).$$

A good starting point is to recall the iterative schemes $\mathbf{x}_{i+1} = (A + I)\mathbf{x}_i + \mathbf{b}$ studied in Chapter 5. Here $\mathbf{F}(\mathbf{x}) = (A + I)\mathbf{x} + \mathbf{b}$, and subtracting $\mathbf{x}_i$ from both sides of this equation, we obtain the *linear* system of difference equations

(6.1.2) $$\mathbf{x}_{i+1} - \mathbf{x}_i = A\mathbf{x}_i + \mathbf{b}.$$

Such systems are studied in Section 6.2 below.

However, in preparation for our study of *nonlinear* systems of difference equations, we will also consider more general iterative schemes of the form

(6.1.3) $$\mathbf{x}_{i+1} = \mathbf{F}(\mathbf{x}_i).$$

225

Here $\mathbf{F}$ associates with a given vector $\mathbf{x}$ the vector value $\mathbf{F}(\mathbf{x})$, and subtraction of $\mathbf{x}_i$ from both sides of this equation yields (6.1.1).

Unfortunately, our efforts to relate (6.1.1) to phenomena considered in Chapters 2 and 3 will encounter some notational problems. Having studied logistic growth by means of the difference equation

$$x_{i+1} - x_i = ax_i - bx_i^2,$$

we now note that the expression bx^2 makes no sense when the scalar variable x is replaced by a vector $\mathbf{x}$. To be sure, we could interpret multiplication in terms of the vector "dot product," replacing $\mathbf{x}^2$ by $\mathbf{x} \cdot \mathbf{x}$. That is, for a row vector $\mathbf{x} = (a, b)$ and a column vector $\mathbf{y} = \binom{c}{d}$, we could use matrix multiplication to define

$$\mathbf{x} \cdot \mathbf{y} = (a, b)\binom{c}{d} = ac + bd.$$

But $\mathbf{x} \cdot \mathbf{x}$ is then a scalar quantity, whereas $\mathbf{f}(\mathbf{x})$ is supposed to be a vector, defeating our original purpose.

We can circumvent this difficulty with a variant of the matrix notation introduced in Chapter 5. With an eye to studying phenomena such as logistic growth, we shall write nonlinear systems of difference equations in the form

(6.1.4) $\mathbf{x}_{i+1} - \mathbf{x}_i = A(\mathbf{x}_i)\mathbf{x}_i + \mathbf{b}$

rather than (6.1.2) That is, the nonlinearities of the system will be manifested in the fact that the elements of the matrix $A(\mathbf{x})$ may themselves be functions of $\mathbf{x}$.

By way of a specific example, consider two populations u and v whose sizes at time i are given by u_i and v_i and are represented in terms of a single vector

$$\mathbf{x}_i = \begin{bmatrix} u_i \\ v_i \end{bmatrix}.$$

Suppose now that these populations are to undergo logistic growth of the form

$$u_{i+1} - u_i = au_i - bu_i^2$$

and

$$v_{i+1} - v_i = cv_i - dv_i^2,$$

respectively. In the context of (6.1.4), we could represent the entire system (i.e., the subsequent sizes of the two populations u and v) in terms of the system

(6.1.5) $\begin{bmatrix} u_{i+1} \\ v_{i+1} \end{bmatrix} - \begin{bmatrix} u_i \\ v_i \end{bmatrix} = \begin{bmatrix} a - bu_i & 0 \\ 0 & c - dv_i \end{bmatrix}\begin{bmatrix} u_i \\ v_i \end{bmatrix}.$

Adding $\mathbf{x}_i = \begin{bmatrix} u_i \\ v_i \end{bmatrix}$ to both sides of this equation leads to the iterative scheme

(6.1.6) $\mathbf{x}_{i+1} = (A(\mathbf{x}_i) + I)\mathbf{x}_i + \mathbf{b},$

in which

$$A(\mathbf{x}_i) + I = \begin{bmatrix} 1 + a - bu_i & 0 \\ 0 & 1 + c - dv_i \end{bmatrix}$$

and $\mathbf{b} = 0$.

An important observation, one that lies at the heart of the present chapter, is the fact that the *interaction* between two such populations can be represented in the form (6.1.5). This calls for appropriate modification of the matrix

(6.1.7) $$A(\mathbf{x}_i) = \begin{bmatrix} a - bu_i & 0 \\ 0 & c - dv_i \end{bmatrix}$$

in ways to be considered in Sections 6.3 and 6.4.

Problem 6.1.1 Describe a modification of the above matrix $A(\mathbf{x})$ that would correspond to "interaction between the populations u and v" in the sense that the size of u affects the growth rate of v, and/or vice versa.

Problem 6.1.2 By proper choice of matrix $A(\mathbf{x})$, write the system

$$\mathbf{x}_{i+1} - \mathbf{x}_i = a\mathbf{x}_i - b\sqrt{(\mathbf{x}_i \cdot \mathbf{x}_i)}\mathbf{x}_i$$

in the form (6.1.4). Using the format of Figure 6.1.1, create a spreadsheet that computes the solution $\mathbf{x}_i = \begin{bmatrix} u_i \\ v_i \end{bmatrix}$ of this system for arbitrary values of a, b, and $\mathbf{x}_1 = \begin{bmatrix} u_1 \\ v_1 \end{bmatrix}$.

	A	B	C	D	E	F	G	H	I	J	K	L
1	For arbitrary values of a, b, u(1) and v(1), this spreadsheet solves the nonlinear system u(i+1)-u(i) = au(i)-b√[(u(i)^2+v(i)^2]u(i); v(i+1)-v(i) = av(i)-b√[(u(i)^2+v(i)^2]v(i).											
2												
3	a=	0.1		b=	0.005							
4												
5	I=	1	2	3	4	5	6	7	8	9	10	
6												
7	u(i)=	40	22.46	17.91	15.39	13.74	12.57	11.7	11.03	10.49	10.1	
8	v(i)=	100	56.15	44.79	38.46	34.34	31.43	29.25	27.57	26.23	25.1	
9												
10												
11		100										
12		80										
13		60							- - - - - u(i)=			
14		60										
15		40							——— v(i)=			
16		20										
17		0										
18			1 2 3 4 5 6 7 8 9 10									
19												

Figure 6.1.1 Modeling a Nonlinear System

6.2. Linear Systems   We begin our study of systems of difference equations with *linear* systems of the form

$$(6.2.1) \qquad \mathbf{x}_{i+1} - \mathbf{x}_i = A\mathbf{x}_i + \mathbf{b},$$

Here $\mathbf{x}_i$ and $\mathbf{b}$ are column vectors of length n and A is taken to be a *constant* matrix of size $n \times n$, where n is a positive integer.

In the case $n = 2$, we can write

$$\mathbf{x}_i = \begin{bmatrix} u_i \\ v_i \end{bmatrix}, \quad A = \begin{bmatrix} a_{11} & a_{12} \\ a_{21} & a_{22} \end{bmatrix}, \quad \text{and} \quad \mathbf{b} = \begin{bmatrix} b_1 \\ b_2 \end{bmatrix}.$$

The linear system (6.2.1) then corresponds to a pair of difference equations of the form

$$(6.2.2) \qquad \begin{aligned} u_{i+1} - u_i &= a_{11}u_i + a_{12}v_i + b_1, \\ v_{i+1} - v_i &= a_{21}u_i + a_{22}v_i + b_2. \end{aligned}$$

Problem 6.2.1 An investor's portfolio consists of a $100 checking account earning 6% interest and a $1000 savings account earning 10%, both compounded annually. At the start of each new year, half of the savings balance is transferred to the checking account, after which $500 new income is added to the savings account. Formulate a pair of difference equations of the form (6.2.2) to represent this system.

Problem 6.2.2 Consider the following variant of the preceding problem. In place of a savings account, the investor places the $1000 in an equity account whose earnings are randomly distributed between -5% and 25% each year. Furthermore, the investor now transfers the larger of $500 and half of the savings balance into the checking account each year. Create a spreadsheet that tracks these two accounts for 10 years.

The difference equation (6.2.1) is equivalent to the iterative scheme

$$(6.2.3) \qquad \mathbf{x}_{i+1} = (A + I)\mathbf{x}_i + \mathbf{b}$$

already encountered in Chapter 5. There we saw that whenever $\||A + I\|| < 1$, (6.2.3) provides a way of bypassing Cramer's rule in solving the algebraic system $A\mathbf{x} + \mathbf{b} = 0$. That is, the sequence of vectors $\mathbf{x}_1, \mathbf{x}_2, \mathbf{x}_3, \ldots$ obtained from (6.2.3) approaches $-A^{-1}\mathbf{b}$.

In the context of Chapters 2 and 5, we would say that the difference equation (6.2.1) "has an equilibrium value at $\mathbf{x}_0 = -A^{-1}\mathbf{b}$." As before, this equilibrium value is obtained by setting the right side of (6.2.1) equal to zero, i.e., by solving $A\mathbf{x} + \mathbf{b} = 0$. Whenever $\||A + I\|| < 1$, this vector behaves like an "attractor" for (6.2.3), in the sense that solutions of the difference equation (6.2.1) approach $\mathbf{x}_0$.

This is in direct analogy to the theory developed for the scalar difference equation

(6.2.4) $$x_{i+1} - x_i = ax_i + b$$

in Section 2.3.

When $\||A + I\|| < 1$ is not satisfied, these conclusions may not follow. This is not surprising, since even in the scalar case, $|a + 1| < 1$ is required to ensure that a solution of (6.2.4) will approach $-b/a$. In such situations we may want to consider a refinement of (6.2.1) corresponding to $\Delta t \mapsto \Delta t/N$. This would lead us to consider

(6.2.5) $$\mathbf{x}_{i+1} - \mathbf{x}_i = (A\mathbf{x}_i + \mathbf{b})/N$$

and the corresponding iterative scheme

(6.2.6) $$\mathbf{x}_{i+1} = (A/N + I)\mathbf{x}_i + \mathbf{b}/N.$$

As the result of such refinement, the condition $\||A + I\|| < 1$ would be replaced by $\||A/N + I\|| < 1$. In this way refinement can provide a useful tool for solving (6.2.1) by iteration.

Problem 6.2.3 Create a spreadsheet that models (6.2.2) with $a_{11} = -3$, $a_{12} = 1$, $a_{21} = 0$, $a_{22} = -5$, $b_1 = 3$, and $b_2 = -2$ and allows for refinement of the form $\Delta t \mapsto \Delta t/N$. Find a value of N for which the spreadsheet solution approximates the solution of $A\mathbf{x} + \mathbf{b} = 0$ obtained by Cramer's rule.

At this point it is interesting to recall the closed-form solution of (6.2.1),

(6.2.7) $$\mathbf{x}_i = (A + I)^{i-1}\mathbf{x}_1 - A^{-1}(I - (A + I)^{i-1})\mathbf{b}.$$

A derivation of (6.2.7) is outlined in Problem 5.2.3. However, as an alternative to recalling this derivation, we can simply verify its correctness.

Problem 6.2.4 For $\mathbf{b} = 0$ (and with a change of notation from $\mathbf{x}_i$ to $\mathbf{y}_i$), (6.2.7) asserts that $\mathbf{y}_i = (A + I)^{i-1}\mathbf{y}_1$ is a closed-form solution of $\mathbf{y}_{i+1} - \mathbf{y}_i = A\mathbf{y}_i$. Given this formula for $\mathbf{y}_i$, verify directly that $\mathbf{y}_{i+1} - \mathbf{y}_i = A\mathbf{y}_i$.

Problem 6.2.5 For $z_1 = 0$ (and with a change of notation from $\mathbf{x}_i$ to $\mathbf{z}_i$), (6.2.7) asserts that $\mathbf{z}_i = -A^{-1}(I - (A + I)^{i-1})\mathbf{b}$ is a closed-form solution of $\mathbf{z}_{i+1} - \mathbf{z}_i = A\mathbf{z}_i + \mathbf{b}$. Given this formula for $\mathbf{z}_i$, verify directly that $\mathbf{z}_{i+1} - \mathbf{z}_i = A\mathbf{z}_i + \mathbf{b}$.

Problem 6.2.6 For $\mathbf{y}_i$ and $\mathbf{z}_i$ as in Problems 6.2.4 and 6.2.5, let $\mathbf{x}_i = \mathbf{y}_i + \mathbf{z}_i$. Make use of the fact that $\mathbf{x}_{i+1} - \mathbf{x}_i = \mathbf{y}_{i+1} - \mathbf{y}_i + \mathbf{z}_{i+1} - \mathbf{z}_i$ to show that (6.2.7) satisfies (6.2.1).

Problem 6.2.6 illustrates a general principle known as *superposition*. The difference equation

$$(6.2.8) \qquad \mathbf{x}_{i+1} - \mathbf{x}_i = A\mathbf{x}_i$$

is called the *homogeneous equation* associated with (6.2.1). If $\mathbf{y}_i$ is any solution of the homogeneous equation (6.2.8) and $\mathbf{z}_i$ is any solution of the nonhomogeneous (original) equation (6.2.1), then $\mathbf{x}_i = \mathbf{y}_i + \mathbf{z}_i$ is also a solution of (6.2.1).

These ideas carry over to a somewhat more general class of systems for which closed-form solutions are generally not available. In place of (6.2.1), we consider

$$(6.2.9) \qquad \mathbf{x}_{i+1} - \mathbf{x}_i = A(i)\mathbf{x}_i + \mathbf{b}(i),$$

where rather than being constant, the matrix A and the vector $\mathbf{b}$ may depend on the number of iterations that have occurred. Since iterations are often assumed to proceed at a specified rate, (6.2.9) can be thought of as a time-dependent generalization of (6.2.1). However, we still refer to (6.2.9) as a linear equation and to

$$(6.2.10) \qquad \mathbf{x}_{i+1} - \mathbf{x}_i = A(i)\mathbf{x}_i$$

as the associated homogeneous equation.

Problem 6.2.7 Show that if $\mathbf{y}_i$ is any solution of the homogeneous equation (6.2.10) and $\mathbf{z}_i$ is any solution of the nonhomogeneous (original) equation (6.2.9), then $\mathbf{x}_i = \mathbf{y}_i + \mathbf{z}_i$ is also a solution of (6.2.9).

In solving (6.2.9) and the associated iterative scheme

$$(6.2.11) \qquad \mathbf{x}_{i+1} = (A(i) + I)\mathbf{x}_i + \mathbf{b}(i)$$

it is necessary to compute the changing values of $A(i)$ and $\mathbf{b}(i)$ together with the solution $\mathbf{x}_i$. Our spreadsheet format for doing this calls for arranging the elements of $A(i)$ in a single column of length n^2 (rather than as a square $n \times n$ array) and aligning both $A(i)$ and $\mathbf{b}(i)$ with the solution $\mathbf{x}(i)$. This is illustrated in Figure 6.2.1 for the case where

$$A(i) = \begin{bmatrix} a - \dfrac{1}{i} & b \\ \dfrac{c}{i} & d\sqrt{i} \end{bmatrix} \quad \text{and} \quad \mathbf{b}(i) = \begin{bmatrix} 1 \\ ei \end{bmatrix}.$$

Finally, we return to the case where A does not depend on i and (6.2.1) has the closed-form solution

$$(6.2.12) \qquad \mathbf{x}_i = (A + I)^{i-1}\mathbf{x}_1 - A^{-1}(I - (A + I)^{i-1})\mathbf{b}.$$

Here we see that the homogeneous equation (6.2.8) has solution

$$(6.2.13) \qquad \mathbf{x}_i = (A + I)^{i-1}\mathbf{x}_1.$$

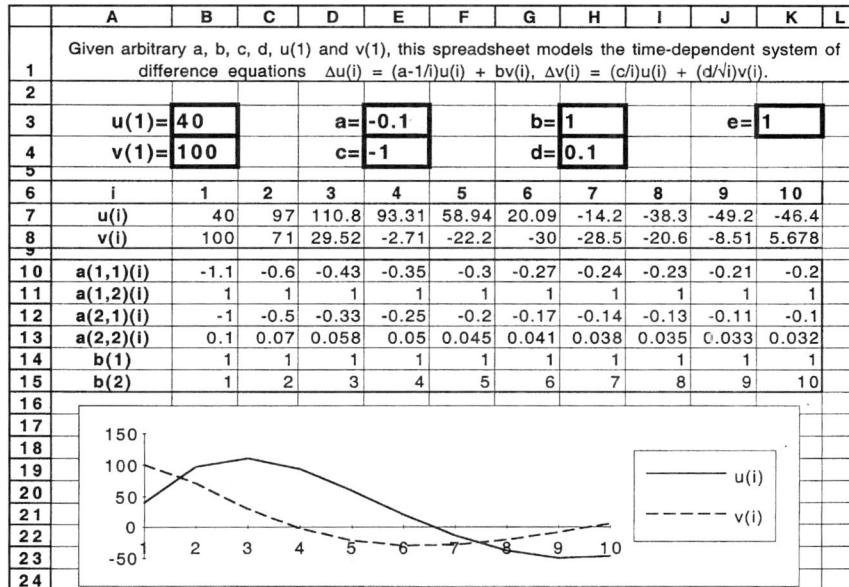

	A	B	C	D	E	F	G	H	I	J	K	L
1	Given arbitrary a, b, c, d, u(1) and v(1), this spreadsheet models the time-dependent system of difference equations Δu(i) = (a-1/i)u(i) + bv(i), Δv(i) = (c/i)u(i) + (d/√i)v(i).											
2												
3	u(1)=	40			a=	-0.1		b=	1		e=	1
4	v(1)=	100			c=	-1		d=	0.1			
5												
6	i	1	2	3	4	5	6	7	8	9	10	
7	u(i)	40	97	110.8	93.31	58.94	20.09	-14.2	-38.3	-49.2	-46.4	
8	v(i)	100	71	29.52	-2.71	-22.2	-30	-28.5	-20.6	-8.51	5.678	
9												
10	a(1,1)(i)	-1.1	-0.6	-0.43	-0.35	-0.3	-0.27	-0.24	-0.23	-0.21	-0.2	
11	a(1,2)(i)	1	1	1	1	1	1	1	1	1	1	
12	a(2,1)(i)	-1	-0.5	-0.33	-0.25	-0.2	-0.17	-0.14	-0.13	-0.11	-0.1	
13	a(2,2)(i)	0.1	0.07	0.058	0.05	0.045	0.041	0.038	0.035	0.033	0.032	
14	b(1)	1	1	1	1	1	1	1	1	1	1	
15	b(2)	1	2	3	4	5	6	7	8	9	10	
16												
17–24	*(chart: 150, 100, 50, 0, -50 vs. 1–10, legend u(i) solid, v(i) dashed)*											

Figure 6.2.1 Solving Time-dependent Linear Systems

This leads us to revisit the homogeneous equation $x_{i+1} - x_i = Ax_i$ in a matrix context.

In place of the homogeneous equation (6.2.8), we now consider a *matrix* difference equation of the form

(6.2.14) $$X_{i+1} - X_i = AX_i.$$

Here $X_1, X_2, X_3, \ldots$ are assumed to be square matrices of the form

$$X = \begin{bmatrix} x_{11} & x_{12} & \cdots & x_{1n} \\ x_{21} & x_{22} & \cdots & x_{2n} \\ x_{n1} & & \cdots & x_{nn} \end{bmatrix},$$

and the difference equation (6.2.14) can again be written as an iterative scheme

(6.2.15) $$X_{i+1} = (A+I)X_i.$$

Problem 6.2.8 Given an initial matrix $X_1 = I$, show that the solution of (6.2.14) is given by $X_i = (A+I)^{i-1}$.

It is readily verified that a sequence of matrices $X_1, X_2, X_3, \ldots$ satisfies (6.2.15) if and only if each individual column, regarded as a vector, is a solution of (6.2.8). In this sense, the matrix equation (6.2.15) is really asking for n (different) solutions of the vector equation (6.2.8), arranged side by side to form a square matrix.

Why might we want n different solutions of (6.2.8) arranged in this format? Well, given such a matrix solution

$$(6.2.16) \qquad\qquad X_1, X_2, X_3, \ldots$$

with $X_1 = I$, we would find that

$$(6.2.17) \qquad \mathbf{x}_1 = X_1 \mathbf{x}_1, \quad \mathbf{x}_2 = X_2 \mathbf{x}_1, \quad \mathbf{x}_3 = X_3 \mathbf{x}_1, \cdots$$

is the corresponding solution of the vector system (6.2.8). In this way (6.2.16) embodies *all* solutions of the homogeneous equation (6.2.8), and for this reason it is sometimes called the *fundamental matrix* solution of the homogeneous equation.

6.3. Nonlinear Systems

6.3. Nonlinear Systems We turn now to more general systems of difference equations represented in the form

$$(6.3.1) \qquad\qquad \mathbf{x}_{i+1} - \mathbf{x}_i = A(\mathbf{x}_i)\mathbf{x}_i + \mathbf{b}.$$

Here the elements of the matrix A may depend on the value of $\mathbf{x}_i$, and as a result, we will have to forgo some of the theoretical tools developed in Section 6.2 for studying linear systems. For example, it is no longer the case that solutions of (6.3.1) consist of a solution of (6.3.1) added to a solution of the associated homogeneous equation

$$(6.3.2) \qquad\qquad \mathbf{x}_{i+1} - \mathbf{x}_i = A(\mathbf{x}_i)\mathbf{x}_i.$$

We also forgo the concept of a "fundamental matrix" as defined at the end of Section 6.2.

On the other hand, (6.3.1) remains equivalent to a simple iterative scheme, namely

$$(6.3.3) \qquad\qquad \mathbf{x}_{i+1} = (I + A(\mathbf{x}_i))\mathbf{x}_i + \mathbf{b}.$$

As a result, we are able to use the spreadsheet format of Figure 6.2.1 to obtain numerical solutions. Indeed, this format also applies to the slightly more general "time-dependent nonlinear system"

$$(6.3.4) \qquad\qquad \mathbf{x}_{i+1} - \mathbf{x}_i = A(i, \mathbf{x}_i)\mathbf{x}_i + \mathbf{b}(i).$$

Many important "rules for change" are expressible in this format, and it is the use of iteration to generate numerical solutions of (6.3.4) that will be of primary interest to us here.

As a first and famous example, we recall the work of a mathematician named Vito Volterra, who during the 1920s pioneered the use of mathematics to study population dynamics. Volterra began by considering a "predator–prey" interaction among two species, say rabbits and foxes, whose populations at time i are given by u_i and v_i, respectively. These populations are assumed confined to an environment that is supportive of one species but not the other, e.g., an island whose vegetation is supportive of rabbits but not of foxes. In the absence of the other species, the rabbit population would grow while the number of foxes would decline. Barring interaction, it would be natural to model these populations by

$$
\begin{aligned}
u_{i+1} - u_i &= au_i, \\
v_{i+1} - v_i &= -cv_i,
\end{aligned}
$$
(6.3.5)

where $a > 0$ and $c > 0$. Here rabbits would grow exponentially while the foxes would decline in accordance with the difference equation for radioactive decay developed in Chapter 2. This system can also be written in the vector form $\mathbf{x}_{i+1} - \mathbf{x}_i = A\mathbf{x}_i$ by setting

$$
\mathbf{x}_i = \begin{bmatrix} u_i \\ v_i \end{bmatrix} \quad \text{and} \quad A = \begin{bmatrix} a & 0 \\ 0 & -c \end{bmatrix}.
$$
(6.3.6)

Let us assume, however, that these populations do interact as the result of chance meetings, events that are bad for the rabbit and good for the fox. Here we shall follow Volterra in assuming that during any given interval of time, the number of meetings depends linearly on the number of rabbits and the number of foxes. For example, a tripling of the rabbit population should triple the number of meetings, while a halving of the fox population should reduce the number of meetings by half. On this basis, Volterra supplemented (6.3.5) by including two "interaction terms" that are proportional to the product $u_i v_i$. This led him to study the system

$$
\begin{aligned}
u_{i+1} - u_i &= au_i - bu_i v_i, \\
v_{i+1} - v_i &= -cv_i + du_i v_i,
\end{aligned}
$$
(6.3.7)

with $b > 0$ and $d > 0$. Here $-bu_i v_i$ represents the reduction of rabbits, and $du_i v_i$ the increase in foxes as the result of meetings during the ith period.

Problem 6.3.1 The matrix A in (6.3.6) has $a_{12} = a_{21} = 0$. Determine formulas for $a_{12}(\mathbf{x}_i)$ and $a_{21}(\mathbf{x}_i)$ that make it possible to write (6.3.7) in the form $\mathbf{x}_{i+1} - \mathbf{x}_i = A(\mathbf{x}_i)\mathbf{x}_i$ (while retaining $a_{11} = a$ and $a_{22} = -c$).

The system (6.3.7) corresponds to an iterative scheme $\mathbf{x}_{i+1} = (A(\mathbf{x}_i) + I)\mathbf{x}_i$, one that can be written

$$
\begin{aligned}
u_{i+1} &= (a + 1)u_i - bu_i v_i, \\
v_{i+1} &= (1 - c)v_i + du_i v_i
\end{aligned}
$$
(6.3.8)

and programmed in the format of Figure 6.2.1. The spreadsheet of Figure 6.3.1 uses this format to solve the predator–prey system (6.3.7) for arbitrary (positive) values of u_1, v_1, a, b, c, and d. Observe that $A(\mathbf{x}_i) = \begin{bmatrix} a_{11} & a_{12} \\ a_{21} & a_{22} \end{bmatrix} = \begin{bmatrix} c & -bu_i \\ dv_i & -c \end{bmatrix}$.

The spreadsheet of Figure 6.3.1 is just a starting point for serious efforts to use mathematics to understand the interaction between two species that exist in a

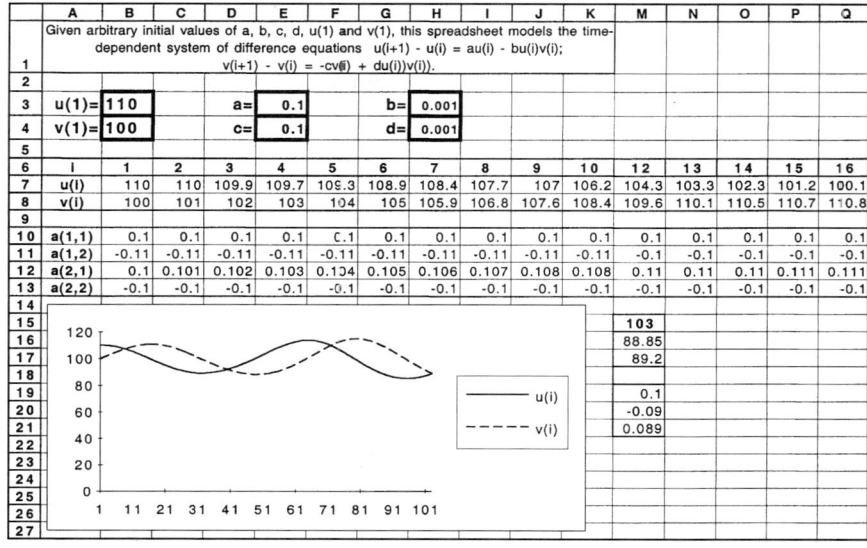

Figure 6.3.1 A Predator–Prey System for Two Species

predator–prey relationship. It is a "simplest case scenario," one that is based on convenient assumptions that happen to give rise to solutions with interesting mathematical properties. However, in trying to adapt this framework to real-world situations, one is likely to modify these assumptions and generate solutions less amenable to theoretical analysis. Examples of such modifications are given in the problems below.

Problem 6.3.2 Create a spreadsheet in the format of Figure 6.3.1 with $u_1 = 110$, $v_1 = 100$, $a = c = 0.1$, $b = d = 0.001$.

Problem 6.3.3 Modify the spreadsheet of Problem 6.3.2 by allowing for a "harvesting of rabbits" that is independent of the size of the populations. This calls for replacing the first equation of (6.3.7) by

$$u_{i+1} - u_i = au_i - bu_iv_i - h,$$

where h denotes the number of rabbits harvested between i and $i + 1$. Repeat Problem 6.3.2 when $h = 1, 2$, and 3.

Problem 6.3.4 Modify the spreadsheet of Problem 6.3.2 by allowing for a "delay" k in the effect that meetings have on the fox population. This calls for replacing the second equation of (6.3.7) by

$$v_{i+1} - v_i = -cv_i + du_{i-k}v_i.$$

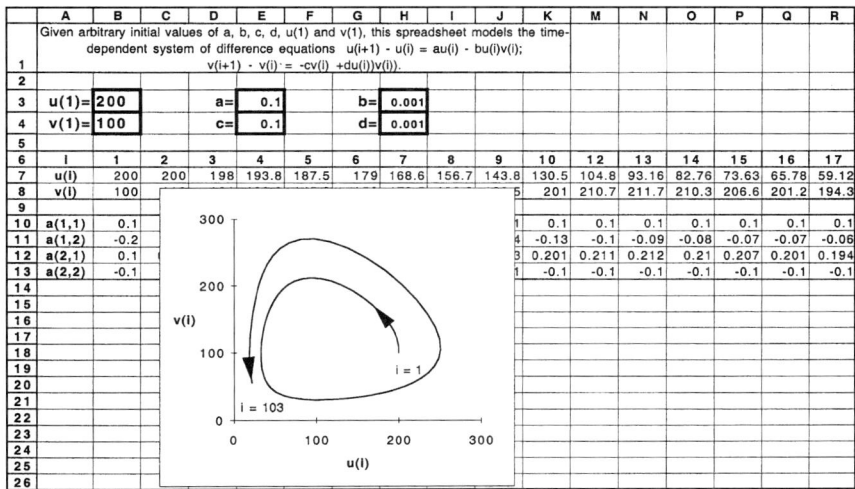

	A	B	C	D	E	F	G	H	I	J	K	M	N	O	P	Q	R
1	Given arbitrary initial values of a, b, c, d, u(1) and v(1), this spreadsheet models the time-dependent system of difference equations u(i+1) - u(i) = au(i) - bu(i)v(i); v(i+1) - v(i) = -cv(i) +du(i))v(i)).																
2																	
3	u(1)=	200			a=	0.1			b=	0.001							
4	v(1)=	100			c=	0.1			d=	0.001							
5																	
6	I	1	2	3	4	5	6	7	8	9	10	12	13	14	15	16	17
7	u(I)	200	200	198	193.8	187.5	179	168.6	156.7	143.8	130.5	104.8	93.16	82.76	73.63	65.78	59.12
8	v(I)	100								5	201	210.7	211.7	210.3	206.6	201.2	194.3
9																	
10	a(1,1)	0.1								1	0.1	0.1	0.1	0.1	0.1	0.1	0.1
11	a(1,2)	-0.2								4	-0.13	-0.1	-0.09	-0.08	-0.07	-0.07	-0.06
12	a(2,1)	0.1								3	0.201	0.211	0.212	0.21	0.207	0.201	0.194
13	a(2,2)	-0.1								1	-0.1	-0.1	-0.1	-0.1	-0.1	-0.1	-0.1
14																	
15																	
16				v(I)													
17																	
18																	
19																	
20																	
21																	
22				i = 103													
23																	
24																	
25					u(I)												
26																	

Figure 6.3.2 An Orbit Representation of Predator–Prey Interaction

Repeat Problem 6.3.2 by modifying the iterative scheme for $i \geq 4$ and determining the effect of delays corresponding to $k = 1$, 2, and 3.

Problem 6.3.5 Following Volterra, modify the spreadsheet of Problem 6.3.2 by assuming a "self-limiting" population of rabbits. This calls for replacing the first equation of (6.3.7) by

$$u_{i+1} - u_i = au_i - bu_i v_i - eu_i^2,$$

where e is a positive constant. What would happen to such a population of rabbits in the absence of foxes?

Figure 6.3.1 provides a graphical representation of u_i (rabbits) and v_i (foxes) as a function of i (time). However, as indicated in Figure 6.3.2, it is also possible to represent these solutions by plotting successive values of (u_i, v_i) in the (u, v)-plane. Rather than telling us the size of the populations at time i, the resulting "orbit" conveys insight into how the size of one species affects the size of the other.

For the original system (6.3.8), the orbits generated in Figure 6.3.2 spiral outward. This corresponds to increasing amplitudes in Figure 6.3.1, whereas a closed orbit would correspond to periodic behavior on the part of these populations. Although closed orbits are not achieved by the system of difference equations (6.3.8), the following problem suggests that refinement may enable us to approach such periodic behavior.

Problem 6.3.6 Create a spreadsheet in the format of Figure 6.3.2, one that allows for refinement of the form $\Delta t \mapsto \Delta t / N$ and contains enough iterations to provide several rotations about $(u, v) = (0, 0)$ when $N = 1$. Describe the effect of successive refinement on such orbits.

Problem 6.3.7 Modify the spreadsheets of Problems 6.3.3–6.3.5 to produce orbit representations of the solutions. Determine the effect of refinement on these orbits.

The populations considered by Volterra were ones for which interaction served to increase one population and decrease the other. There are, however, situations in which two populations are in direct competition with each other. While each of these populations is assumed to grow in isolation from the other, both populations are now assumed to be adversely affected as a result of interaction.

A simple model for such a "competitive system" can be obtained by making two sign changes in the predator–prey system (6.3.7). Writing

$$(6.3.9) \qquad \begin{aligned} u_{i-1} - u_i &= au_i - bu_i v_i, \\ v_{i+1} - v_i &= cv_i - du_i v_i, \end{aligned}$$

with a, b, c, and d positive, it is clear that either u_i or v_i would grow exponentially in the absence of the other species. However, the question of what happens when these species interact is far from obvious.

Before modeling (6.3.9) on a spreadsheet, it is of interest to note that the concept of "equilibrium value" carries over to nonlinear systems such as

$$(6.3.10) \qquad \begin{bmatrix} u_{i+1} \\ v_{i+1} \end{bmatrix} - \begin{bmatrix} u_i \\ v_i \end{bmatrix} = \begin{bmatrix} a & -bu_i \\ -dv_i & c \end{bmatrix} \begin{bmatrix} u_i \\ v_i \end{bmatrix}.$$

For if there exists a value of $\mathbf{x}_2 = \begin{bmatrix} u \\ v \end{bmatrix}$ at which

$$(6.3.11) \qquad \begin{bmatrix} a & -bu \\ -dv & c \end{bmatrix} \begin{bmatrix} u \\ v \end{bmatrix} = 0,$$

then we would have no change in a population that satisfies (6.3.10).

Problem 6.3.8 By writing (6.3.11) as a system of two (nonlinear) equations in two unknowns, verify that (6.3.10) has the two equilibrium values $\begin{bmatrix} 0 \\ 0 \end{bmatrix}$ and $\begin{bmatrix} c/d \\ a/b \end{bmatrix}$.

The big question is whether either of the equilibrium values so defined serves as an "attractor" as defined Chapter 5. Unfortunately, a theoretical answer to this question is beyond the scope of the present discussion. We have, however, been able to determine the location of such equilibrium values and can now call on spreadsheets to help us determine whether they have properties akin to the attractors of Chapter 5. In the context of population dynamics, such attracting fixed points will correspond to a stable form of coexistence between the species being modeled.

Problem 6.3.9 Construct a spreadsheet that models (6.3.10) for $a = 0.1, b = 0.01, c = 0.2$, and $d = 0.03$. Find three initial values of $\mathbf{x}_1 = \begin{bmatrix} u_1 \\ v_1 \end{bmatrix}$ for which u_i approaches 0 and v_i grows without bound. Then find three initial values of $\mathbf{x}_1$ for which v_i approaches 0 and u_i grows without bound. In the (u, v)-plane, estimate the location of the "separatrix" that separates these two categories of initial values.

Problem 6.3.10 Construct a spreadsheet that models (6.3.10) for arbitrary positive values of a, b, c, and d. Test whether $\mathbf{x}_0 = \begin{bmatrix} c/d \\ a/b \end{bmatrix}$ ever behaves like an "attractor" for nearby values of $\mathbf{x}_1$.

As the above problems suggest, while $\mathbf{x}_0 = \begin{bmatrix} c/d \\ a/b \end{bmatrix}$ is an equilibrium value for (6.3.10), it does not play the role of "attractor." This absence of an attracting equilibrium value with $u > 0$ and $v > 0$ suggests that there is no (mathematical) basis for coexistence among two such species. Depending on the initial value $\mathbf{x}_0$, one species will out-compete the other and go on to grow exponentially, at least until factors outside the present model come into play.

These observations lead us to consider a more general class of competitive systems, namely

$$(6.3.12) \qquad \begin{aligned} u_{i+1} - u_i &= a u_i - e u_i^2 - b u_i v_i, \\ v_{i+1} - v_i &= c v_i - f v_i^2 - d u_i v_i, \end{aligned}$$

or

$$(6.3.13) \qquad \begin{bmatrix} u_{i+1} \\ v_{i+1} \end{bmatrix} - \begin{bmatrix} u_i \\ v_i \end{bmatrix} = \begin{bmatrix} a - e u_i & -b u_i \\ -d v_i & c - f v_i \end{bmatrix} \begin{bmatrix} u_i \\ v_i \end{bmatrix}$$

Here we have also allowed for "self-limiting" terms $-e u_i^2$ and $f v_i^2$ of the kind that are associated with logistic growth. If now there were no interaction (i.e., if $b = 0$ and $d = 0$), then we would have a pair of logistic equations for which u_i approaches a/e and v_i approaches c/f. In other words, in the special case $b = d = 0$,

$$\mathbf{x}_0 = \begin{bmatrix} a/e \\ c/f \end{bmatrix}$$

is an attracting equilibrium value for (6.3.12).

But what about the general case? The equilibrium values of (6.3.13) are found by solving the system

$$(6.3.14) \qquad \begin{aligned} a u_i - e u_i^2 - b u_i v_i &= 0, \\ c v_i - f v_i^2 - d u_i v_i &= 0, \end{aligned}$$

and (6.3.14) turns out to have four solutions. The first three are

$$\begin{bmatrix} 0 \\ 0 \end{bmatrix}, \quad \begin{bmatrix} 0 \\ c/f \end{bmatrix}, \quad \begin{bmatrix} a/e \\ 0 \end{bmatrix},$$

while the fourth is obtained by solving

(6.3.15)
$$\begin{bmatrix} e & b \\ d & f \end{bmatrix} \begin{bmatrix} u \\ v \end{bmatrix} = \begin{bmatrix} a \\ c \end{bmatrix}.$$

Whenever this "fourth equilibrium value" has $u > 0$ and $v > 0$, it corresponds to coexistence among the two species modeled by (6.3.12). For this reason, situations in which this equilibrium value behaves like an attractor are of special interest.

We have already seen that in one extreme case ($e = f = 0$) the equilibrium value $\begin{bmatrix} c/d \\ a/b \end{bmatrix}$ fails to attract nearby values of x_1. In another extreme case ($b = d = 0$) the equilibrium value $\begin{bmatrix} a/e \\ c/f \end{bmatrix}$ does attract values of x_1 with $u_1 > 0$ and $v_1 > 0$. Thus, for the general case of (6.3.12), we would expect that either situation might arise.

Problem 6.3.11 Create a spreadsheet that models the system (6.3.12) for arbitrary (positive) values of a, b, c, d, e, and f and also displays the equilibrium value x_0 defined by (6.3.15). Show that if b and d are "small" relative to e and f and if x_1 is chosen sufficiently near x_0, then the subsequent values of x_i approach x_0. If, however, b and d are "large" relative to e and f, then no matter how close x_1 is chosen to x_0, the subsequent values of x_i will *not* approach x_0.

By way of giving this problem physical meaning, we can think of b and d as representing the level of competition between the two species, whereas e and f represent the tendencies of the populations to stabilize at self-imposed limits. If competition is small relative to self-limitation, then coexistence is possible. However, if competition is large relative to self-limitation, then one or the other of the species will be eliminated.

What is the meaning to be attached to "small" and "large"? On the basis of ideas developed in Problem 6.7.9, it can be shown that $bd > ef$ will ensure the elimination of one or the other species.

Problem 6.3.12 Use the spreadsheet of Problem 6.3.11 to test the impact of the inequality $bd > ef$ on the prospects for coexistence.

Problem 6.3.12 raises the question of whether $bd < ef$ always allows for coexistence. This is a question we return to in Figure 6.5.1 with an answer of "essentially yes."

What happens when we try to adapt the mathematically tractable system (6.3.12) to real-world situations? Well, as we depart from this special format the underlying mathematics quickly becomes more challenging. On the other hand, computational techniques such as those used in Problem 6.3.11 generalize quite readily. The following problems suggest some of the ways in which (6.3.12) might be made more realistic.

Problem 6.3.13 Modify (6.3.12) by replacing the terms $-eu_i^2$ and $-fv_i^2$ by $-eu_i^{3/2}$ and $-fv_i^{3/2}$, respectively. Will such a change enhance or reduce the self-limiting tendencies of the two species? Try to anticipate the effect such a change will have on coexistence. Adapt the spreadsheet of Problem 6.3.11 to test your intuition when $a = 0.1, b = 0.01, c = 0.2, d = 0.03, e = 0.02, f = 0.02$.

Problem 6.3.14 Modify (6.3.12) by replacing b and d with random numbers lying in the intervals $[b - 0.005, b + 0.005]$ and $[d - 0.01, d + 0.01]$, respectively. Choose values of e and f such that successive runs of this model give rise to both coexistence and to the annihilation of one of the species.

Problem 6.3.15 With $a = 0.1, b = 0.01, c = 0.2, d = 0.03, e = 0.02, f = 0.02$, modify (6.3.12) to allow for delays of $i = 1, 2,$ and 3 in the effect of competition on changes in the size of the first species u.

The formalisms we have used to study interacting populations find applications in other areas as well. An interesting example is found in "consumer side economics" as developed by John Maynard Keynes during the 1930s. Here we adopt some notation, using capital letters to represent a national economy whose output Y has three components:

Consumption, to be denoted by C;
Investment, to be denoted by I;
Government spending, to be denoted by G.

This classification leads us to measure the level of economic activity in the i th year as

$$(6.3.16) \qquad Y_i = C_i + I_i + G_i,$$

In modeling the dynamics of such an economy, we begin by disregarding government spending (i.e., by setting $G_i = 0$). We also assume that it is the previous year's level of economic activity that determines this year's level of consumption. This last assumption is often formulated as

$$(6.3.17) \qquad C_i = cY_{i-1},$$

where c is a constant $(0 < c < 1)$ called "the marginal propensity to consume." Going on to assume that investment is constant (i.e., setting $I_i = I_0$), (6.3.16) reduces to

$$(6.3.18) \qquad Y_i = C_i + I_0.$$

Problem 6.3.16 By eliminating C_i, combine (6.3.17) and (6.3.18) into a single difference equation. For this equation, derive the closed-form solution

$$Y_i = c^{i-1}Y_1 + \left[\frac{1 - c^{i-1}}{1 - c}\right] I_0.$$

Problem 6.3.17 Given $0 < c < 1$, what happens to Y_i as i becomes large? Explain why $1/(1 - c)$ might be termed the "Keynesian multiplier" for such an economy.

An interesting variation of this model is to replace the assumption "investment is constant" with a link between investment and national income. However, rather than making I_i proportional to the *level* of investment, we will assume that such "induced investment" depends on the *change* in Y_i, e.g., that

(6.3.19) $I_i = d(Y_i - Y_{i-1}) + e,$

where d and e are constant and $0 < d < 1$. Combining (6.3.16), (6.3.17), and (6.3.19) now yields

(6.3.20) $Y_i = cY_{i-1} + d(Y_i - Y_{i-1}) + G_i + e.$

Problem 6.3.18 Assuming $G_i = 0$, find a closed-form solution of (6.3.20). [Hint: Write the equation in the form $Y_i = aY_{i-1} + b$.]

While the models considered so far have allowed for closed-form solution, more realistic models for national income tend to require numerical solution. For example, in many situations there is a delay in the impact of economic growth on induced investment. Assuming a one-year delay would lead us to replace (6.3.19) by

(6.3.21) $I_i = d(Y_{i-1} - Y_{i-2}) + e.$

As indicated by the spreadsheet below, this modification can lead to economic phenomena very different from those encountered in Problem 6.3.17. In particular, the second-order difference equation

(6.3.22) $Y_i = cY_{i-1} + d(Y_{i-1} - Y_{i-2}) + G_i + e$

allows for an "acceleration effect" that accounts for economic overshoot and various forms of cyclic behavior. The model in Figure 6.3.3 again assumes $G_i = 0$.

Problem 6.3.19 With $G_i = 0$, use the spreadsheet format of Figure 6.3.3 to solve (6.3.22) for $c = 0.3$, $d = 0.1$, and $e = 10$.

One way of moderating the "boom and bust" that can arise in such accelerator models is to reintroduce government spending. In particular, one can try to establish rules for G_i that moderate cycles of the kind that Figure 6.3.3 can generate.

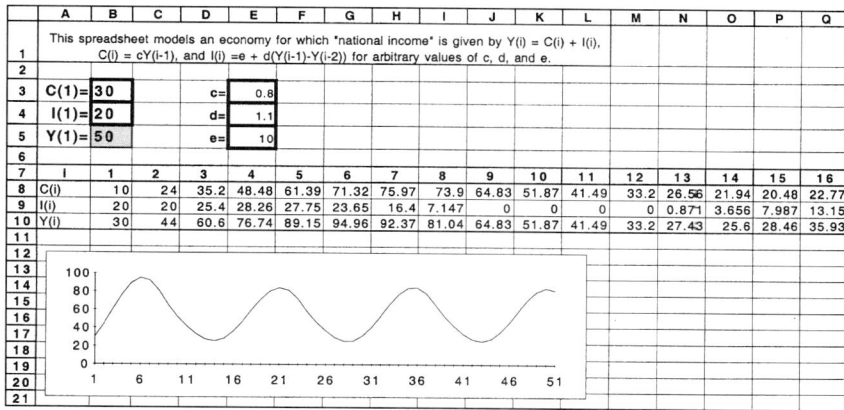

	A	B	C	D	E	F	G	H	I	J	K	L	M	N	O	P	Q
1	This spreadsheet models an economy for which "national income" is given by Y(i) = C(i) + I(i), C(i) = cY(i-1), and I(i) =e + d(Y(i-1)-Y(i-2)) for arbitrary values of c, d, and e.																
2																	
3	C(1)=	30			c=	0.8											
4	I(1)=	20			d=	1.1											
5	Y(1)=	50			e=	10											
6																	
7	I	1	2	3	4	5	6	7	8	9	10	11	12	13	14	15	16
8	C(i)	10	24	35.2	48.48	61.39	71.32	75.97	73.9	64.83	51.87	41.49	33.2	26.56	21.94	20.48	22.77
9	I(i)	20	20	25.4	28.26	27.75	23.65	16.4	7.147	0	0	0	0	0.871	3.656	7.987	13.15
10	Y(i)	30	44	60.6	76.74	89.15	94.96	92.37	81.04	64.83	51.87	41.49	33.2	27.43	25.6	28.46	35.93
11																	
12																	

Figure 6.3.3 A Keynesian Accelerator Model

Problem 6.3.20 Modify the spreadsheet of Figure 6.3.3 to try your hand at prudent government spending. That is, after creating a spreadsheet for the pair of difference equations

$$Y_i = cY_{i-1} + I_i + G_i,$$
$$I_i = d(Y_{i-1} - Y_{i-2}) + e,$$

define G_i as a function of Y_{i-1}, Y_{i-2}, and I_{i-1} so as to moderate the business cycles that occur for various values of c, d, and e. Do your policies remain effective if the second equation is replaced by $I_i = d(Y_{i-2} - Y_{i-3}) + e$?

6.4. Higher-Order Equations

In the previous section we were led to systems of difference equations through models for the interaction of several animal species or economic phenomena. In the present section we will arrive at such systems by a different route, namely that of transforming a single difference equation of order n ($n \geq 2$) into a system of n first-order equations. It turns out that such system representations of higher-order difference equations can provide new and interesting insights into the phenomena that underlie such equations.

The second-order difference equation

(6.4.1) $$\Delta\Delta x_i = -32$$

arose in Section 2.5, where it was used to model the vertical motion of a body subject to a constant gravitational force. Here Δx_i approximates the *velocity* of such a body, while the "second difference"

(6.4.2) $$\Delta\Delta x_i = \Delta(x_{i+1} - x_i) = (x_{i+2} - x_{i+1}) - (x_{i+1} - x_i)$$
$$= x_{i+2} - 2x_{i+1} + x_i,$$

approximates its *acceleration*.

Problem 6.4.1 Near the surface of the earth, a body of mass m is subject to a downward force of approximately $-32m$ pounds. Show that (6.4.1) corresponds to Newton's second law, asserting that "force equals mass times acceleration," or $F = ma$.

In fact, Newton's second law allows for nonconstant force fields and applies to bodies whose mass may change with time (e.g., a rocket that expels its burned fuel). Allowing for such generalizations calls for replacing (6.4.1) by

$$(6.4.3) \qquad\qquad \Delta[m(i)\Delta x_i] = f(i, x_i, \Delta x_i).$$

Here $m(i)$ denotes the mass of the body at time i, and $f(i, x_i, \Delta x_i)$ denotes a force whose magnitude may depend on time (i) as well as the position (x_i) and velocity (Δx_i) of the body. In this case $m(i)\Delta x_i$ corresponds to the body's *momentum* at time i, and Newton's second law asserts that "force equals rate of change of momentum."

But before embodying such ideas in spreadsheets, it is important to emphasize that Newton's second law corresponds to a *differential* equation

$$(6.4.4) \qquad\qquad \frac{d}{dt}\left(m(t)\frac{dx}{dt}\right) = f\left(t, x, \frac{dx}{dt}\right).$$

While it is possible to use difference equations and iteration to *approximate* solutions of such differential equations, the question of the closeness of such approximations is always lurking in the background. In Chapter 2 we emphasized the role of *refinement* in ensuring that (6.4.1) provides a good approximation to the corresponding differential equation $\frac{d^2x}{dt^2} = -32$, and refinement will play a similar role in the generalizations to be considered here.

For this reason we will be well advised to build the capacity for refinement into the spreadsheets used to represent such differential equations. It is only when further refinement fails to have significant effect that the solution of the difference equation provides a meaningful approximation to the solution of the corresponding differential equation.

With such qualifications in mind, we proceed to write (6.4.3) as a system. Here we will denote momentum at time i by

$$(6.4.5) \qquad\qquad p_i = m(i)\Delta x_i$$

and use this notation to write (6.4.3) as a system of difference equations

$$(6.4.6) \qquad\qquad \begin{aligned} \Delta x_i &= \frac{1}{m(i)}p_i, \\ \Delta p_i &= f(i, x_i, \Delta x_i). \end{aligned}$$

An important application of (6.4.6) arises in modeling mechanical oscillations. Our first example is that of a mass attached to a spring that obeys "Hooke's law,"

or

$$f(i, x_i, \Delta x_i) = -kx_i$$

with $k > 0$. Here a positive displacement $x_i > 0$ gives rise to a negative restoring force toward $x = 0$, whereas a negative displacement $x_i < 0$ gives rise to a positive restoring force, also toward $x = 0$. The subsequent motion of such a "simple harmonic oscillator" is readily modeled in spreadsheet format.

Problem 6.4.2 Construct a spreadsheet for a simple harmonic oscillator that allows for arbitrary (positive) values of the mass m and "spring constant" k and for refinement of the form $\Delta t \mapsto \Delta t / N$. For $m = 10, k = 0.1, x_1 = 0, p_1 = 20$, and $N = 1$, use this spreadsheet to estimate the amount of time required for a given mass to return to its initial position $x = 0$. Repeat for $N = 10$ and $N = 20$.

Problem 6.4.3 In Problem 6.4.2, estimate the mass's maximum displacement from $x = 0$ when $m = 10, k = 0.1, x_1 = 0, p_1 = 20$, and $N = 1$. Repeat for $N = 10$ and $N = 20$.

While spreadsheet approximations do not quite succeed in generating the periodic solution that is characteristic of such a simple harmonic oscillator, refinement does enable us to approach such periodicity.

Problem 6.4.4 Compare the answers obtained in Problem 6.4.2 with those generated by the "exact" solution $x_i = \frac{p_1}{\sqrt{mk}} \sin(\frac{\sqrt{\frac{k}{m}}(i-1)}{N})$. Repeat for Problem 6.4.3.

Problem 6.4.5 Make use of the spreadsheet from Problem 6.4.2 to estimate the effect of "a doubling of the spring constant k" on the mass's maximum displacement and the amount of time required for a given mass to return to its initial position.

The general format of equations (6.4.6) enables us to consider interesting variations of Problem 6.4.2. The spreadsheet of Figure 6.4.1 represents a harmonic oscillator whose mass is subject to air resistance that is proportional to the speed of the mass. It also allows for "metal fatigue" in the spring, replacing the spring constant k by $k/(1 + c(i - 1))$, where c is a positive constant.

Problem 6.4.6 Generalize the spreadsheet of Figure 6.4.1 by allowing for a mass that changes with time according to the rule $m(i) = h/i$, where h is a positive constant.

Our graphical representation of data generated by (6.4.6) has until now depicted position (x) and momentum (p) as functions of time. It is, however, also of interest to view the solution of (6.4.6) in the (x, p)-plane, also known as the "phase

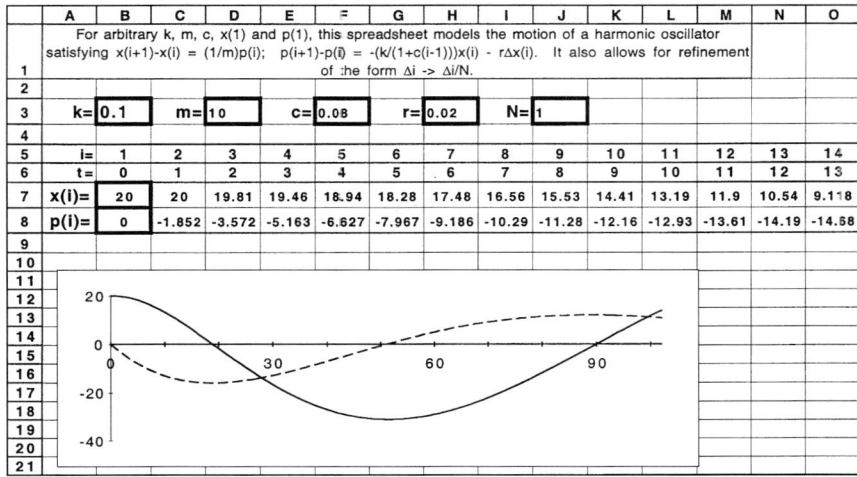

	A	B	C	D	E	F	G	H	I	J	K	L	M	N	O
	For arbitrary k, m, c, x(1) and p(1), this spreadsheet models the motion of a harmonic oscillator														
	satisfying x(i+1)-x(i) = (1/m)p(i); p(i+1)-p(i) = -(k/(1+c(i-1)))x(i) - rΔx(i). It also allows for refinement														
1						of the form Δi -> Δi/N.									
2															
3	k=	0.1		m=	10		c=	0.08		r=	0.02		N=	1			
4															
5	i=	1	2	3	4	5	6	7	8	9	10	11	12	13	14
6	t=	0	1	2	3	4	5	6	7	8	9	10	11	12	13
7	x(i)=	20	20	19.81	19.46	18.94	18.28	17.48	16.56	15.53	14.41	13.19	11.9	10.54	9.118
8	p(i)=	0	-1.852	-3.572	-5.163	-6.627	-7.967	-9.186	-10.29	-11.28	-12.16	-12.93	-13.61	-14.19	-14.68
9															
10															

Figure 6.4.1 A Damped Harmonic Oscillator with Metal Fatigue

plane." This is analogous to the "orbit representation" for interacting species given in Figure 6.3.2.

The spreadsheet of Figure 6.4.2 is identical to that of Figure 6.4.1, except that it relies on a "scatter graph" of points with coordinates (x_i, p_i) to represent its data. Given $c = 0$ and $r = 0$, refinement will enable us to approximate the elliptical orbit that corresponds to the periodic motion of a "simple harmonic oscillator."

Another interesting use of the phase plane is to represent the motion of a pendulum in difference equation form. Here we consider a mass m at the end of a string of length l. The mass is subject to a vertical gravitational force, and its displacement is now measured in terms of the angle θ between the string and the vertical axis. If θ_i denotes such angular displacement at time i, it can be shown by a "resolution of forces" diagram (see Problem 6.7.12) that the motion of the pendulum is approximated by

(6.4.7)
$$\Delta\theta_i = \frac{p_i}{ml}$$
$$\Delta p_i = k\sin(\theta_i)$$

Problem 6.4.7 Use the built-in function "Sin" to create a spreadsheet that models (6.4.7) for arbitrary values of l, m, k, θ_1, and p_1 and graphs the resulting values of θ_i and p_i in a phase plane. For $\theta_1 = 0$ and $p_1 > 0$, give a physical interpretation for the qualitative differences in "phase portraits" obtained for small vs. large values of p_1.

Problem 6.4.8 Modify the preceding model to allow for "damping due to air resistance" that is encountered by the mass at the end of the pendulum.

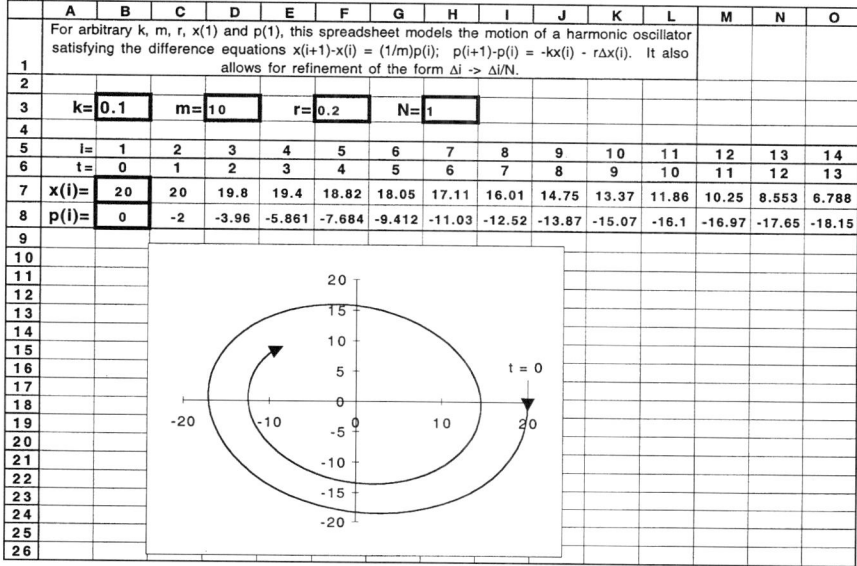

	A	B	C	D	E	F	G	H	I	J	K	L	M	N	O
1		For arbitrary k, m, r, x(1) and p(1), this spreadsheet models the motion of a harmonic oscillator satisfying the difference equations x(i+1)-x(i) = (1/m)p(i); p(i+1)-p(i) = -kx(i) - rΔx(i). It also allows for refinement of the form Δi -> Δi/N.													
2															
3	k=	0.1		m=	10		r=	0.2		N=	1				
4															
5	i=	1	2	3	4	5	6	7	8	9	10	11	12	13	14
6	t=	0	1	2	3	4	5	6	7	8	9	10	11	12	13
7	x(i)=	20	20	19.8	19.4	18.82	18.05	17.11	16.01	14.75	13.37	11.86	10.25	8.553	6.788
8	p(i)=	0	-2	-3.96	-5.861	-7.684	-9.412	-11.03	-12.52	-13.87	-15.07	-16.1	-16.97	-17.65	-18.15

Figure 6.4.2 A Phase-Plane Representation of the Harmonic Oscillator

Having introduced angular coordinates, it becomes possible also to consider the "two-body problem" corresponding to a sun and a single planet. Here we regard the sun as fixed at the origin of a coordinate system and use Newton's second law to determine the motion of a planet of mass m whose position is described by its polar coordinates (r, θ). It can be shown that the resulting motion takes place in a plane that is determined by the planet's initial velocity.

According to the "inverse square law," at time i the sun's gravitational force on the planet is given by

$$F = -\frac{m\,MG}{r_i^2}.$$

Here r_i is the distance from the planet to the sun at time i, G is the *universal gravitational constant*, and M is the mass of the sun. The minus sign reflects the fact that the force on the planet is directed toward the sun. (See Figure 6.4.3.)

In the absence of such a gravitational force, the planet would move along a straight line. The fact that the planet moves along a curved (in fact, elliptical) path is the result of its acceleration toward the sun, i.e., its "radial acceleration." As shown in problems at the end of the chapter, if the planet's path has polar-coordinate representation $r = r(\theta)$, then its radial acceleration is approximated by $\Delta\Delta r_i - r_i(\Delta\theta_i)^2$. Setting $F = ma$ and transposing now yields

(6.4.8)
$$\Delta\Delta r_i = r_i(\Delta\theta_i)^2 - \frac{C}{r_i^2},$$

where $C = MG$.

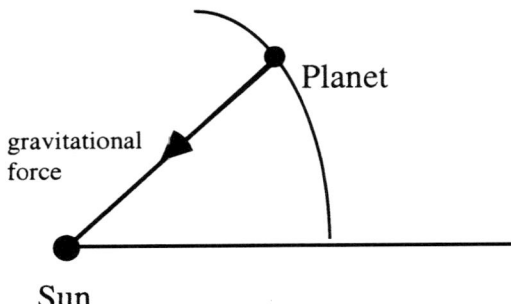

Figure 6.4.3 A Planet in a Central Force Field

Because all forces act toward the sun, we say that the planet is subject to a "central force field." This corresponds to an absence of angular forces (or "torques"), and as a result, the planet's "angular acceleration" is zero. As shown in the problems at the end of the chapter, this leads to $\Delta(r_i^2(\Delta\theta_i)) = 0$ or

$$(6.4.9) \qquad\qquad r_i^2(\Delta\theta_i) = k,$$

where k does not change with i. Squaring both sides of (6.4.9) and dividing by r_i^3 now yields

$$(6.4.10) \qquad\qquad \frac{k^2}{r_i^3} = r_i(\Delta\theta_i)^2.$$

Problem 6.4.9 On the basis of (6.4.8) and (6.4.10), establish the equation

$$(6.4.11) \qquad\qquad \Delta\Delta r_i = \frac{k^2}{r_i^3} - \frac{C}{r_i^2}.$$

Show that $r_i < k^2/C$ implies $\Delta\Delta r_i > 0$, while $r_i > k^2/C$ implies $\Delta\Delta r_i < 0$.

One conclusion of Problem 6.4.6 is that a planet whose motion is governed by the inverse square law could pursue, among others, a circular orbit at a distance k^2/C from the sun. Such a circular orbit corresponds to a solution of the system (6.4.8), (6.4.9) with initial conditions

$$(6.4.12) \qquad\qquad r_1 = \frac{k^2}{C}, \quad \Delta r_1 = 0, \quad \text{and} \quad \Delta\theta_1 = \frac{k}{r_1^2}.$$

These observations suggest that a solution of (6.4.8) with initial values close to those prescribed by (6.4.12) would lead to a nearly circular orbit at a distance from the sun that is sometimes greater and sometimes less than k^2/C. The radial component of such an orbit is approximated in the spreadsheet of Figure 6.4.4.

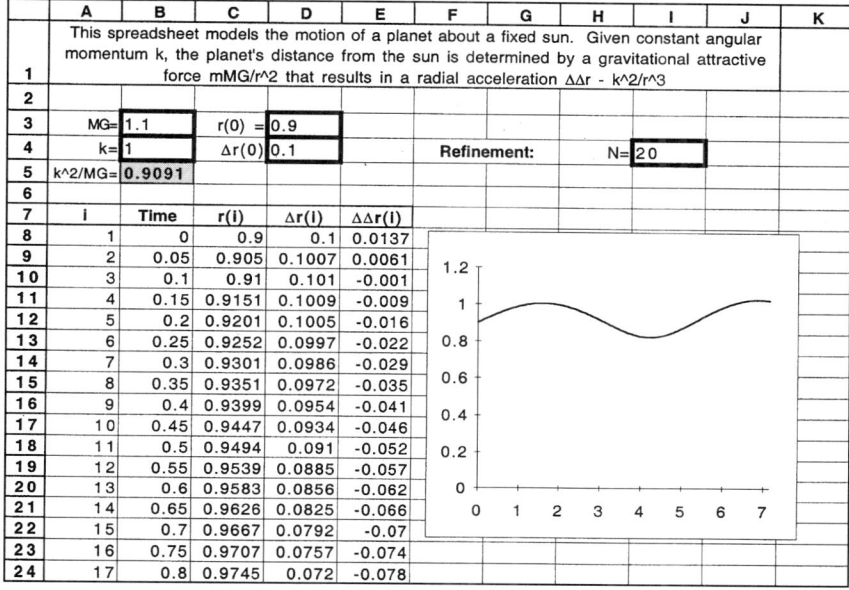

Figure 6.4.4 The Radial Component of the 2-Body Problem

6.5. Chaos Revisited

In Chapter 3 we first encountered chaos in connection with the logistic equation

$$(6.5.1) \qquad x_{i+1} - x_i = ax_i - bx_i^2,$$

which has equilibrium values at $x = 0$ and $x = a/b$. Here we found that $x = 0$ is always a repeller for the corresponding iterative scheme

$$(6.5.2) \qquad x_{i+1} = (1 + a)x_i - bx_i^2,$$

whereas $x = a/b$ is an attractor when $0 < a < 2$ and a repeller whenever $a > 2$. For $a > 2$, the iterates produced by (6.5.2) can find themselves "trapped" between two repellers at $x = 0$ and $x = a/b$. This can lead to solutions $x_1, x_2, x_3, \ldots$ exhibiting a variety of behaviors, including cycles of length 2, 3, 4, $\ldots$, and eventually chaos.

We also showed that the phenomena of cycles and chaos represent a tendency for iterates to "overshoot" the equilibrium value $x = a/b$. That is, in the case of both cycles and chaos the solution $x_1, x_2, x_3, \ldots$ includes values that are both larger and smaller than the equilibrium value a/b.

Given $a > 2$, there is, however, a way of "restoring order" in the face of such overshoot behavior. Refinement of the form $\Delta i \mapsto \Delta i/N$ serves to replace (6.5.1) by

$$(6.5.3) \qquad x_{i+1} - x_i = \left(ax_i - bx_i^2\right)/N,$$

leading to the iterative scheme

(6.5.4)
$$x_{i+1} = \left(1 + \frac{a}{N}\right)x_i - \frac{b}{N}x_i^2,$$

which has the same fixed points as (6.5.2). By choosing $N > a/2$, we are able to transform a/b into an attractor, thereby ensuring that iterations produced by the refined scheme (6.5.4) do approach the equilibrium value a/b. Conversely, when the solution of the original logistic equation does approach a/b, sufficiently large antirefinements of the form $\Delta t \mapsto M\Delta t$ will serve to disrupt this process and produce phenomena such as cycles and chaos.

In this context we saw that cycles and chaos can be regarded as "typical behavior" for solutions of nonlinear equations of the form

(6.5.5)
$$x_{i+1} - x_i = f(x_i).$$

Should such a difference equation have an equilibrium value x_0 satisfying $f(x) = 0$, and should x_0 serve as attractor for the corresponding iterative scheme $x_{i+1} = x_i + f(x_i)$, then appropriate antirefinements can be expected to generate both cycles and chaos.

Problem 6.5.1 Confirm that when x is measured in radians, the difference equation

$$x_{i+1} - x_i = \sin(x_i) - 0.1x_i$$

has an equilibrium value x_0 that lies between $x = 8.42$ and $x = 8.43$. Using $x_1 = 8$, create a spreadsheet that calculates x_0 to five decimal places. Can you find another equilibrium value?

Problem 6.5.2 Use the above spreadsheet to determine a value of M for which the antirefinement $\Delta t \mapsto M\Delta t$ generates cycles of length 2. Determine a value of M that corresponds to cycles of length 4 and a value of M that corresponds to chaos.

Given that such overshoot behavior is typical of solutions of nonlinear difference equations, it becomes natural to ask whether similar phenomena arise in connection with nonlinear *systems* of the form

(6.5.6)
$$\mathbf{x}_{i+1} - \mathbf{x}_i = \mathbf{f}(\mathbf{x}_i).$$

Here we have seen that solutions of $\mathbf{f}(\mathbf{x}) = 0$ play the role of equilibrium values for solutions of nonlinear systems of the form (6.5.6) and that an equilibrium value $\mathbf{x}_0$ can also play the role of attractor, i.e., ensuring that a solution $\mathbf{x}_1, \mathbf{x}_2, \mathbf{x}_3, \ldots$ will approach $\mathbf{x}_0$ whenever $\mathbf{x}_1$ is sufficiently close to $\mathbf{x}_0$. While the formulation of conditions under which such an $\mathbf{x}_0$ plays the role of attractor is beyond the scope of the present treatment, we can use spreadsheets to

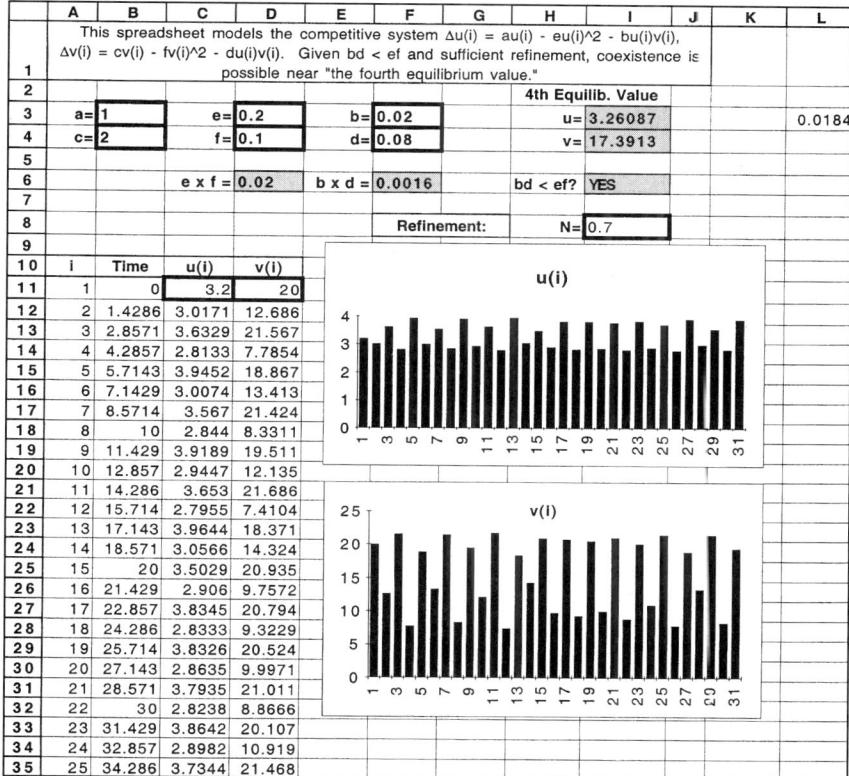

Figure 6.5.1 The Effect of Refinement on a Nonlinear System

confirm such behavior, e.g., in connection with our model for competing species (6.3.13).

A spreadsheet in the format of Figure 6.5.1 will allow the reader to explore the effect of refinement of the form $\Delta t \mapsto \Delta t / N$ on the prospects for achieving a stable form of coexistence at what was called "the fourth equilibrium value" of (6.3.13).

Problem 6.5.1 For (6.3.13), with $a = 0.1, b = 0.01, c = 0.2, d = 0.03, e = 0.02$, and $f = 0.02$, determine a value $M = 1/N$ for which both $u_1, u_2, u_3, \ldots$ and $v_1, v_2, v_3, \ldots$ display cycles of length 2. Repeat for cycles of length 4. Repeat for chaotic behavior.

Such examples extend our earlier observations about chaotic behavior to solutions of nonlinear systems of difference equations of the form $\mathbf{x}_{i+1} - \mathbf{x}_i = \mathbf{f}(\mathbf{x}_i)$. Specifically, *in the presence of an attracting equilibrium value x_0, cycles and chaos are* *the result of "sloppiness" in the form of "inadequate refinement."* In such situations it is possible to transform cycles and chaos into an "approach toward equilibrium" by implementing a refinement of the form $\Delta t \mapsto \Delta t / N$ for sufficiently large values of N.

It is at this point that some profoundly important qualifications arise. There are examples of nonlinear systems of difference equations of the form $\mathbf{x}_{i+1} - \mathbf{x}_i = \mathbf{f}(\mathbf{x}_i)$ for which chaotic behavior *cannot* be eliminated by reductions in the size of Δt. While an explanation of this more persistent form of chaos is beyond the scope of our presentation, we will be able to explore some well-known examples of such "real chaos" with techniques developed in Section 6.3 above.

Perhaps the most famous example of a system for which chaos persists as $\Delta t \to 0$ is that of the Lorenz equations

$$
\begin{aligned}
u_{i+1} - u_i &= -a u_i + a v_i, \\
v_{i+1} - v_i &= -u_i w_i + b u_i - v_i, \\
w_{i+1} - w_i &= u_i v_i - c w_i,
\end{aligned}
$$

(6.5.7)

where a, b, and c are constants yet to be assigned. At first glance, this system may suggest a model for the interaction of three populations, rather than the two populations considered by Volterra. In fact, however, these equations arose from efforts to represent meteorological, rather than demographic, change.

Problem 6.5.2 Assuming $a > 0$, $b > 1$, and $c > 0$, solve the nonlinear system of algebraic equations

$$
\begin{aligned}
-a u + a v &= 0, \\
b u - v - u w &= 0, \\
u v - c w &= 0,
\end{aligned}
$$

(6.5.8)

to find an equilibrium value $x_0 = (u_0, v_0, w_0)$ for (6.5.7) with $v_0 > 0$ and $w_0 > 0$.

Problem 6.5.3 Confirm that $a = 10$, $b = 28$, and $c = \frac{8}{3}$ correspond to an equilibrium value at $(\sqrt{72}, \sqrt{72}, \sqrt{27})$ and $(-\sqrt{72}, \sqrt{72}, 27)$.

Deciding whether such an equilibrium value is an attractor is beyond the theoretical base we have developed here. However, we *are* able to construct a spreadsheet to explore such questions. To prevent an "escape to infinity" in the iterative process associated with (6.5.7), considerable refinement will be needed. To prepare for such refinement, a spreadsheet with about 1000 iterations will be required. Figure 6.5.2 presents a graphical representation of these iterations by projecting the resulting 3-dimensional orbit onto the (u, w)-plane.

As indicated in Figure 6.5.2, with $a = 10$, $b = 28$, and $c = \frac{8}{3}$ the Lorenz system starts out as if it has every intention of approaching one of its nonzero equilibrium

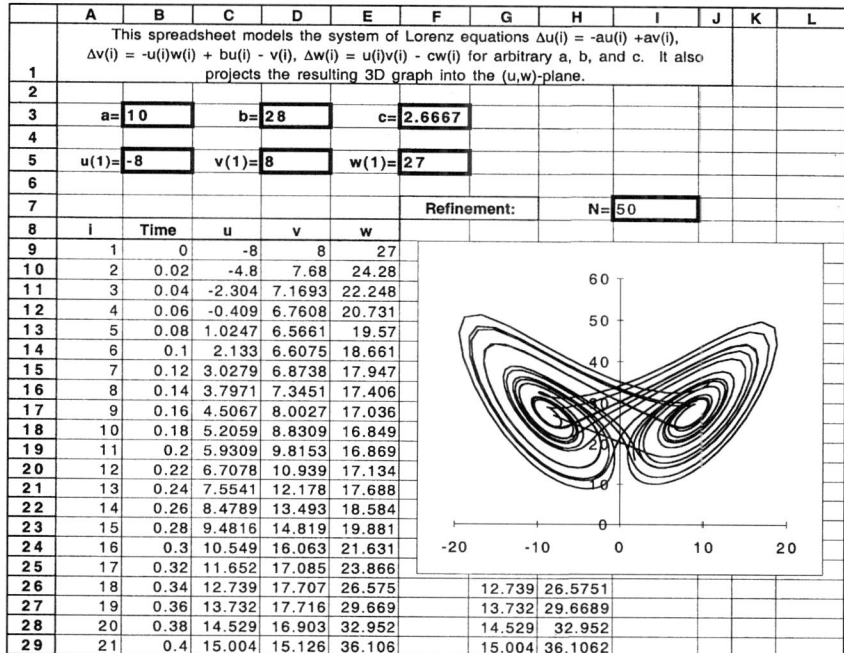

This spreadsheet models the system of Lorenz equations Δu(i) = -au(i) +av(i), Δv(i) = -u(i)w(i) + bu(i) - v(i), Δw(i) = u(i)v(i) - cw(i) for arbitrary a, b, and c. It also projects the resulting 3D graph into the (u,w)-plane.

a= 10   b= 28   c= 2.6667

u(1)= -8   v(1)= 8   w(1)= 27

Refinement: N= 50

i	Time	u	v	w		
1	0	-8	8	27		
2	0.02	-4.8	7.68	24.28		
3	0.04	-2.304	7.1693	22.248		
4	0.06	-0.409	6.7608	20.731		
5	0.08	1.0247	6.5661	19.57		
6	0.1	2.133	6.6075	18.661		
7	0.12	3.0279	6.8738	17.947		
8	0.14	3.7971	7.3451	17.406		
9	0.16	4.5067	8.0027	17.036		
10	0.18	5.2059	8.8309	16.849		
11	0.2	5.9309	9.8153	16.869		
12	0.22	6.7078	10.939	17.134		
13	0.24	7.5541	12.178	17.688		
14	0.26	8.4789	13.493	18.584		
15	0.28	9.4816	14.819	19.881		
16	0.3	10.549	16.063	21.631		
17	0.32	11.652	17.085	23.866		
18	0.34	12.739	17.707	26.575	12.739	26.5751
19	0.36	13.732	17.716	29.669	13.732	29.6689
20	0.38	14.529	16.903	32.952	14.529	32.952
21	0.4	15.004	15.126	36.106	15.004	36.1062

Figure 6.5.2 The "Strange" Lorenz Attractor

values. However, before its orbit "closes in" on $u_0 = \sqrt{72}$ and $w_0 = 72$, this search for equilibrium is abandoned for one near $u_0 = -\sqrt{72}$ and $w_0 = 72$. After a while this search for equilibrium is abandoned for $u_0 = \sqrt{72}$ and $w_0 = 72$, etc. What is also important here is the fact that *this strange behavior cannot be eliminated by further refinement*. That is, such "strange attractors" arise as the solution of the system of *differential* equations

(6.5.9)
$$\frac{du}{dt} = -au + av,$$
$$\frac{dv}{dt} = bu - v - uw,$$
$$\frac{dw}{dt} = uv - cw,$$

as well as (6.5.7)

How does one account for such bizarre behavior? Well, the above description is based on a *projection* of the system's 3-dimensional orbits onto the (u, w)-plane. The two-dimensional orbits described above are, in fact, accompanied by concurrent displacement in the v-direction. What looks like an attractor for one value of v may look very different for others. As observed by Edward Abbott over 100 years ago,[1] transforming a 2-dimensional image of the world into a 3-dimensional realization can be a profound and challenging exercise.

This example leaves open the question of whether for systems of *two* difference equations we can continue to count on refinement to eliminate cycles and chaos. An example based on a nonlinear oscillating system indicates that the answer is "no."

In Section 6.3 we saw that the motion of a unit mass ($m = 1$) subject to a force $f(x, \Delta x, t)$ can be approximated by the system

$$(6.5.10) \qquad \begin{aligned} \Delta x_i &= p_i, \\ \Delta p_i &= f(x_i, \Delta x_i, i). \end{aligned}$$

When $f(x, \Delta x, t) = -kx$, we obtain an approximation to the "simple harmonic motion" associated with a particle attracted to $x = 0$ by a spring with spring constant k.

We now consider a more complicated situation associated with the "forced Duffing equation." Here

$$(6.5.11) \qquad f(x, \Delta x, t) = x - \Delta x - x^3 + k \cos t,$$

and the terms defining f have the following physical interpretations:

> x corresponds to a spring that repels the mass from $x = 0$.
> $-\Delta x$ corresponds to air resistance that is proportional to velocity.
> $-x^3$ corresponds to a nonlinear spring that attracts the mass to $x = 0$.
> $k \cos t$ corresponds to a periodic force being applied to the mass.

The corresponding system of difference equations is

$$(6.5.12) \qquad \begin{aligned} \Delta x_i &= p_i, \\ \Delta p_i &= x_i - p_i - x_i^3 + k \cos i, \end{aligned}$$

where k is a constant that controls the strength of the "periodic driving force."

Problem 6.5.4 For $k = 0$ and $\mathbf{x}_i = \begin{bmatrix} x_i \\ p_i \end{bmatrix}$, write (6.5.12) in the form

$$(6.5.13) \qquad \mathbf{x}_{i+1} - \mathbf{x}_i = A(\mathbf{x}_i)\mathbf{x}_i.$$

Find the three equilibrium values of this system.

Problem 6.5.5 Using the matrix $A(\mathbf{x}_i)$ from Problem 6.5.4, create a spreadsheet that models (6.5.13) near $\mathbf{x}_0 = \begin{bmatrix} 1 \\ 0 \end{bmatrix}$, allows for refinement of the form

---

[1] Edward Abbott was the author of a remarkable book called *Flatland*. It provides a delightful account of a mathematician's effort to understand a 3-dimensional world based on two-dimensional perceptions.

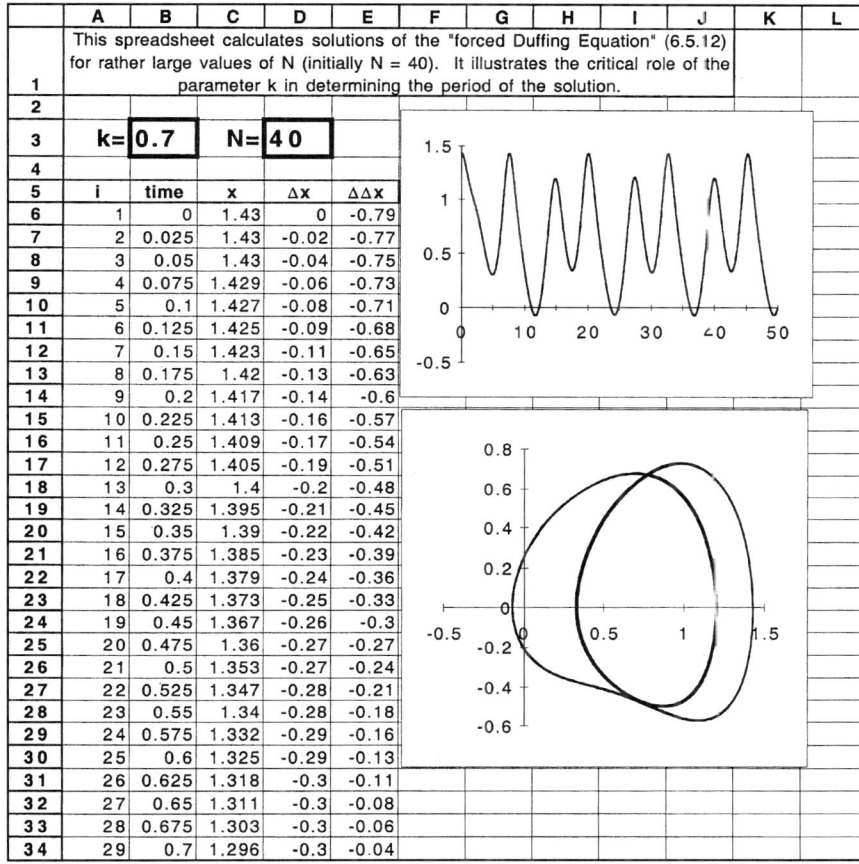

	A	B	C	D	E	F	G	H	I	J	K	L
1	This spreadsheet calculates solutions of the "forced Duffing Equation" (6.5.12) for rather large values of N (initially N = 40). It illustrates the critical role of the parameter k in determining the period of the solution.											
2												
3	k=	0.7		N=	40							
4												
5	i	time	x	Δx	ΔΔx							
6	1	0	1.43	0	-0.79							
7	2	0.025	1.43	-0.02	-0.77							
8	3	0.05	1.43	-0.04	-0.75							
9	4	0.075	1.429	-0.06	-0.73							
10	5	0.1	1.427	-0.08	-0.71							
11	6	0.125	1.425	-0.09	-0.68							
12	7	0.15	1.423	-0.11	-0.65							
13	8	0.175	1.42	-0.13	-0.63							
14	9	0.2	1.417	-0.14	-0.6							
15	10	0.225	1.413	-0.16	-0.57							
16	11	0.25	1.409	-0.17	-0.54							
17	12	0.275	1.405	-0.19	-0.51							
18	13	0.3	1.4	-0.2	-0.48							
19	14	0.325	1.395	-0.21	-0.45							
20	15	0.35	1.39	-0.22	-0.42							
21	16	0.375	1.385	-0.23	-0.39							
22	17	0.4	1.379	-0.24	-0.36							
23	18	0.425	1.373	-0.25	-0.33							
24	19	0.45	1.367	-0.26	-0.3							
25	20	0.475	1.36	-0.27	-0.27							
26	21	0.5	1.353	-0.27	-0.24							
27	22	0.525	1.347	-0.28	-0.21							
28	23	0.55	1.34	-0.28	-0.18							
29	24	0.575	1.332	-0.29	-0.16							
30	25	0.6	1.325	-0.29	-0.13							
31	26	0.625	1.318	-0.3	-0.11							
32	27	0.65	1.311	-0.3	-0.08							
33	28	0.675	1.303	-0.3	-0.06							
34	29	0.7	1.296	-0.3	-0.04							

Figure 6.5.3 Cycles in the Duffing Equation When $k = 0.7$

$\Delta t \mapsto \Delta t / N$, and represents the solution as an orbit in the (x, p)-plane. Confirm that for x_1 near x_0 these orbits approach a closed curve about x_0 as N grows increasingly large.

Given the spreadsheet from Problem 6.5.5, we are now in a position to determine the effect of periodic driving forces of the form $k \cos t$ for varying values of k.

Problem 6.5.6 Using the format of Figure 6.5.3 create a spreadsheet that graphs x_i as a function of i.

Problem 6.5.7 Find a value of k that corresponds to cycles of length 4. Find a value of k that corresponds to chaos.

6.6. Multivariate Systems Until now we have considered difference equations for a variable x that depends on a single index i. Here increases in i

i =	1	2	3	4	5	6	7	8
x_{i1} =	$160	$30	$80	$110	$200	$80	$60	$40

Figure 6.6.1 An Unequal Initial Distribution of Wealth

corresponded to the passage of time, and x_i approximated a function of the form $x(t)$ that allows for a graphical representation in the "space–time coordinates" (x, t).

We now consider difference equations involving two or more indices. Our first example is formulated in terms of a variable x_{ij} for which i corresponds to a spatial displacement while j continues to represent passage of time. Here x_{ij} will provide an approximation to a function $x(s, t)$ that allows for a graph in three dimensions.

By way of example in the financial realm, we consider a rule for change aimed at the "equalization of wealth" among individuals arranged in a row. Starting with eight such individuals, we denote their respective locations by $i = 1, i = 2, \ldots, i = 8$. The variable x_{ij} will denote the assets of the individual located at position i when time equals j. For $j = 1$, the assets of these eight individuals are rather arbitrarily given as shown in Figure 6.6.1.

Our rule for "sharing of wealth" is based on the differences in assets between these individuals and their immediate neighbors. For example, the individual at $i = 3$ has $80, which is $50 more than $i = 2$ and $30 less than $i = 4$. The individual at $i = 5$ has $90 more than $i = 4$ and $120 more than $i = 6$.

Having recorded such pairs of differences for $i = 1, 2, \ldots, 8$, we impose a "rule for change" that calls for a transfer of funds among neighbors, from richer to poorer, equal to 10% of the differences just recorded.

According to this rule, the individual at $i = 3$ is to receive $3 from $i = 4$ while paying $5 to $i = 2$. The resulting net loss of $2 can be calculated as

$$x_{32} - x_{31} = 0.1(x_{41} - x_{31}) + 0.1(x_{21} - x_{31}) = 0.1(x_{41} - 2x_{31} + x_{21}).$$

Analogously, the fact that the individual at $i = 5$ pays out a total of $21 follows from

$$x_{52} - x_{51} = 0.1(x_{61} - x_{51}) + 0.1(x_{41} - x_{51}) = 0.1(x_{61} - 2x_{51} + x_{41}).$$

A useful notation for representing such processes is to define symbols Δ_i and Δ_j by

(6.6.1a) $\Delta_i x_{ij} = x_{i+1\,j} - x_{ij},$

(6.6.1b) $\Delta_i \Delta_i x_{ij} = \Delta_i(x_{i+1,j} - x_{ij}) = x_{i+2,j} - 2x_{i+1,j} - x_{ij},$

i =	1	2	3	4	5	6	7	8
$x_{i1} =$	$160	$30	$80	$110	$200	$80	$60	$40
$x_{i2} =$	$147	$48	$78	$116	$179	$90	$60	$42

Figure 6.6.2 A Partial Sharing of Wealth

and

(6.6.2) $$\Delta_j x_{ij} = x_{i,j+1} - x_{ij}.$$

That is, Δ_i denotes change with respect to the first (spatial) index i, while Δ_j denotes change with respect to the second (temporal) index j. In this context, the above rule for "sharing the wealth" can be written

$$\Delta_j x_{ij} = 0.1 \Delta_i x_{ij} - 0.1 \Delta_i x_{i-1,j},$$

or

(6.6.3) $$\Delta_j x_{ij} = 0.1 \Delta_i \Delta_i x_{i-1,j}.$$

One application of this rule to the data in Figure 6.6.1 now yields the results shown in Figure 6.6.2.

At first glance, this method of "sharing of wealth" leaves much to be desired. Individual 4, who started out with assets of $110, grew wealthier, while Individual 3, who started out with $80, lost money. Individual 7, who started out with only $60, saw no change. Although these observations may raise questions about the fairness of this particular "rule for change," we will be well advised to implement several more iterations with respect to j before drawing conclusions.

But before continuing this iterative process, let us consider the special situation that exists at the ends of the row. Having only one neighbor, the individuals at $i = 1$ and at $i = 8$ engaged in only a single funds transfer—despite the fact that (6.6.3) calls for two such transfers. We could deal with this situation by acknowledging a need for special rules at the ends of the row but choose instead to introduce two "fictional players" located at $i = 0$ and at $i = 9$.

By defining

$$x_{0j} = x_{1j} \quad \text{and} \quad x_{8j} = x_{9j}$$

for $j = 1, 2, \ldots$, we are ascribing to these fictional individuals the same wealth as that of their immediate (less fictional) neighbor. This convention leads to $\Delta_i x_{0j} = 0$ and $\Delta_i x_{8j} = 0$ for all j and enables us to apply (6.6.3) for $1 \le i \le 8$ without special rules.

$i =$	0	1	2	3	4	5	6	7	8	9
$x_{i1} =$	$160	$160	$30	$80	$110	$200	$80	$60	$40	$40
$x_{i2} =$	$147	$147	$48	$78	$116	$179	$90	$60	$42	$42

Figure 6.6.3 A Partial Sharing of Wealth, Including Fictional Players

Inclusion of these fictional individuals calls for replacing Figure 6.6.2 by Figure 6.6.3.

Problem 6.6.1 Use a hand calculator to obtain the values of x_{i3} for $1 \le i \le 9$. Confirm that the combined wealth of individuals 1 through 8 has remained constant.

While carrying out such calculations manually may build character, technology provides us with more convenient ways of continuing this process and representing its outcome in graphical form. The spreadsheet of Figure 6.6.4 continues Figure 6.6.3 another 19 places. It also allows for exchanges based on rules other than "10% of the difference."

Many spreadsheets also provide a 3-dimensional graphing capability, enabling one to represent the data of Figure 6.6.4 in graphical format (See Figure 6.6.5).

	A	B	C	D	E	F	G	H	I	J	K
1	This spreadsheet models a "sharing of wealth" among 8 individuals in a row. It calls for repeated exchanges of 100c% of $\Delta x(i)$ among neighbors, from the richer neighbor to the poorer, including the fictional individuals at the ends of the row.										
2											
3	c=	0.3									
4											
5	j \ i	0	1	2	3	4	5	6	7	8	9
6	1	160.0	160	30	80	110	200	80	60	40	40.0
7	2	121.0	121.0	84.0	74.0	128.0	137.0	110.0	60.0	46.0	46.0
8	3	109.9	109.9	92.1	93.2	114.5	126.2	103.1	70.8	50.2	50.2
9	4	104.6	104.6	97.8	99.3	111.6	115.8	100.3	74.3	56.4	56.4
10	5	102.5	102.5	103.3	102.5	109.2	109.9	97.2	76.7	61.8	61.8
11	6	101.8	101.8	101.6	103.8	107.4	105.9	94.9	78.4	66.3	66.3
12	7	101.8	101.8	102.3	104.2	105.9	103.0	93.2	79.7	69.9	69.9
13	8	101.9	101.9	102.7	104.2	104.5	100.9	92.1	80.8	72.8	72.8
14	9	102.2	102.2	102.9	103.8	103.3	99.4	91.4	81.8	75.2	75.2
15	10	102.4	102.4	103.0	103.4	102.3	98.1	90.9	82.7	77.2	77.2
16	11	102.6	102.6	102.9	102.9	101.4	97.2	90.6	83.5	78.8	78.8
17	12	102.7	102.7	102.8	102.5	100.6	96.5	90.5	84.2	80.2	80.2
18	13	102.7	102.7	102.7	102.0	99.9	95.9	90.4	84.9	81.4	81.4
19	14	102.7	102.7	102.5	101.6	99.3	95.5	90.4	85.5	82.5	82.5
20	15	102.6	102.6	102.3	101.2	98.9	95.1	90.5	86.1	83.4	83.4
21	16	102.5	102.5	102.1	100.8	98.4	94.8	90.5	86.6	84.2	84.2
22	17	102.4	102.4	101.8	100.5	98.1	94.6	90.6	87.0	84.9	84.9
23	18	102.2	102.2	101.6	100.2	97.8	94.5	90.8	87.5	85.5	85.5
24	19	102.0	102.0	101.4	99.9	97.5	94.3	90.9	87.9	86.1	86.1
25	20	101.8	101.8	101.1	99.6	97.3	94.2	91.0	88.3	86.7	86.7

Figure 6.6.4 A Scheme for Sharing Wealth

	A	B	C	D	E	F	G	H	I	J	K
1	This spreadsheet models a "sharing of wealth" among 8 individuals in a row. It calls for repeated exchanges of 100c% of Δx(i) among neighbors, from the richer neighbor to the poorer, including the fictional individuals at the ends of the row.										
2											
3	c=	0.3									
4											
5	j \ i	0	1	2	3	4	5	6	7	8	9
6	1	160.0	160	30	80	110	200	80	60	40	40.0
7	2	121.0									46.0
8	3	109.9									50.2
9	4	104.6									56.4
10	5	102.5									61.8
11	6	101.8									66.3
12	7	101.8									69.9
13	8	101.9									72.8
14	9	102.2									75.2
15	10	102.4									77.2
16	11	102.6									78.8
17	12	102.7									80.2
18	13	102.7									81.4
19	14	102.7									82.5
20	15	102.6									83.4
21	16	102.5									84.2
22	17	102.4									84.9
23	18	102.2									85.5
24	19	102.0									86.1
25	20	101.8	101.8	101.1	99.6	97.3	94.2	91.0	88.3	86.7	86.7

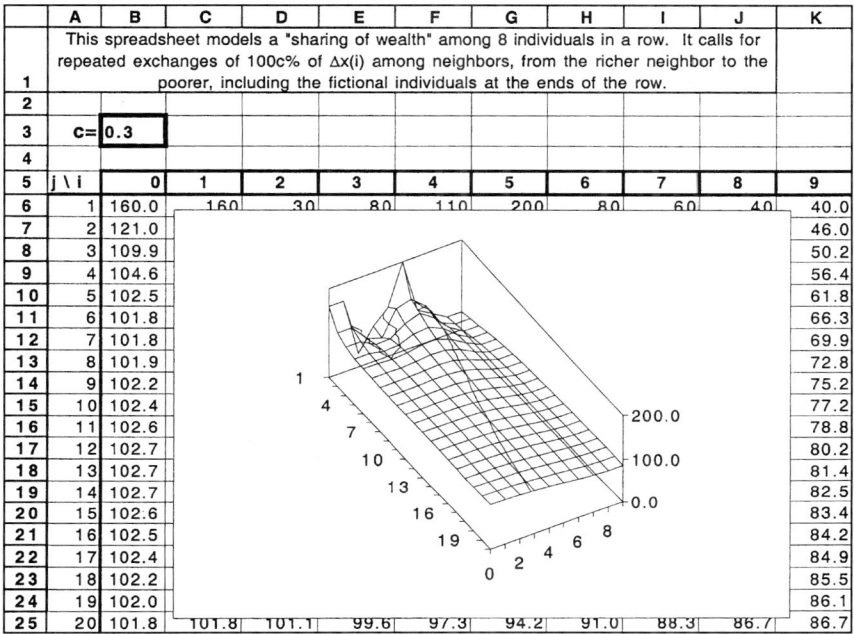

Figure 6.6.5 A 3-D Representation of Sharing of Wealth

Problem 6.6.2 Calculate the total initial wealth of the eight individuals represented in Figure 6.6.1. Confirm that as j becomes large, the system represented in the spread-sheet of Figure 6.6.4 is tending toward an equitable division of this initial wealth.

While we have until now interpreted (6.6.3) in monetary terms, this multivariate difference equation has other interpretations as well. In place of individuals of varying wealth, we can think of masses at varying temperatures arranged in a row. Assuming that neighboring masses are in contact with each other, energy will flow from hotter to cooler. According to Newton's law of cooling, the rate of energy flow is proportional to the difference in temperature (or temperature "gradient") between neighboring masses. The "10% of the difference rule" that governed the flow of money is now interpreted as "conductivity" and denoted by c. In the latter context,

$$(6.6.4) \qquad \Delta_j x_{ij} = c \Delta_i \Delta_i x_{i-1,j}$$

provides a basis for calculating heat flow in a rod of uniform conductivity c.

But what about the fictional individuals positioned at the end of the row and whose wealth was assumed to satisfy $x_{0j} = x_{1j}$ and $x_{8j} = x_{9j}$ for $j = 1, 2, \ldots$? In the context of heat flow, we can interpret their role as ensuring that the temperature gradient remains zero at both ends of the rod. This in turn precludes energy flow to the left of $i = 1$ and to the right of $i = 8$, corresponding to "insulated boundary conditions" at the ends of the rod. Among the consequences of these particular boundary conditions is that the total energy in the rod remains constant (just as the combined wealth of individuals 1 through 8 remained constant in Problem 6.6.1).

From a physical point of view, other boundary conditions are also of interest. For example, we may want to consider a situation where the left end of a rod is placed in ice water (0 °C) while the right end is immersed in boiling water (100 °C). In monetary terms, this calls for placing a "tax collector" at the left end who with every increase in j takes 10% of the assets of the unfortunate individual at $i = 1$. At the right end we have a Robin Hood who with every increase in j, takes or gives money to $i = 8$ depending on whether x_{ij} is greater or less than \$100.

Implementing these new boundary conditions on a spreadsheet simply requires new rules in the columns labeled $i = 0$ and $i = 9$. For the initial data given in Figure 6.6.1, these boundary conditions lead to the graphical representation of Figure 6.6.6.

Problem 6.6.3 Modify the spreadsheet from Problem 6.6.2 to allow for boundary conditions of the form $x_{0j} = x_{1j}$ and $x_{8j} = b$, where b is an arbitrary constant. Generate a graph like that of Figure 6.6.6 when $b = 20$ and interpret the outcome in terms of heat flow.

	A	B	C	D	E	F	G	H	I	J	K
1			This spreadsheet models a "sharing of wealth" among 8 individuals in a row. It calls for repeated exchanges of 100c% of Δx(i) among neighbors, from the richer neighbor to the poorer, including the fictional individuals at the ends of the row.								
2											
3	c=	0.3									
4											
5	j \ i		0	1	2	3	4	5	6	7	8	9
6	1	0	160	30	80	110	200	80	60	40	100
7	2	0	73.0							64.0	100
8	3	0	54.4							73.6	100
9	4	0	45.1							82.3	100
10	5	0	40.6							88.0	100
11	6	0	37.9							91.8	100
12	7	0	35.9							94.3	100
13	8	0	34.3							96.0	100
14	9	0	33.0							97.2	100
15	10	0	31.7							98.0	100
16	11	0	30.5							98.5	100
17	12	0	29.4							98.8	100
18	13	0	28.4							99.0	100
19	14	0	27.5							99.1	100
20	15	0	26.6							99.1	100
21	16	0	25.8							99.0	100
22	17	0	25.1							98.9	100
23	18	0	24.4							98.8	100
24	19	0	23.7							98.6	100
25	20	0	23.1	44.4	62.4	76.3	86.0	92.4	96.2	98.4	100

Figure 6.6.6 Heat Flow in a Rod with Ends in Cold and Hot Water

Given such new boundary conditions, we may no longer be approaching an equitable distribution among $x_{1j}, x_{2j}, \ldots, x_{8j}$. Rather, the x_{ij} will now be approaching a *linear* distribution, one for which $\Delta_i x_{ij}$ has a constant value—but not necessarily zero. This can be interpreted in the context of Chapters 2 and 3 by thinking of $\Delta_i \Delta_i x_{ij} = 0$ as defining an "equilibrium value" for the equation

$$(6.6.5) \qquad \Delta_j x_{ij} = c \Delta_i \Delta_i x_{ij}.$$

When the right side of (6.6.5) is zero, increases in j have no effect on x_{ij}, and this occurs whenever $\Delta_i x_{ij}$ is constant for all i. The nature of this equilibrium solution is determined by boundary conditions such as those we have considered above.

Problem 6.6.4 Given $x_{0j} = 50$ and $x_{9j} = 140$ for all j, find values of x_{1j}, $x_{2j}, \ldots, x_{8j}$ such that $\Delta_i \Delta_i x_{i-1,j} = 0$ for $0 \le i \le 8$.

In the spreadsheets constructed above, one can readily alter the original "10% of the difference" rule by assigning values other than 0.1 to c. Increasing from $c = 0.1$ to $c = 0.2$ increases the rate at which a solution of (6.6.5) approaches such an equilibrium value, and this strategy seems to work up to $c = 0.5$. At this point, however, we encounter a form of "overshoot," much like that encountered in solving $\Delta x_i = -c x_i$ when $c > 1$. This analogy suggests that refinements of the form $\Delta t \mapsto \Delta t / N$ can be useful in analyzing phenomena associated with (6.6.5)

Problem 6.6.5 Modify the spreadsheet in Problem 6.6.3 to allow for solutions of $\Delta_j x_{ij} = (c/N) \Delta_i \Delta_i x_{ij}$ for arbitrary values of N. Given $c = 2$, find the smallest integer value of N for which the corresponding refinement of (6.6.5) approximates the equilibrium solution determined by these boundary conditions.

To motivate another application of multivariate difference equations, we replace "equalization of wealth" with a "bucket brigade" problem. Here we again consider individuals arranged in a row. The first three individuals have assets of $50, $30, and $80, respectively, and our goal is to pass these assets along to individuals to their right whose initial assets are $0. With each increase in j, the assets of the individual at i are to be passed to the individual at $i + 1$, as illustrated in Figure 6.6.7.

Such a process is readily implemented on a spreadsheet. If we associate the index i with horizontal displacement to the right and the index j with vertical displacement downward, the spreadsheet is simply implementing the rule

$$(6.6.6) \qquad x_{i+1,j+1} = x_{ij}.$$

In order to express (6.6.6) as a difference equation that utilizes the symbols Δ_i and Δ_j defined earlier, we add and subtract the expression $x_{i+1,j}$ to obtain

$$x_{i+1,j+1} - x_{i+1,j} + x_{i+1,j} - x_{ij} = 0,$$

i=	1	2	3	4	5	6	7	8
$x_{i1}=$	$50	$30	$80	$0	$0	$0	$0	$0
$x_{i2}=$	$0	$50	$30	$80	$0	$0	$0	$0
$x_{i3}=$	$0	$0	$50	$30	$80	$0	$0	$0

Figure 6.6.7 Assets Being Passed to the Right

or

$$\Delta_j(x_{i+1,j}) + \Delta_i(x_{ij}) = 0.$$

Replacing i by $i - 1$ and transposing the second term finally yields

(6.6.7) $$\Delta_j(x_{ij}) = -\Delta_i(x_{i-1,j}).$$

Problem 6.6.6 Continue the table of Figure 6.6.7 by programming the above rule into the second row of a spreadsheet and then continuing down 10 more rows.

An important modification of (6.6.7) occurs when such a monetary bucket brigade is provided with leaky buckets. That is, what happens if each individual passes only a fraction of his or her assets to the next person in line? This corresponds to replacing (6.6.7) by

(6.6.8) $$\Delta_j(x_{ij}) = -c\Delta_i(x_{i-1,j}), \qquad \text{with} \quad 0 < c < 1.$$

Problem 6.6.7 Modify the spreadsheet of Problem 6.6.6 to the case where $100c\%$ of the assets of the individual at i are passed to the individual at $i + 1$. Create a three-dimensional graph of the resulting data when $c = 0.8$.

As suggested by Figure 6.6.8, for $c < 1$ we no longer have a well-defined "wave packet" moving to the right. While the leading edge of the wave continues to advance one cell to the right with each increase in j, the trailing edge is becoming blurred as money is retained by individuals who would previously have passed along all their assets.

An interesting way to build on these observations is to give a different interpretation to (6.6.8). We now consider cars moving from left to right along a stretch of highway that has signposts at 100-foot intervals. These signposts are labeled $i = 1, 2, 3, \ldots$, and x_{ij} now denotes the number of cars in the ith interval, i.e., between signposts labeled i and $i + 1$, at time j.

Under ideal conditions, $x_{i+1,j+1} = x_{ij}$, and every car traverses about 100 foot in the time interval corresponding to j.

Problem 6.6.8 Suppose j measures time in seconds. If $x_{i+1,j+1} = x_{ij}$, find the speed at which traffic is moving in miles/hour.

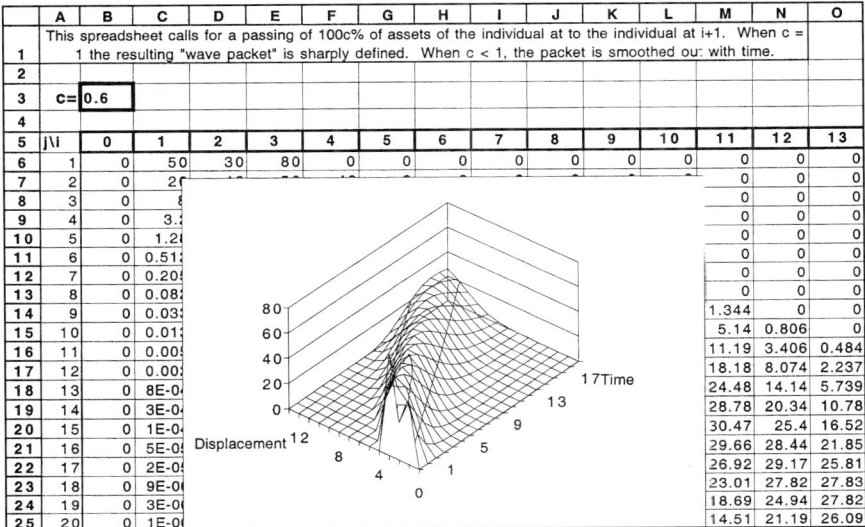

	A	B	C	D	E	F	G	H	I	J	K	L	M	N	O
1	This spreadsheet calls for a passing of 100c% of assets of the individual at to the individual at i+1. When c = 1 the resulting "wave packet" is sharply defined. When c < 1, the packet is smoothed out with time.														
2															
3	c=	0.6													
4															
5	j\i	0	1	2	3	4	5	6	7	8	9	10	11	12	13
6	1	0	50	30	80	0	0	0	0	0	0	0	0	0	0
7	2	0	20										0	0	0
8	3	0	8										0	0	0
9	4	0	3.2										0	0	0
10	5	0	1.28										0	0	0
11	6	0	0.512										0	0	0
12	7	0	0.205										0	0	0
13	8	0	0.082										0	0	0
14	9	0	0.033										1.344	0	0
15	10	0	0.013										5.14	0.806	0
16	11	0	0.005										11.19	3.406	0.484
17	12	0	0.002										18.18	8.074	2.237
18	13	0	8E-04										24.48	14.14	5.739
19	14	0	3E-04										28.78	20.34	10.78
20	15	0	1E-04										30.47	25.4	16.52
21	16	0	5E-05										29.66	28.44	21.85
22	17	0	2E-05										26.92	29.17	25.81
23	18	0	9E-06										23.01	27.82	27.83
24	19	0	3E-06										18.69	24.94	27.82
25	20	0	1E-06										14.51	21.19	26.09

Figure 6.6.8 An Inefficient Bucket Brigade

If, however, $c < 1$ and $x_{i+1,j+1} = x_{ij}$ is replaced by (6.6.8), the situation becomes more complicated. Cars at the head of the pack will continue to move at the speed determined in Problem 6.6.8, but cars at the rear of the pack may be sitting still, waiting for the traffic jam to clear!

What is it that determines the value of c? Well, experience suggests that it is the *density* of traffic that determines whether cars move at the maximum speed ($c = 1$) or whether some cars slow down as they approach congestion ($c < 1$). If so, c should *not* be taken as a constant but be allowed to vary from cell to cell in our spreadsheet.

In Figure 6.6.9, we assume that $c(x_{ij})$ is the smaller of 1 and $4/x_{i+1,j-1}$. That is, if there are more than 4 cars in the $(i + 1)$st interval at time j, then not all cars will advance from i to $i + 1$ as time increases to $j + 1$. If, however, the interval has 4 or fewer cars, then traffic will flow freely according to $x_{i+1,j+1} = x_{ij}$. The spreadsheet of Figure 6.6.9 enables us to visualize the resulting traffic patterns.

As the trailing edge begins to move toward larger values of i, the density of the overall traffic distribution also begins to even out. However, the unevenness of this density suggests why large numbers of cars result in stop-and-go traffic at reduced speed.

6.7. Problems, Exercises, and Projects
This chapter has dealt with rules for change not for a single variable, but for a *system* of variables, each of which can affect the changes that other variables undergo. Such "dynamical systems" are at the heart of mathematical modeling, a topic of central importance in

	A	B	C	D	E	F	G	H	I	J	K	L	M	N	O	P	Q	R
1	This spreadsheet models traffic flow when the number of cars in cell i+1 determines the percentage of the cars in cell i that will advance. It allows us to assign the largest number of cars/cell that allows traffic to flow freely.																	
2																		
3	N=	6																
4																		
5	j\i	0	1	2	3	4	5	6	7	8	9	10	11	12	13	14	15	16
6	1	0	19	31	35	12	17	27	25	14	30	36	22	13	26	30	25	25
7	2	0	15														26.2	25
8	3	0	12														26.9	25.2
9	4	0	9.3														27.1	25.6
10	5	0	7.0														26.9	25.9
11	6	0	5.0														26.5	26.1
12	7	0	3.3														26	26.2
13	8	0	2.0														25.5	26.2
14	9	0	1.0														25.1	26
15	10	0	0														24.8	25.8
16	11	0	0.0														24.6	25.6
17	12	0															24.7	25.4
18	13	0															24.8	25.2
19	14	0															25	25.1
20	15	0															25.3	25.1
21	16	0															25.6	25.1
22	17	0															25.9	25.3
23	18	0															26.2	25.4
24	19	0															26.5	25.6
25	20	0															26.6	25.8
26	21	0															26.8	26
27	22	0															26.9	26.2
28	23	0															26.9	26.3
29	24	0															26.9	26.5
30	25	0															26.8	26.6
31	26	0															26.7	26.6
32	27	0															26.5	26.6
33	28	0															26.4	26.6
34	29	0															26.2	26.5
35	30	0															26	26.5
36	31	0	0	0	0	0	0	0	1.01	7.44	15.2	20.1	22.4	23.3	23.9	24.8	25.7	26.3
37	32	0	0	0	0	0	0	0	0.19	4.89	12.9	18.8	21.8	23	23.7	24.6	25.5	26.2
38	33	0	0	0	0	0	0	0	0	2.7	10.3	17.1	21	22.7	23.6	24.4	25.3	26.1

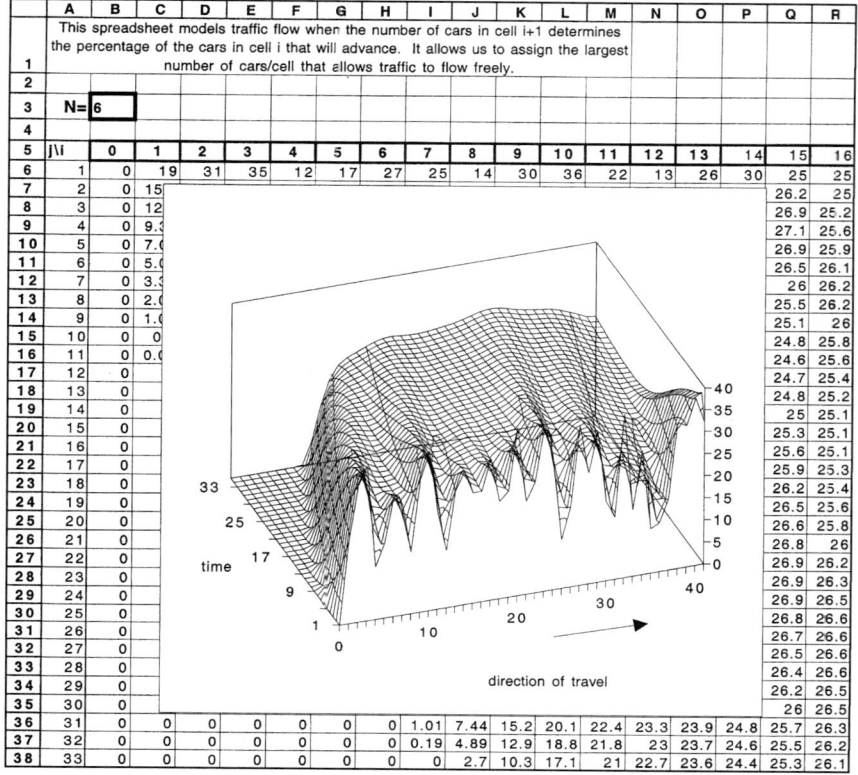

Figure 6.6.9 Density-Dependent Traffic Flow

diverse fields such as economics, ecology, and physics. The problems that follow extend some of the ideas developed in this chapter. It is hoped they will provide the reader with a basis for further efforts to use mathematics as a tool for understanding the world and its dynamics.

A Financial System with Randomness and Delays

Problem 6.7.1 An investor's portfolio consists of a checking account earning 5% monthly interest into which $1000 is deposited at the beginning of each month. This is linked to an investment account with an initial balance of $10,000 whose earnings are randomly distributed between −1% and 3% each month. After 12 months of accumulating assets in both accounts, this investor makes a monthly transfer of one-half of the amount in the checking account into the investment account.

(a) Create a spreadsheet to track these two accounts for 60 months.

(b) Compare the output from part (a) with a situation in which there is a delay of 6 months in the transfer process. That is, rather than transferring half of the checking balance at the beginning of the ith month ($i = 13, 14, \ldots$), the investor transfers half of the balance that existed 6 months earlier (at $i = 7, 8, \ldots$).

(c) Repeat part (b) in the context of a "bear market" in which the earnings in the investment account are randomly distributed between -3% and 1% each month.

Calculating A^n for a Matrix A

Problem 6.7.2 In the spreadsheet of Figure 6.2.1, a matrix $A = \begin{bmatrix} a_{11} & a_{12} \\ a_{21} & a_{22} \end{bmatrix}$ is represented in terms of a single column

$$\begin{bmatrix} a_{11} \\ a_{12} \\ a_{21} \\ a_{22} \end{bmatrix}$$

in order to facilitate matrix iteration. Using this format,

(a) Create a spreadsheet that calculates A^2, A^3, A^4, $\ldots$, A^{10} for an arbitrary 2×2 matrix A.

(b) For $A = \begin{bmatrix} 0.3 & 0.4 \\ 0.7 & 0.6 \end{bmatrix}$, find the 2×2 matrix form of A^{10}.

Problem 6.7.3 Using a computer to determine matrix powers may involve large numbers of arithmetic computations. Even if carried out to many decimal places (e.g., 15 on some spreadsheets), such computations can involve "round-off errors" that create uncertainty about the precise value of A^n. These considerations lead us to seek a precise representation of A^n for matrices having the special form

$$A = \begin{bmatrix} p & q \\ 1 - p & 1 - q \end{bmatrix}.$$

Note that the matrix in part (b) of Problem 6.7.2 above is of this form.

(a) Show that if A is of the above type and B is any matrix of the form $\begin{bmatrix} a & a \\ b & b \end{bmatrix}$, then $BA = B$. [Note that we cannot assume $BA = AB$.]

(b) For A as above, show that $A^2 = B + \lambda I$, where $\lambda = (p - q)^2$ and B is a matrix of the form considered in part (a), $-$i.e., a matrix whose columns are equal.

(c) From part (b) deduce that:

   (i) $A^3 = B + \lambda A$.

   (ii) $A^4 = B + \lambda A^2 = B + \lambda(B + \lambda I) = (1 + \lambda)B + \lambda^2 I$.

   (iii) $A^5 = (1 + \lambda)B + \lambda^2 A$.

   (iv) $A^6 = (1 + \lambda + \lambda^2)B + \lambda^3 I$.

(d) Observing the pattern in part (c), explain why

$$A^{2n} = (1 + \lambda + \lambda^2 + \cdots + \lambda^{n-1})B + \lambda^n I,$$
$$A^{2n+1} = (1 + \lambda + \lambda^2 + \cdots + \lambda^{n-1})B + \lambda^n A$$

for $n = 1, 2, \ldots$.

(e) Assuming $0 < p < 1$ and $0 < q < 1$, show that $\lambda = (p - q)^2 < 1$ and that A^m approaches $\frac{1}{1-\lambda} B$ as m grows without bound. [Hint: Recall the formula for summing a geometric series.]

(f) Use part (d) to find the exact answer to Problem 6.7.2(b).

(g) In the case $A = \begin{bmatrix} 0.3 & 0.4 \\ 0.7 & 0.6 \end{bmatrix}$, show that A^m approaches

$$\begin{bmatrix} \dfrac{36}{99} & \dfrac{36}{99} \\[2mm] \dfrac{63}{99} & \dfrac{63}{99} \end{bmatrix}$$

as m grows without bound.

Problem 6.7.4 Another case in which powers of a matrix can be calculated precisely is when the matrix is diagonal, i.e., of the form $D = \begin{bmatrix} d_1 & 0 \\ 0 & d_2 \end{bmatrix}$.

(a) Show that for the above diagonal matrix D we have $D^n = \begin{bmatrix} d_1^n & 0 \\ 0 & d_2^n \end{bmatrix}$.

(b) Show that if A is a matrix of the form $T^{-1}BT$ for some matrices B and T, then $A^n = T^{-1}B^n T$. [Hint: $(T^{-1}BT)(T^{-1}BT) = (T^{-1}B)(TT^{-1})(BT) = (T^{-1}B)(BT).$]

(c) Let $\alpha = (1 + \sqrt{5})/2$ and $\beta = (1 - \sqrt{5})/2$ be the two roots of $x^2 - x - 1 = 0$. Show that the inverse of the matrix $T = \begin{bmatrix} 1 & 1 \\ \beta & \alpha \end{bmatrix}$ is given by

$$T^{-1} = \begin{bmatrix} \dfrac{\alpha}{\alpha - \beta} & -\dfrac{1}{\alpha - \beta} \\[3mm] -\dfrac{\beta}{\alpha - \beta} & \dfrac{1}{\alpha - \beta} \end{bmatrix}.$$

(d) Letting $D = \begin{bmatrix} \beta & 0 \\ 0 & \alpha \end{bmatrix}$ and $A = \begin{bmatrix} 0 & 1 \\ 1 & 1 \end{bmatrix}$, show that $A = TDT^{-1}$ and therefore that $A^n = TD^n T^{-1}$.

(e) Noting that $A = \begin{bmatrix} 0 & 1 \\ 1 & 1 \end{bmatrix}$ is the Fibonacci matrix arising in (5.6.7) and that according to Problem 5.8.31, $A^n = \begin{bmatrix} x_{n-1} & x_n \\ x_n & x_{n+1} \end{bmatrix}$, where x_i is the ith Fibonacci number, use part (d) to prove that $x_n = \frac{\alpha^n - \beta^n}{\alpha - \beta}$. [Hint: After calculating $TD^n T^{-1}$, recall that since $\alpha\beta = -1$, we have $\alpha\beta^n = -\beta^{n-1}$, etc.]

(f) Problem 5.6.2 provides one basis for showing that x_{n+1}/x_n approaches α as n grows without bound. Show that this result also follows from the characterization $x_n = \frac{\alpha^n - \beta^n}{\alpha - \beta}$ established in part (e).

A Fundamental Matrix for Buffaloes

Problem 6.7.5 In Chapter 5, the growth of a 3-cohort buffalo population was described in terms of a vector

$$\mathbf{y}_i = \begin{bmatrix} b_i \\ t_i \\ a_i \end{bmatrix}$$

and a difference equation $\mathbf{y}_{i+1} - \mathbf{y}_i = A\mathbf{y}_i$, in which

$$A = \begin{bmatrix} -1 & \tau & q \\ \alpha & -1 & 0 \\ 0 & \beta & \gamma-1 \end{bmatrix} \quad \text{and} \quad A+I = \begin{bmatrix} 0 & \tau & q \\ \alpha & 0 & 0 \\ 0 & \beta & \gamma \end{bmatrix}.$$

(a) For $A+I$ as above, calculate $(A+I)^2$ and $(A+I)^3$.

(b) For $\tau = 0.1, q = 0.5, \alpha = 0.9, \beta = 0.8$ and $\gamma = 0.7$, use the column format of Problem 6.7.2 to calculate $(A+I)^2, (A+I)^3, \ldots, (A+I)^{10}$.

(c) For

$$\mathbf{y}_1 = \begin{bmatrix} 30 \\ 60 \\ 50 \end{bmatrix},$$

use the results of part (b) to calculate $\mathbf{y}_9$.

Superposition of Solutions

Problem 6.7.6 On a plot of land adjacent to that occupied by the buffaloes in the previous problem, a commercial buffalo operation is launched that is subject to the same fertility and survival rates τ, q, α, β, and γ. However, this operation imports 10 babies at the beginning of each year while also harvesting 8 adults. Letting the vector $\mathbf{z}_i$ denote its population of babies, teens, and adults at time i,

(a) Show that this population is described by $\mathbf{z}_{i+1} - \mathbf{z}_i = A\mathbf{z}_i + \mathbf{b}$, where

$$\mathbf{b} = \begin{bmatrix} 10 \\ 0 \\ -8 \end{bmatrix}.$$

(b) For

$$\mathbf{z}_1 = \begin{bmatrix} 0 \\ 80 \\ 20 \end{bmatrix},$$

calculate $\mathbf{z}_9$. [Hint: Recall the closed form (6.2.7).]

(c) For $\mathbf{y}_i$ as in Problem 6.7.5 and letting $\mathbf{x}_i$ describe the total population of buffaloes on both plots (i.e., $\mathbf{x}_i = \mathbf{y}_i + \mathbf{z}_i$,), find $\mathbf{x}_9$.

More on Predator–Prey and Competition

Problem 6.7.7 A variant of the predator–prey model (6.3.7) arises when the environment provides a safe haven for a given number k of prey. In this case meetings occur only when $u > k$, and then in proportion to $\text{Max}[0, (u_i - k)v_i]$. Create a spreadsheet that incorporates this modification into Figure 6.3.1 and compare the results obtained when $k = 50$ with those obtained when $k = 0$.

Problem 6.7.8 Consider a modification of (6.3.7) in which rather than depending on random meetings, the predator population harvests the smaller of one prey per predator and 5% of the remaining prey, - i.e., a system in which we replace the term $-bu_i v_i$ by

$$\text{Min}[v_i, 0.05u_i].$$

Modify the spreadsheet of Figure 6.3.1 to reflect this modification and find values of a, c, and d for which both populations survive for 100 time periods.

Competing Species—Heading Toward Equilibrium or Extinction?

Problem 6.7.9 Recall the system of difference equations

$$(*) \qquad \begin{aligned} u_{i+1} - u_i &= au_i - eu_i^2 - bu_i v_i, \\ v_{i+1} - v_i &= cv_i - fv_i^2 - du_i v_i \end{aligned}$$

for two competiting species (see (6.3.12) of Section 6.3).

(a) Confirm the assertion in Section 6.3 that three of the equilibrium values of $(*)$ are given by

$$\begin{bmatrix} 0 \\ 0 \end{bmatrix}, \quad \begin{bmatrix} 0 \\ c/f \end{bmatrix}, \quad \begin{bmatrix} a/e \\ 0 \end{bmatrix},$$

while the fourth is obtained by solving

$$\begin{bmatrix} e & b \\ d & f \end{bmatrix} \begin{bmatrix} u \\ v \end{bmatrix} = \begin{bmatrix} a \\ c \end{bmatrix}.$$

(b) The fourth equilibrium value is the intersection of the lines $eu + bv = a$ and $du + fv = c$. Keeping in mind that the coefficients in $(*)$ are all assumed positive, show that this intersection is in the first quadrant if and only if one of the following conditions holds:

$$\frac{c}{d} > \frac{a}{e} \quad \text{and} \quad \frac{c}{f} < \frac{a}{b}$$

or

$$\frac{c}{d} < \frac{a}{e} \quad \text{and} \quad \frac{c}{f} > \frac{a}{b}.$$

[Hint: Note the intersection points of each line with the u-axis and the v-axis.]

(c) Show that if the first of the conditions in part (b) is satisfied, then $bd < ef$.

(d) Explain why parts (b) and (c) imply the following: If the fourth equilibrium value $\mathbf{x}_0 = (u_0, v_0)$ lies in the first quadrant and $bd > ef$, then

$$\frac{c}{d} < \frac{a}{e} \quad \text{and} \quad \frac{c}{f} > \frac{a}{b}.$$

(e) The figure below represents the situation analyzed in part (d) when $bd > ef$ and A, B, C, D are the four regions into which the first quadrant is partitioned by the two lines.

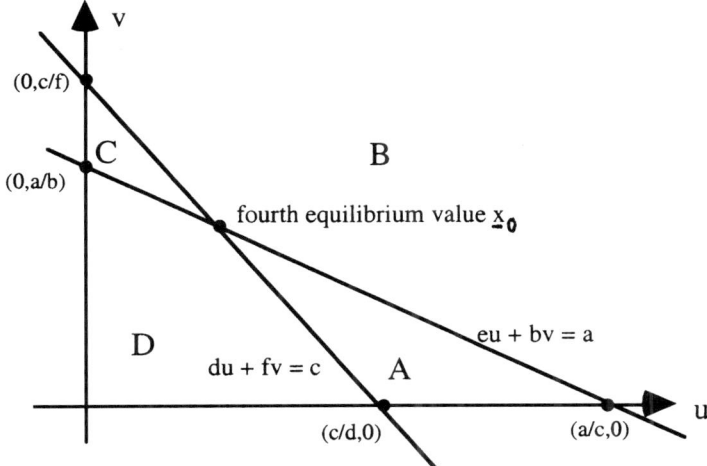

Rewriting (*) in the form

(**)
$$\Delta u_i = (a - eu_i - bv_i)u_i,$$
$$\Delta v_i = (c - du_i - fv_i)v_i,$$

explain why

$(u_i, v_i) \in A$ implies that $\Delta u_i > 0$ and $\Delta v_i < 0$.

$(u_i, v_i) \in B$ implies that $\Delta u_i < 0$ and $\Delta v_i < 0$.

$(u_i, v_i) \in C$ implies that $\Delta u_i < 0$ and $\Delta v_i > 0$.

$(u_i, v_i) \in D$ implies that $\Delta u_i > 0$ and $\Delta v_i > 0$.

[Hint: The line $eu + bv = a$ separates the (u, v)-plane into two half-planes, one in which $a - eu - bv < 0$ and the other in which $a - eu - bv > 0$.]

(f) From part (e), deduce that if $\mathbf{x}_i = (u_i, v_i)$ is in either region B or D, then $\Delta \mathbf{x}_i = (\Delta u_i, \Delta v_i)$ points in the direction of the region A or C. This means that succeeding iterates follow a path that enters A or C. (In very exceptional cases they might follow a path approaching $\mathbf{x}_0$, but without ever reaching it. If they move ever so slightly off this path, they will then be on a path approaching A or C.)

(g) From part (e), deduce that if $\mathbf{x}_i = (u_i, v_i)$ is in either region A or C, then $\Delta \mathbf{x}_i = (\Delta u_i, \Delta v_i)$ points *away* from the equilibrium value $\mathbf{x}_0$. Since succeeding iterates cannot enter (or reenter) B or D (why?), they follow a path approaching a point that is either on the u-axis or the v-axis, corresponding to the extinction of one or the other species.

(h) Now suppose $bd < ef$ and use reasoning analogous to that in part (d) to show that $c/d > a/e$ and $c/f < a/b$. Referring to the figure in part (e), show that in this case $\Delta \mathbf{x}_i$ points *toward* $\mathbf{x}_0$ whenever $\mathbf{x}_i = (u_i, v_i)$ belongs to region A or C. Explain why this corresponds to a form of equilibrium between the species represented by u and v.

Epidemics

Problem 6.7.10 Consider a contagious disease whose duration is one week and that infects only those who have not been infected previously. Letting x_i denote the number of people who have the disease at the beginning of the ith week, suppose that the number infected during the ith week is the result of random meetings between individuals who are infected and those who are susceptible.

(a) Show that the above assumptions lead to

$$x_{i+1} = m x_i s_i,$$

where s_i denotes the number of people susceptible at the beginning of the ith week and m is a constant of proportionality.

(b) In the absence of births, deaths, or migration, show that

$$s_{i+1} - s_i = -m x_i s_i$$

describes the number of susceptible individuals at $i = 2, 3, \ldots$.

(c) Create a spreadsheet that models the course over 5 years of an epidemic caused by the infection of 2000 individuals in a population of 30,000 susceptible individuals when $m = 0.00003$.

(d) Modify part (c) to apply to a population that gains 300 susceptible members each week.

"Y Gotta Eat!"

Problem 6.7.11 As a modification of the basic Keynesian model of Section 6.3, suppose there is an level of consumption k that will occur regardless of last year's level of economic activity. One way of incorporating such "essential consumption" into the model is to replace (6.3.17) by $C_i = cY_{i-1} + k$, where $k > 0$. Modify the spreadsheet of Figure 6.3.3 to incorporate this assumption. Does such a modification seem to enhance or undermine the stability of the system?

Motion in Two Dimensions

In Section 6.4 we began by approximating Newton's differential equation (6.4.4) by means of a pair of difference equations (6.4.6). This approximation enabled us to study the motion of a body whose displacement is measured by a single variable x. Implicit in such a model is that the body is constrained to move in a one-dimensional space. Such considerations were followed by examples of "two-dimensional motion" such as the swing of a pendulum and the motion of a planet about its sun. Here we again used difference equations with "first and second differences" to approximate differential equations that call for "first and second derivatives." Without providing a full calculus-based explanation of how these differential equations are arrived at, the final problems of this chapter provide some further insight into such two-dimensional motion.

The Pendulum Swings

Problem 6.7.12 The figure below represents a bob of mass m at the end of a string of constant length ℓ. Gravity produces a vertical (downward) force of magnitude mg. It is the component of this force that is tangent to the direction of motion that governs the motion of the pendulum.

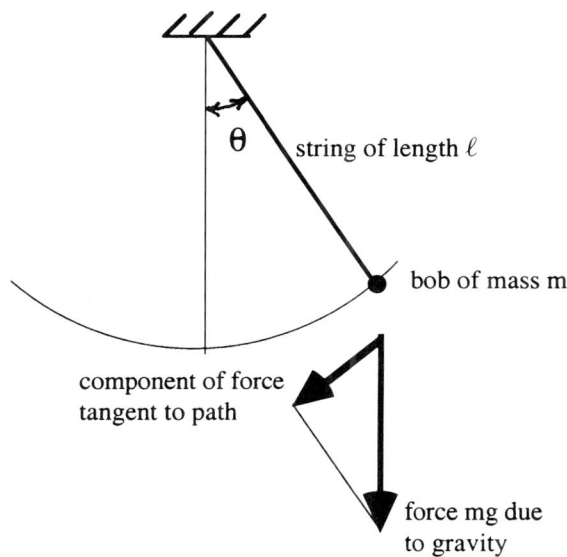

θ

string of length ℓ

bob of mass m

component of force
tangent to path

force mg due
to gravity

(a) Use similar triangles to show that the component of the gravitational force tangent to the path of motion of the bob is $mg \sin \theta$

(b) The bob moves along the arc of a circle of radius ℓ centered at the point of suspension. Show that the speed with which the bob moves along this circle is approximated by $v = \ell \frac{\Delta \theta}{\Delta t}$, where $\Delta \theta$ is the change in angle in the time interval Δt. Show that the acceleration a along this circle is approximated by $a = \frac{\Delta v}{\Delta t} = \ell \frac{\Delta \Delta \theta}{(\Delta t)^2}$.

(c) Use the discrete approximation of Newton's second law (6.4.3) to show that $m\ell \frac{\Delta \Delta \theta}{(\Delta t)^2} = -mg \sin \theta$ and, assuming small *unit* time steps, establish the difference equation $\Delta \Delta \theta_i = -(g/\ell) \sin \theta_i$.

(d) Introducing the momentum term $p_i = mv_i$, show that the second-order difference equation in part (c) can be written in the form (6.4.7) with $k = -mg$.

Planetary Motion and Circular Orbits

The motion of a planet moving about a stationary sun can be described in terms of the distance r between the bodies and an angle θ that measures rotation relative to a fixed ray emanating from the sun. Expressing Newton's second law relative to such a system of "polar coordinates" and letting F denote the force with which the planet is attracted to the sun, the laws of calculus lead to a pair of differential equations,

$$m \left(\frac{d^2 r}{dt^2} - r \left(\frac{d\theta}{dt} \right)^2 \right) = F \quad \text{and} \quad r^2 \frac{d\theta}{dt} = \text{constant} = k.$$

It is important to note that these equations correspond to the assumption that the force is a "central force," i.e., one that always attracts the planet directly toward the sun. Here $\frac{d^2 r}{dt^2} - r(\frac{d\theta}{dt})^2$ represents the "radial component of acceleration," while $\frac{d}{dt}(r^2 \frac{d\theta}{dt})$ denotes the "tangential component," which in the absence of angular forces is zero. The inverse square law of gravity corresponds to $F = -\frac{GMm}{r^2}$, where G is the universal gravitational constant, M is the mass of the sun, and m is the mass of the planet.

The spreadsheet of Figure 6.4.4 is based on a pair of difference equations that correspond to the above, namely

$$\Delta \Delta r_i = r_i (\Delta \theta_i)^2 - \frac{MG}{r_i^2} \quad \text{and} \quad r_i^2 (\Delta \theta_i) = k.$$

Problem 6.7.13 A planet is traveling at a constant speed v in a circular path of radius R about the sun. Show that (6.4.9) and (6.4.10) imply

$$R = \frac{k^2}{MG}, \quad v^2 = \frac{MG}{R}, \quad \text{and} \quad R = \left(\frac{MG}{4\pi^2} \right)^{1/3} T^{2/3},$$

where T is a "year" in the life of the planet, i.e., the time required to traverse the circle once.

Another Universe

Problem 6.7.14 Consider a universe in which our "inverse square law of gravitation" is replaced by $F = -\frac{GMm}{r^k}$, where k is a constant, $2 < k < 3$.

  (a) For $k = \frac{5}{2}$ and a given time T for a single rotation, find a value of r at which a circular orbit is possible.
  (b) Given our earthly year $T = 1$, the value $k = 2$ places the earth about 93,000,000 miles from the sun. Assuming that M, G, and $T = 1$ remain unchanged, determine the effect of a transition from $k = 2$ to $k = 2.5$.
  (c) Modify the spreadsheet in Figure 6.4.4 to allow for varying values of k. Setting $k = 2.5$, choose r_1 and Δr_1 to obtain a nearly circular orbit.

Another Approach to Equalizing Wealth

Problem 6.7.15 The "rule for change" embodied in the diffusion equation (6.6.3) is based on the differences in wealth among individuals arranged in a row. At times $i = 1, 2, 3, \ldots$ it calls for a transfer of 10% of such differences, from richer to poorer, among adjacent individuals. Consider now the following rule: At times $j = 1, 2, \ldots$ each individual conveys 10% of his or her assets to each immediate neighbor. Show that (with due modification for individuals at the end of the row) this is equivalent to the rule embodied in Figure 6.6.3.

Problem 6.7.16 The spreadsheet in Figure 6.6.4 generalizes (6.6.3) by allowing for transfers of 100c% of the differences in wealth among adjacent individuals.

  (a) Describe the effect of increasing c from 0.1 to 0.4 in our efforts to effect an equalization of wealth.
  (b) Describe the effect of allowing c to equal 0.6.
  (c) Modify the spreadsheet in Figure 6.6.4 to allow for refinements of the form "$\Delta t \mapsto \Delta t/N$." [Hint: Enter a value for N in cell D3 and then insert a column for "Time $= (j - 1)/N$" to the left of the values for j. Now, instead of requiring transfers of 100c% of the difference at the beginning of each unit of time, implement N successive transfers of $(100c/N)$% in the course of each unit of time.]
  (d) Use the spreadsheet of part (c) to implement an equalization of wealth when $c = 0.6$.

Problem 6.7.17 Consider the problem of equalizing wealth among individuals arranged in a rectangular grid. Here individuals are located in terms of coordinates (h, i), and the wealth of the individual at (h, i) at time $j - 1$ is denoted by x_{hij}.

(a) Suppose that at $j = 1, 2, \ldots$ there is a comparison of wealth between an individual located at (h, i) and such immediate neighbors as may be located at $(h - 1, i)$, $(h + 1, i)$, $(h, i - 1)$, and $(h, i + 1)$. This comparison is followed by a transfer of 10% of the difference in the wealth of these individuals, from richer to poorer. Write down the difference equation describing such a process for an "interior individual" (i.e., one having four immediate neighbors located at $(h - 1, i)$, $(h + 1, i)$, $(h, i - 1)$, and $(h, i + 1)$.

(b) Show that the process described in part (a) is equivalent to one in which each individual passes 10% of his or her assets to each immediate neighbor.

(c) We will say that an individual whose assets remain constant under part (a) is "in harmony" with his or her neighbors. Show that for an individual with four neighbors, this condition is satisfied whenever

$$x_{hij} = \frac{x_{h-1,ij} + x_{h+1,ij} + x_{h,i-i,j} + x_{h,i+1,j}}{4}$$

(d) Suppose that a noninterior individual at (h, i) has only three neighbors (say, at $(h + 1, i)$, $(h, i - 1)$, and $(h, i + 1)$). Under what conditions is this individual in harmony with his or her neighbors? Repeat for individuals with only two neighbors. With only one.

(e) Construct a 3×3 grid of individuals whose wealth is not identical but such that all 9 individuals are in harmony with their neighbors.

(f) Consider a grid of individuals $x_{h,i}$ (the subscript j is no longer relevant) who are in harmony with their neighbors. Show that for interior individuals with four neighbors this is equivalent to

$$\Delta_h \Delta_h x_{h+1,i} + \Delta_i \Delta_i x_{h,i+1} = 0.$$

7
STELLA MODELING

7.1. Introduction While iterative algebra has a life and interest of its own, our primary focus has been its role in modeling "rules for change." Such rules have tended to be difference equations of the form

$$(7.1.1) \qquad\qquad x_{i+1} - x_i = f(x_i)$$

and various generalizations thereof.

As a starting point we considered "the dynamics of $100 invested at 10% interest compounded quarterly." This gave rise to the difference equation $x_{i+1} - x_i = 0.1x_i$ and the concept of refinement. The spreadsheet format for implementing the underlying procedures is restated in Figure 7.1.1. (See also Figure 2.2.3.) This format has since been extended to many other difference equations as well.

Given such a "template" for solving and refining difference equations, it becomes natural to ask whether we can call on technology to simplify the process. In particular, can technology help us model "rules for change" without confronting the iterative procedures that underlie their solution. This chapter deals with one such technology, namely a software simulation package called Stella.

As we soon shall see, Stella provides an environment in which one can generate both tabular and graphical outputs for problems such as "determine the dynamics of $100 invested at 10% interest compounded quarterly." What is different here is that Stella enables us to formulate such problems in an iconic or "picture-based" environment. Aside from making the solution process more convenient, these icons provide a visual representation of the system being modeled and of the functional relationships on which it is based.

While such convenience is surely to be welcomed, there are dangers in relying on a technology whose solution techniques we do not understand. For this reason,

		N=	4
i	**Time**	**x$_i$**	**f(x$_i$)**
1	0	100	10
2	.25	102.5	10.25
3	.5	105.0625	10.5625
etc.			

Figure 7.1.1 A Format for Solving $x_{i+1} - x_i = f(x_i)$

the present chapter focuses on the process by which Stella goes about calculating solutions to its icon-based rules for change. In particular, Section 7.2 focuses on the relationship between Stella's icons and the spreadsheet techniques we have developed for solving difference equations.

With this relationship understood, it will be possible to take fuller advantage of Stella's icon-based modeling environment. In particular, we will be able to study phenomena that, while also accessible in terms of spreadsheets, are more readily modeled in this engaging new format.

7.2. Flows, Reservoirs, and Difference Equations

As preparation for relating Stella modeling to spreadsheets, it will be useful to develop a difference equation that corresponds to "water flowing into a reservoir at a prescribed rate." The simplest case, that of a constant flow, does not require such iterative machinery. If the reservoir starts with 10 gallons and the rate of flow is 3 gallons/minute, then the amount of water in the reservoir at time t is given by $10 + 3t$. Underlying this assertion is the rule

(7.2.1)   total accumulation = initial accumulation + rate of flow × time.

Here it is essential that the same units of volume (e.g., gallons) and time (minutes) be used consistently throughout. Denoting accumulation at time t by $x(t)$ and the (constant) rate of flow by f, (7.2.1) becomes

(7.2.2) $x(t) = x(0) + f \times t.$

Such problems become more interesting when the rate of flow is not constant but changes continuously with time. Here, we can think of a variable flow, one whose rate is denoted by $f(t)$, being monitored at one-minute time intervals. Denoting these intervals by $i = 1, 2, 3, \ldots$, let x_i denote the amount accumulated at the beginning of the ith interval. If f_i denotes the rate of flow at the beginning of the ith interval, an application of (7.2.1) yields $x_{i+1} \approx x_i + f_i \times 1$. It is on this basis that we approximate "water flowing into a reservoir at a prescribed rate" with the difference equation

(7.2.3) $$x_{i+1} - x_i = f_i.$$

To reconcile difference equations with Stella modeling, it will be important to make explicit the unit of time associated with (7.2.3). This issue already arose in Chapter 2, where we included a column for "Time" in representing refinements of the form $\Delta t \mapsto \Delta t/N$ on a spreadsheet. If t denotes time and each iteration in i corresponds to one unit of time, then $t = i - 1$. However, under refinements of the form $\Delta t \mapsto \Delta t/N$, this relationship becomes $t = (i - 1)/N$.

Suppose now that a reservoir has an initial accumulation $x(0)$ and that the rate of flow into the reservoir is given by a specific function such as

$$f(t) = 2 + t.$$

Taking note of the fact that "$t = i - 1$," we would approximate $x(t)$ by solving

(7.2.4) $$x_{i+1} - x_i = 1 + i; \quad x_1 = x(0).$$

See Figure 7.2.1.

The above approximations to $x(t)$ can usually be improved by more frequent monitoring of the flow. If we choose to monitor $f(t)$ every 15 seconds but continue to express flow in terms of gallons per *minute*, then in place of (7.2.3) we

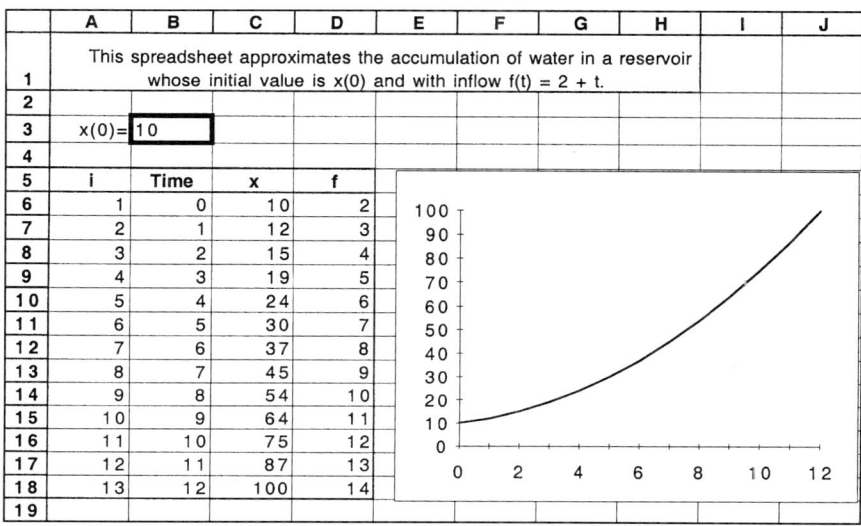

	A	B	C	D	E	F	G	H	I	J
1	This spreadsheet approximates the accumulation of water in a reservoir whose initial value is x(0) and with inflow f(t) = 2 + t.									
2										
3	x(0)=	10								
4										
5	i	Time	x	f						
6	1	0	10	2		100				
7	2	1	12	3		90				
8	3	2	15	4		80				
9	4	3	19	5		70				
10	5	4	24	6		60				
11	6	5	30	7		50				
12	7	6	37	8		40				
13	8	7	45	9		30				
14	9	8	54	10		20				
15	10	9	64	11		10				
16	11	10	75	12		0				
17	12	11	87	13		0 2 4 6 8 10 12				
18	13	12	100	14						
19										

Figure 7.2.1 Accumulation in a Reservoir with Time-dependent Flow I

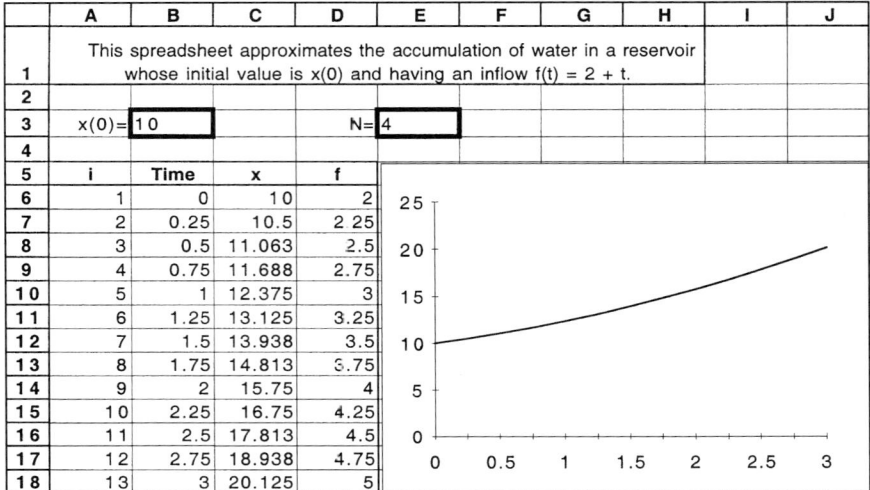

Figure 7.2.2 Accumulation in a Reservoir with Time-dependent Flow II

would arrive at the difference equation[1]

$$(7.2.5) \qquad\qquad y_{j+1} - y_j \approx \frac{f_j}{4}.$$

Such improvements in accuracy correspond to refinements of the form $\Delta t \mapsto \Delta t/N$ that we have already learned to build into our spreadsheet models. They lead to the spreadsheet format of Figure 7.2.2 for approximating the accumulation of water in a tank when a variable flow rate $f(t)$ and initial accumulation $x(0)$ have been specified.

Problem 7.2.1 The calculus-based solution of the problem underlying (7.2.4) can be shown to be $x(t) = x(0) + 2t + t^2/2$. Use a spreadsheet in the format of Figure 7.2.2 to compare $x(8)$ with x_9 which is generated by (7.2.4), for $N = 1$; with x_{33} for $N = 4$; and with x_{81} for $N = 10$.

Problem 7.2.2 A tank that is initially empty has an inflow of $f(t) = 1 + t^2$ and an outflow of $g(t) = 2t$ gallons/minute. Create a spreadsheet that allows for refinement and approximates the accumulation in the tank for $0 \le t \le 8$.

Problem 7.2.3 In a cylindrical tank, the height h of water is proportional to the volume accumulated. A drain at the bottom of the tank leads to an outflow whose rate is proportional to h. Suppose a tank with $x(0) = 200$ satisfies $h = x/10$ and that water is being drained at the rate $g(t) = 2h(t)$. Given an inflow

---

[1] Recall that once the refined equation is established, we revert to the old variable x and index i

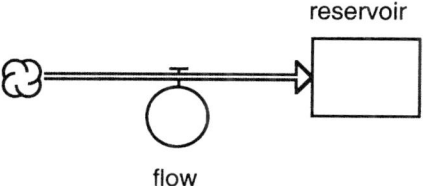

Figure 7.2.3 The First Two Stella Icons

satisfying $f(t) = 1 + t^2$, approximate the accumulation of water in the tank for $0 \le t \le 8$.

In light of the concepts developed in Chapter 2, Figure 7.2.2 should be a familiar one. Our reason for revisiting these iterative procedures now is to establish their close connection to Stella modeling. This is because Stella relies on "flows into reservoirs" to provide a generic framework for representing a wide range of iterative processes, and it will be important to have a clear and explicit understanding of how such representations relate to the difference equations and systems of difference equations studied in Chapters 2 and 6.

Against this background, we note that Stella modeling is based on four icons. One icon is a rectangle that corresponds to a *reservoir*. A second icon is a pipe with a circular manometer[2] that represents a measured *flow*. These two icons enable us to pose the problem of "water flowing into a reservoir at a prescribed rate" as shown in Figure 7.2.3.

While this Figure conveys the structure, or "gestalt," of the phenomenon being modeled, it does not contain any of the detail that has gone into programming the spreadsheet of Figure 7.2.2. This is done in the "functional phase" of Stella modeling, one that is signalled by the appearance of question marks within these icons (Figure 7.2.4).

To remove the question mark in the reservoir, one double-clicks the icon to obtain a dialog box. This dialog box asks us to specify the value of $x(0)$, and an entry such as "$x(0) = 10$" causes the question mark to disappear. To remove the question mark in the flow icon, one double-clicks the circular manometer to obtain a dialog box calling for a rate of flow. Here Stella's syntax for specifying "$f(t) = 2 + t$"

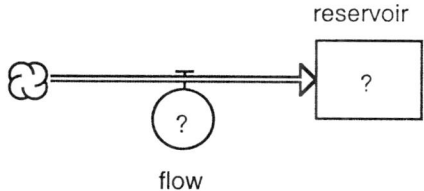

Figure 7.2.4 The 'Functional Phase" of Stella Modeling

---

[2] A device that measures rate of flow in prescribed units, such as gallons/minute.

Time	reservoir	flow
0	10.00	2.00
1	12.00	3.00
2	15.00	4.00
3	19.00	5.00
4	24.00	6.00
5	30.00	7.00
6	37.00	8.00
7	45.00	9.00
8	54.00	10.00
9	64.00	11.00
10	75.00	12.00
11	87.00	13.00
Final	100.00	

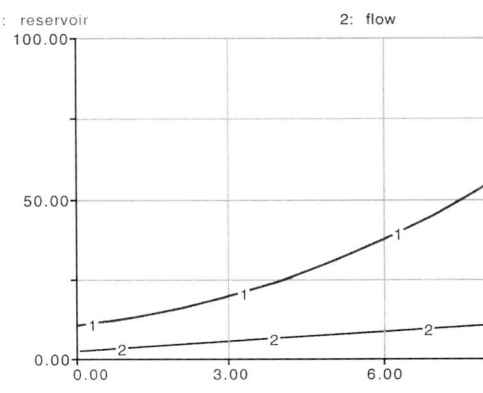

Figure 7.2.5 Stella's Rendition of Figure 7.2.1

calls for entering "2 + TIME" in the flow's dialog box. With these operations performed, Stella is prepared to implement (7.2.4) and display the result for $0 \le t \le T$ in both tabular and graphical format (Figure 7.2.5).

But what if we also seek to implement refinements of the form $\Delta t \mapsto \Delta t/N$? It is here that Stella offers considerable advantage over a spreadsheet. Under Stella's Time Specs menu we are able to specify a "Time Interval" (e.g., $T = 8$) for the duration being modeled. This menu also allows us to implement refinements by specifying a value for "DT." Here DT is a parameter equal to the reciprocal of the integer N we have come to associate with refinement. In order to implement $\Delta t \mapsto \Delta t/N$, we now specify DT $= 1/N$. For example, specifying DT $= 0.25$ produces an output identical to that of Figure 7.2.2 when $N = 4$. (See Figure 7.2.6.)

What makes Stella so convenient in this regard is that we no longer have to "make the spreadsheet N times as long" to obtain a solution for a given time interval $0 \le t \le T$. Once we have specified a time interval by assigning a value to T, such adjustments are automatically performed by the software.

While reservoirs and flows provide an engaging context for thinking about rules for change, Stella's icons can be used to model processes other than "water flowing into a tank." A reservoir can be thought of as the balance in a bank account whose inflow corresponds to periodic payments of interest. Alternatively, a reservoir can be thought of as a population, with an inflow corresponding to births and outflow corresponding to deaths. A closer look at such situations will provide a basis for introducing Stella's other two icons.

In modeling "$100 deposited in a bank at 10% interest" we begin with a reservoir labeled "Balance" and an inflow labeled "interest." (See Figure 7.2.7.) The fact that the rate of inflow now depends on the size of the balance creates a need for a third Stella icon, namely a *connector* leading from Balance to interest. In order to accommodate interest rates other than 10%, we have called on Stella's fourth

Time	reservoir	flow
.00	10.00	2.00
.25	10.50	2.25
.50	11.06	2.50
.75	11.69	2.75
1.00	12.38	3.00
1.25	13.12	3.25
1.50	13.94	3.50
1.75	14.81	3.75
2.00	15.75	4.00
2.25	16.75	4.25
2.50	17.81	4.50
2.75	18.94	4.75

Figure 7.2.6 Stella's Approach to Refinement

icon, a circle that is called a *converter*. Labeling this converter "r ," we also draw a connector from r to the inflow labeled "interest." With these two connectors in place, we are able to specify the product "r * Balance" in the inflow's dialog box.

Here Stella's converter plays a role very similar to the "cell with heavy boundary" used in spreadsheet models. It contains a numerical parameter that is subject to change.

The compounding of interest N times/year is now accomplished by setting $DT = 1/N$ under Stella's Time Specs menu.

While Stella has many more "bells and whistles" that are both engaging and useful, it is ideas from "iterative algebra" that are central to understanding how its tabular and graphical outputs are generated. An ability to go back and forth

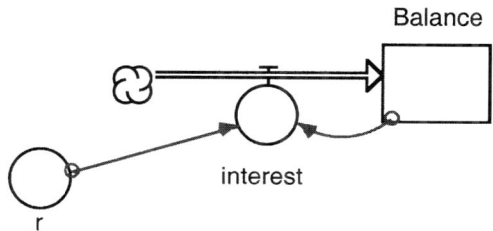

Figure 7.2.7 The Gestalt of Money Earning Interest

between Stella and spreadsheet formulations of "rules for change" will provide us with an ability to use both types of software with an understanding of their uses and limitations.

Problem 7.2.4 A reservoir labeled "x" has inflow "ax" and outflow "bx-squared." Use this structure to create a Stella model for the logistic equation $x_{i+1} - x_i = ax_i - bx_i^2$. With $DT = 1$, confirm that the x_i approach a/b whenever $|a| < 1$.

Problem 7.2.5 Setting $a = 6$ and $b = 2$, find a value of DT for which the x_i defined in the above problem approach 3. How do you explain the model's behavior when $DT = 5$?

7.3. Systems   In Chapter 6 we made the transition from a single difference equation

$$(7.3.1) \qquad\qquad x_{i+1} - x_i = f(i, x_i)$$

to systems of the form

$$(7.3.2) \qquad\qquad \mathbf{x}_{i+1} - \mathbf{x}_i = \mathbf{f}(i, \mathbf{x}_i).$$

In many cases such systems were also written as

$$(7.3.3) \qquad\qquad \mathbf{x}_{i+1} - \mathbf{x}_i = A(i, \mathbf{x}_i)\mathbf{x}_i + \mathbf{b}_i.$$

An important feature of Stella is the ease with which it enables us to make the transition from a single equation to a system and the extent to which it facilitates generalizations of models such as those considered in Chapter 6.

By way of illustration, our discussion of predator–prey interaction began with the system

$$(7.3.4) \qquad\qquad \begin{aligned} u_{i+1} - u_i &= au_i, \\ v_{i+1} - v_i &= -cv_i, \end{aligned}$$

corresponding to the growth of rabbits and the decline of foxes. In a Stella context (See Figure 7.3.1), we might begin with two reservoirs labeled u and v. Reservoir u has an inflow labeled au with a connector from u, while reservoir v has an outflow labeled cv with a connector from v.

In order to include the interaction terms contained in

$$(7.3.5) \qquad\qquad \begin{aligned} u_{i+1} - u_i &= au_i - bu_iv_i, \\ v_{i+1} - v_i &= -cv_i + du_iv_i, \end{aligned}$$

we simply append an outflow buv to u and an inflow duv to v, both of which are the objects of connectors from both u and v (Figure 7.3.2). To facilitate changes

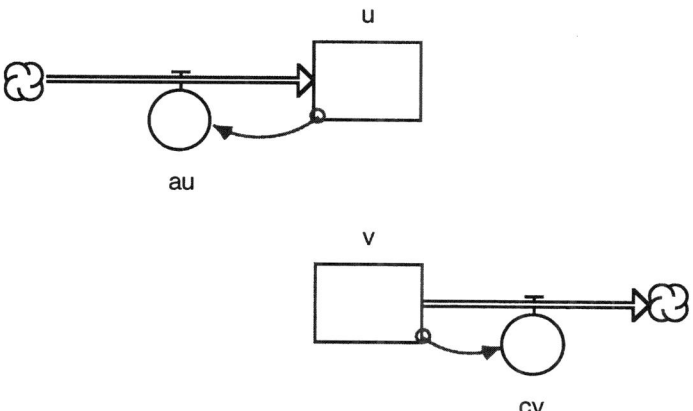

Figure 7.3.1 Rabbits Grow while Foxes Decline

in the parameters a, b, c, and d, we also include four converters that embody their values.

Setting $a = c = 0.1, b = d = 0.0001, u(0) = 110$ and $v(0) = 100$, and specifying the size of the four flows to correspond to (7.3.5), this Stella model provides us with both time-dependent and orbit graphs of the solution of (7.3.5). See Figure 7.3.3.

Here it is easy to change from $DT = 1$ to $DT = 0.5$ and observe the approach to periodic solutions for both u and v. See Figure 7.3.4.

Here it is interesting to recall the vector representation $\mathbf{x}_{i+1} - \mathbf{x}_i = A\mathbf{x}_i$ for such systems. With $\mathbf{x}_i = \begin{bmatrix} u_i \\ v_i \end{bmatrix}$, the Stella diagrams of Figures 7.3.1 and 7.3.2

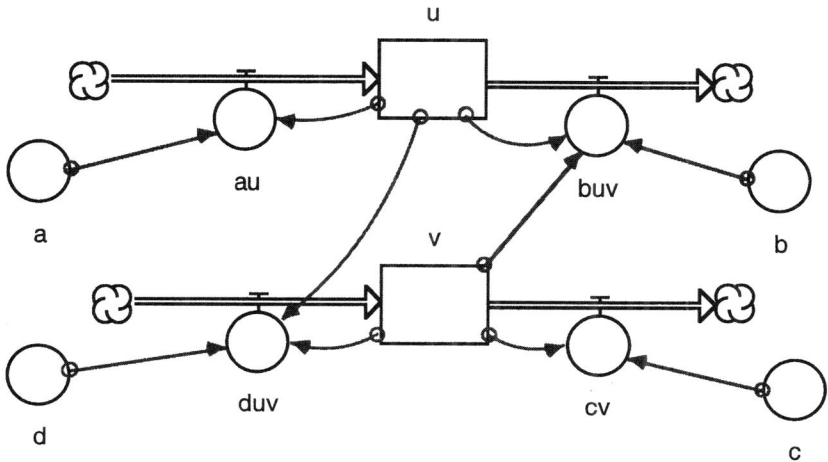

Figure 7.3.2 Gestalt of a Predator–Prey System

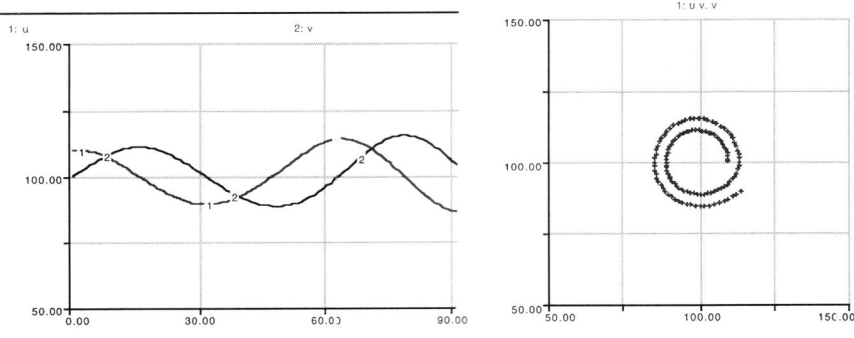

Figure 7.3.3 Stella Representations of Predator-Prey Interaction

correspond to solutions of (7.3.3) with $A = \begin{bmatrix} a & 0 \\ 0 & -c \end{bmatrix}$ and $\begin{bmatrix} a & -bu_i \\ -dv_i & c \end{bmatrix}$, respectively. The following problem explores the relationship between connectors in Stella modeling and the "off-diagonal" terms in the corresponding matrix.

Problem 7.3.1 The dynamics of three noninteracting species can be represented by $\mathbf{x}_{i+1} - \mathbf{x}_i = A\mathbf{x}_i$, where $\mathbf{x}_i$ is a column vector $\text{col}[u_i, v_i, w_i]$ and A is a "diagonal matrix" whose elements satisfy $a_{ij} = 0$ whenever $i \neq j$. Their dynamics can also be modeled by a Stella diagram with three reservoirs labeled u, v, and w whose connectors go only from a reservoir to flows in or out of that reservoir (see Figure 7.3.1). If A is now augmented with an entry $a_{32} \neq 0$, describe the additional connector required in the corresponding Stella diagram. Repeat for general $a_{ij} \neq 0$ when $i \neq j$.

Having created a Stella model for the basic predator–prey system, it is easy to extend it to more complex and realistic situations. For example, the "harvesting of rabbits" considered in Problem 6.3.3 would simply call for a connector from a new converter labelled h to the outflow from u in Figure 7.3.2.

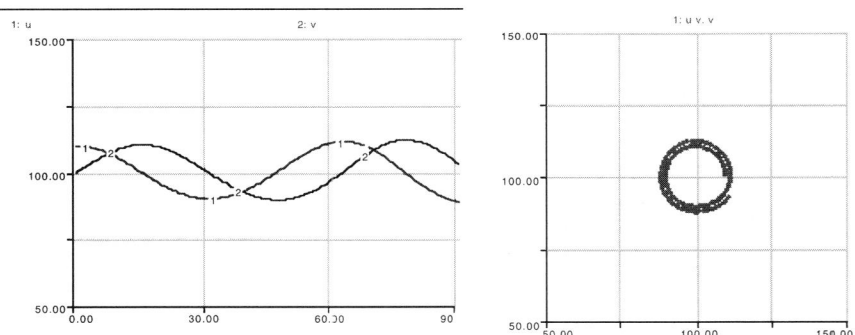

Figure 7.3.4 Refined Stella Representations of Predator–Prey Interaction

Problem 7.3.2 Use Stella to redo Problem 6.3.3.

Problem 7.3.3 Use Stella to redo Problem 6.3.5.

Rather than just redoing Problem 6.3.6, let us use Stella's convenient programming format to embed it in a more complete system. As already noted, Volterra extended the basic predator–prey format (7.3.5) to the interaction of two species in which the prey is "self-limiting." That is, he considered systems of the form

(7.3.6)
$$u_{i+1} - u_i = au_i - bu_i v_i - eu_i^2$$
$$v_{i+1} - v_i = -cv_i + du_i v_i.$$

Such systems continue to fall within the structure of Figure 7.3.2, requiring only a reprogramming of the outflow from the reservoir labeled u to accommodate the additional term $-eu_i^2$.

But suppose we want to go beyond the concept of "superlinear damping" to inquire into the mechanism for self-limitation. The fact that rabbits on an island are dependent on a finite source of vegetation might lead us to expand Figure 7.3.2 with a reservoir labeled "Grass" (Figure 7.3.5). The dependence of the rabbit population on vegetation now calls for a connector from Grass to the

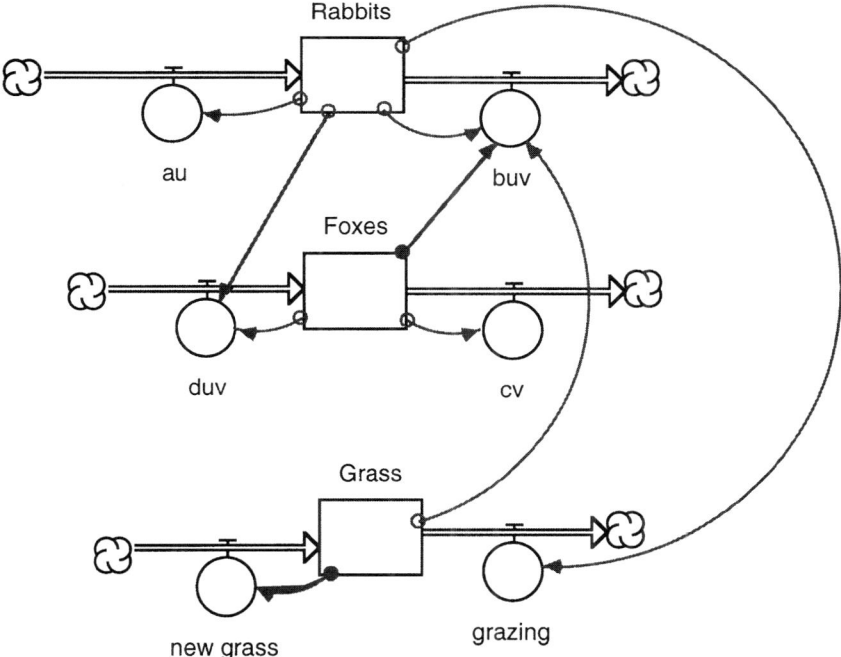

Figure 7.3.5 A Mechanism for Self-limitation

outflow corresponding to rabbit deaths. In place of relying on a superlinear damping term $-eu_i^2$, we might instead build in a mechanism that goes into effect whenever Grass falls below some level, e.g., replace $-eu_i^2$ by

$$\text{Min}(\text{Grass} - 100, 0)$$

in defining the outflow from u.

Since grass is self-regenerating, we will need a connector from "Grass" to the inflow "new grass." And finally, since the rate at which grass is eaten depends on the size of the rabbit population, we need a connector from the reservoir u representing rabbits to the outflow labeled "grazing."

Such an approach to self-limitation may lead us to consider mechanisms that were overlooked in (7.3.6). While the rate at which grass regenerates is proportional to the amount of grass present, there is a substantial delay between the time that grass produces seeds and the time that these seeds contribute to the rabbits' food supply. This observation provides a basis for introducing a delay into the flow "new grass" and, implicitly, into the damping mechanism that makes the rabbit population self-limiting.

Problem 7.3.4 Modify the predator–prey system (7.3.5) along the lines of Figure 7.3.5. Determine the effect of delaying "new grass" by 1, 2, and 3 units of time.

Problem 7.3.5 Elaborate on Problem 7.3.4 by including a converter that corresponds to the harvesting of rabbits. Rather than setting a constant harvest rate h, use a connector to make h dependent on the number of foxes, and then try to devise a rule that stabilizes the predator–prey interactions observed in Problem 7.3.4.

By way of another example expressible in Stella's iconic format, we return to the realm of economics and the broccoli market considered in Section 2.3. The Stella diagram of Figure 7.3.6 may in itself serve as a reminder of some of the concepts developed there.

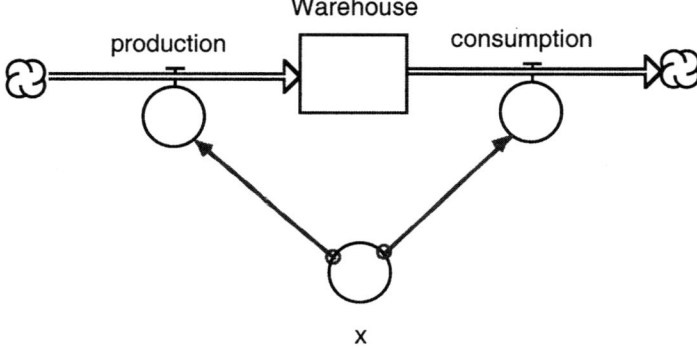

x

Figure 7.3.6 The Broccoli Market

As indicated by Figure 7.3.6, the flow of broccoli in and out of a "Warehouse" is controlled by price, where price is denoted by x. This dependence is reflected by connectors from the converter labeled x to the flows "production" and "consumption." In Stella's functional phase we are asked describe the way in which these flows depend on x in explicit terms. Here we may impose artificially simple rules such as "production equals $2x$" and "consumption equals $4 - x$." Alternatively, Stella allows us to use data based on observation to define these functional relationships (see Stella's "make graph" capability). It is these rules, or "elasticities," that define the marketplace within which broccoli is bought and sold. Since production increases with price while consumption decreases, these rules should enable us to determine an equilibrium price at which production equals consumption.

The dynamics of this market can be based on the concept of "excess demand," which was defined to be "consumption minus production." To introduce this concept into Figure 7.3.6, we simply add a converter labeled E, draw connectors from the two flows to E, and program E's dialog box accordingly.

This raises the question, "How does excess demand affect price?" The "Walrasian assumption" introduced in Chapter 2 asserted that it is price *change*, rather than price itself, that is determined by excess demand. Since converters do not allow for in- or outflows (only reservoirs allow flows), the Walrasian assumption is difficult to build directly into Figure 7.3.6. We can, however, replace the converter labeled x by a reservoir, one that *does* allow for an inflow called "price change." On this basis, a Stella framework for studying the broccoli market might be one like that of Figure 7.3.7.

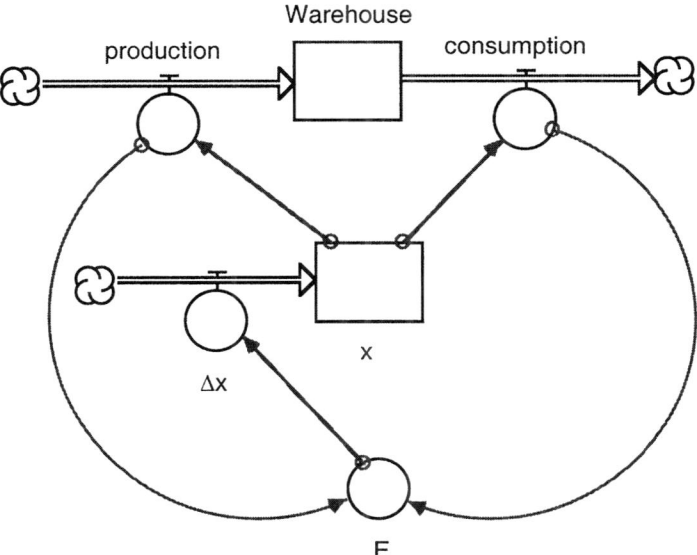

Figure 7.3.7 The Walrasian Broccoli Market

Problem 7.3.6 In Chapter 2 the Walrasian assumption for the broccoli market was embodied in the difference equation $x_{i+1} - x_i = m(e - c) - m(d + f)x_i$. Use Stella to reproduce the model underlying Figure 2.3.3 and use this Stella model to address the questions raised in Problem 2.3.12.

Another approach to microeconomic modeling is to let accumulation in a warehouse determine price. Here we may want to replace broccoli by a more storable commodity, such as petroleum. When storage tanks are full of petroleum, distributors can encourage consumption by lowering the price. Conversely, when levels are low, an increase in price or taxes can be used to reduce demand. Such mechanisms are readily accommodated in Figure 7.3.6 by "dynamiting" the icon representing excess demand and establishing a connector from the reservoir labeled "Warehouse" to the flow labeled "price increase."

However, one may prefer to use the amount of petroleum in storage to determine *price* rather than price *change*. Since one cannot draw a connector to a reservoir (only flows are allowed to affect the accumulation in a reservoir), using storage to determine price would require a converter for x.

Problem 7.3.7 Modify Figure 7.3.7 so that accumulation in the Warehouse can be used to determine price. Setting the initial accumulation in the Warehouse at 100 units, formulate a pricing rule that seeks to stabilize this accumulation at 50 units. [Hint: You may want to take into account the equilibrium price determined by the ways in which price affects production and consumption.]

Problem 7.3.8 Modify Figure 7.3.7 so that accumulation in the Warehouse can be used to determine price change. Setting the initial accumulation in the Warehouse at 100 units, formulate a price change rule that seeks to stabilize this accumulation at 50 units.

Problem 7.3.9 Modify Problem 7.3.8 so that *change* in accumulation in the Warehouse is taken into account in determining price change.

As another example of using Stella to elaborate on previously considered models, we return to the logistic equation $x_{i+1} - x_i = a x_i - b x_i^2$ as developed in Problem 7.2.4. In a population growth context, the fact that solutions of this equation (or a refinement thereof) approach the equilibrium value a/b can be interpreted in terms of "carrying capacity." That is, an unfettered growth rate ax combines with a nonlinear damping term $-bx^2$ to define the equilibrium value $x_0 = a/b$ at which such a population can be expected to stabilize.

But there are situations in which populations themselves affect the carrying capacity of their environment. If agricultural practices degrade the land upon which a population depends for food, then population growth would contribute to a decline in carrying capacity. If, on the other hand, a growing population is able to develop more efficient and sustainable forms of agriculture, then population growth could be linked to an increase in carrying capacity.

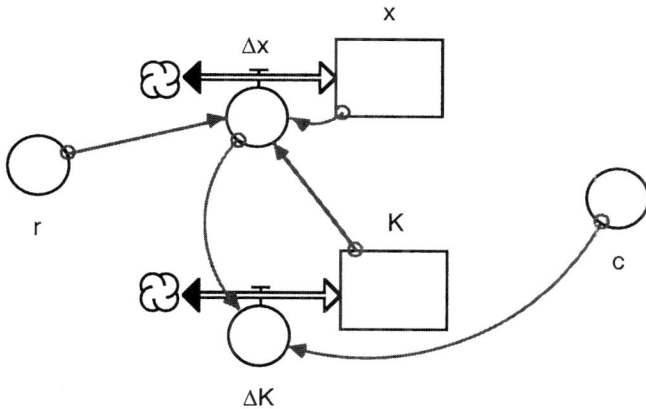

Figure 7.3.8 Population Growth Affects Carrying Capacity

Such considerations suggest replacing the logistic equation with a system such as

$$\Delta x_i = r\,x_i\,(1 - K_i\,x_i),$$
$$(7.3.7) \qquad \Delta K_i = c\,\Delta x_i,$$

one in which changes in carrying capacity are linked to population growth. These ideas are embodied in Figure 7.3.8, where connectors enable us to specify a change in carrying capacity that is proportional to the change in the population itself. The degree of impact of population change on carrying capacity is determined by the converter labeled c.

The distinction between optimists and pessimists can now be related to the sign of the constant c. Optimists may point out that during the past 200 years food production has outpaced population growth, indicating a correlation between the growth of human population and an increase in carrying capacity. Pessimists, on the other hand, may point to the fact that in this process, food production has become heavily dependent on nonrenewable resources, notably the fossil fuels used to sustain such high levels of food production. The finiteness of fossil fuel reserves can be cited as a basis for arguing that the demands of extra mouths now exceeds the contributions of extra hands as they affect the earth's carrying capacity.

Problem 7.3.10 Modify Figure 7.3.8 to allow for the possibility that c is *not* constant but dependent on the size of the population denoted by x. Taking the world's population as 5.8 billion, hypothesize a level at which human population begins (or began) to degrade, rather than enhance, the earth's carrying capacity. [Remark: Be sure that changes in population and carrying capacity are represented by *biflows*, rather than flows that are unidirectional.]

Finally, we note that Stella can also be used to model mechanical systems, such as those considered in Section 6.4. Here we have to adopt a somewhat broader view of the concept of "reservoir." Rather than representing an accumulation, such as water in a tank or individuals in a population, reservoirs can be used to represent

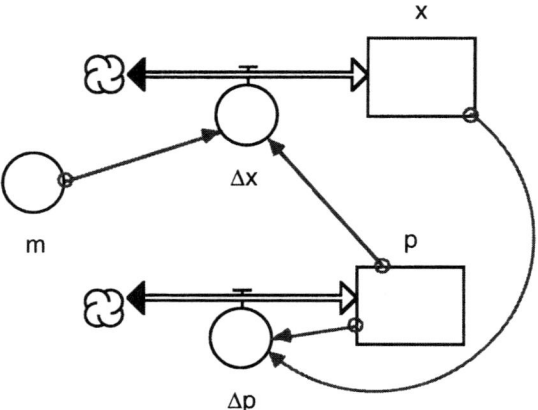

Figure 7.3.9 Newton's Second Law

any variable that is subject to change.[1] The rule for change to which that variable responds can then be programmed into the in- or outflows associated with such a reservoir.

In this context, the systems form of Newton's second law,

$$(7.3.8) \qquad \begin{aligned} x_{i+1} - x_i &= \frac{p_i}{m}, \\ p_{i+1} - p_i &= f(i, x_i, p_i), \end{aligned}$$

can readily be modeled in terms of Stella's icons.

The reservoirs in Figure 7.3.9 correspond to x and p, and their biflows correspond to Δx_i and Δp_i, respectively. The first of these biflows is the object of a connector from the reservoir for p, while the latter biflow has connectors from both x and p.

Problem 7.3.11 Use the Stella format of Figure 7.3.9 to solve Problem 6.4.2.

Problem 7.3.12 Use the Stella format of Figure 7.3.9 to solve Problem 6.4.3.

7.4. Iteration, Reservoirs, and Flows Our development of difference equations was based on the equivalence of the iterative scheme

$$(7.4.1) \qquad\qquad x_{i+1} = F(x_i)$$

and difference equations of the form

$$(7.4.2) \qquad\qquad x_{i+1} - x_i = f(x_i).$$

---

[1] This point of view has already arisen in using a reservoir to represent price in Figure 7.3.6.

Starting with (7.4.1), we were able to extend the iterative techniques used to solve polynomial equations to the solution of difference equations as well.

Stella, on the other hand, starts with difference equations. Its most fundamental system of icons, that of a flow into a reservoir, corresponds to difference equations of the form

(7.4.3) $$x_{i+1} - x_i = f(i),$$

while the addition of a connector from the reservoir to the flow enables us to model (7.4.2). This raises the question of whether and how Stella can be used to perform iterations, such as those represented by (7.4.1).

Perhaps the simplest answer is to write (7.4.1) as

(7.4.4) $$x_{i+1} - x_i = F(x_i) - x_i$$

and then to program the inflow to the reservoir labeled x as $F(x) - x$. But this approach fails to take advantage of Stella's icons to convey a visual representation of the iterative process.

To develop an icon-based representation of iteration we may be tempted to use a circular flow, one that feeds the same reservoir from which it emanates.

Unfortunately, the circular flow in Figure 7.4.1 feeds into an iconic "cloud" rather than the reservoir from which it emanates. This cloud indicates that the flow did not really "connect" to the reservoir, i.e., that the flow will not affect the accumulation in x and that Stella has rejected this particular effort to represent the iterative process.

We can, however, call on another Stella icon in this regard. In programming the dialog box for x, we can call for a "conveyor" x in place of a reservoir. In practice, this means that x no longer "accumulates" its inflow but simply "conveys" it to another destination (in our case to an iconic cloud whose contents are outside the system being modeled).

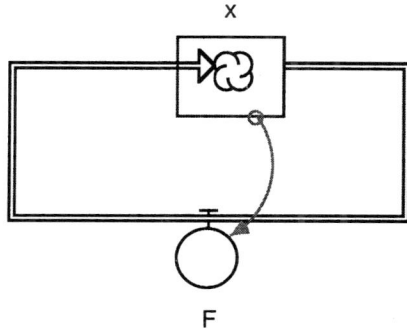

Figure 7.4.1 A Circular Flow

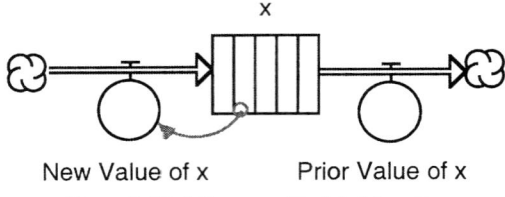

New Value of x   Prior Value of x

Figure 7.4.2 A Conveyor Model of Iteration

In Figure 7.4.2 we have specified $DT = 1$, and the conveyor has been programmed with a "transit time $= 1$." That is, at the end of the ith unit of time,

$$x_i \text{ is expelled into a cloud, and}$$
$$x_{i+1} = F(x_i) \text{ enters the conveyor.}$$

This provides an iconic model in which the conveyor both suggests and models the iterative scheme $x_{i+1} = F(x_i)$.

Problem 7.4.1 Recalling the Babylonian iterator $F(x) = (x^2 + k)/(2x)$ for approximating $\sqrt{k}$, use the Stella format of Figure 7.4.2 to calculate $\sqrt{5}$.

Problem 7.4.2 Use Stella to construct a quadratic equation solver.

Problem 7.4.3 Use Stella to construct a cubic equation solver.

Having used Stella to implement iterations based on reservoirs, we turn to a more challenging problem, that of implementing iterations based on flows. In working with spreadsheets, such distinctions did not arise. Here we found ourselves iterating abstract variables such as x or u or v without regard to the role they play in the model that underlies them. However, when we use Stella's icons to model the same situation, iterating a flow labeled x is not the same as iterating a reservoir.

This situation is illustrated by the Keynesian multiplier model considered in Section 6.3. Here $C(i)$ was a flow used to represent consumption during the ith year, while $I(i)$ was another flow used to denote investment. Representing these processes in terms of Stella might begin with two reservoirs labeled Households and Industry, respectively. Consumption would now be a flow into Households, while Investment would be a flow into Industry. In the absence of government spending (i.e., with $G(i) = 0$), national income is defined as $Y(i) = C(i) + I(i)$. This leads to the Stella representation of Figure 7.4.3.

The challenge now (see (6.3.17)) is to build the iterative rule

(7.4.5) $C(i) = cY(i - 1)$

into Figure 7.4.3. Including a converter labeled c to represent "the marginal propensity to consume" causes no difficulty. However, in order to implement

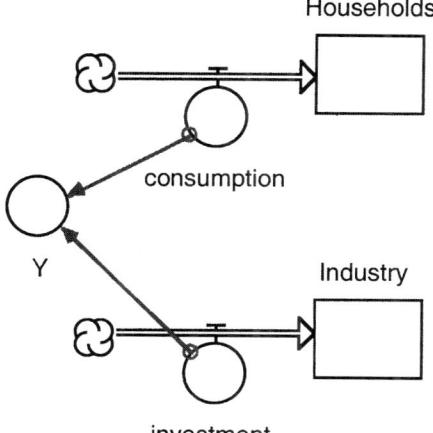

Figure 7.4.3 A Definition of National Income

(7.4.5) it would be natural to introduce a converter labeled "prior Y" that is dependent on Y by the rule 1DELAY(Y,1)" and then to define "consumption" as "c× prior Y." (See Figure 7.4.4.)

The problem here is that Stella will not allow us to draw the final connector from "prior Y" to "consumption." Rather, we receive a message pointing out that such a rule would constitute a circular set of definitions of (Figure 7.4.5).

In the face of such rules it becomes necessary to "trick" Stella into performing the desired iterations. This can be done by elaborating on Figure 7.4.4.

In the representation of the Keynes multiplier effect of Figure 7.4.6 we note that in the course of a year, households transform consumption into waste. Accordingly,

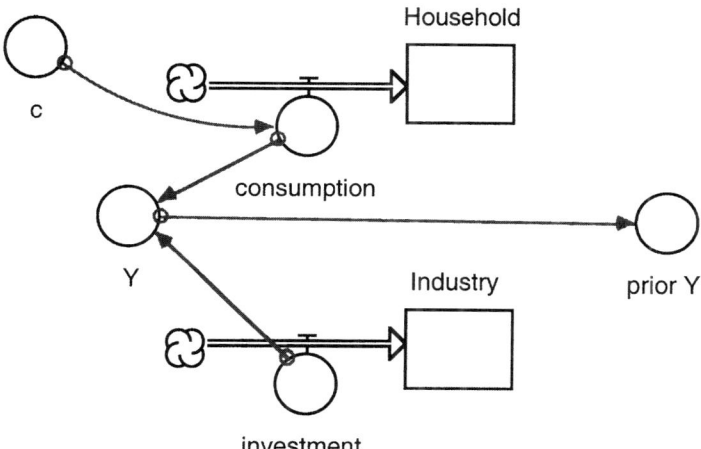

Figure 7.4.4 An Attempt at Iteration of Flows

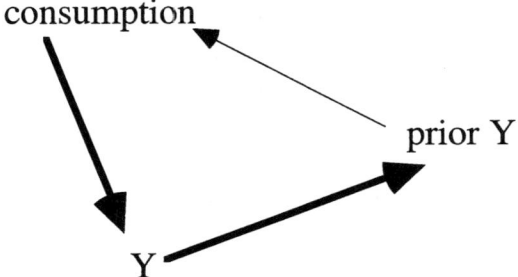

Figure 7.4.5 An Unacceptable Connection

"Households" are now represented by a conveyor with "transit time = 1" whose output is the same size as would be "DELAY(consumption,1)." In place of a connector from "prior Y" (rejected by Stella), we now employ *two* connectors (from "waste" and from "prior I") to consumption. Stella *does* allow us to make these two connections without rejecting them as a circular.

In this way, Figure 7.4.6 provides a useful iconic representation of the system underlying Keynes's "consumer side" model of an economy. Assuming that $I(i)$ has the constant value 20, and setting $c = 0.6$, we again obtain the previously observed multiplier effect in which $Y(i)$ approaches $20/(1-c)$. (See Figure 7.4.7.)

Problem 7.4.4 Create a Stella model that implements the "induced investment rule" (6.3.20). Use this model to solve Problems 6.3.19 and 6.3.20.

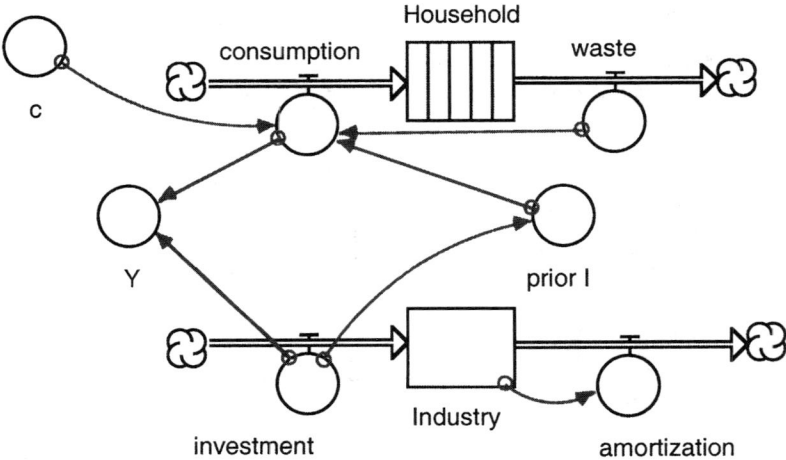

Figure 7.4.6 A Keynesian Multiplier Model

7.5. Submodels
In modeling predator–prey systems, we started with a rabbit population that in the absence of foxes undergoes exponential growth in accordance with the difference equation

(7.5.1) $$x_{i+1} - x_i = ax_i.$$

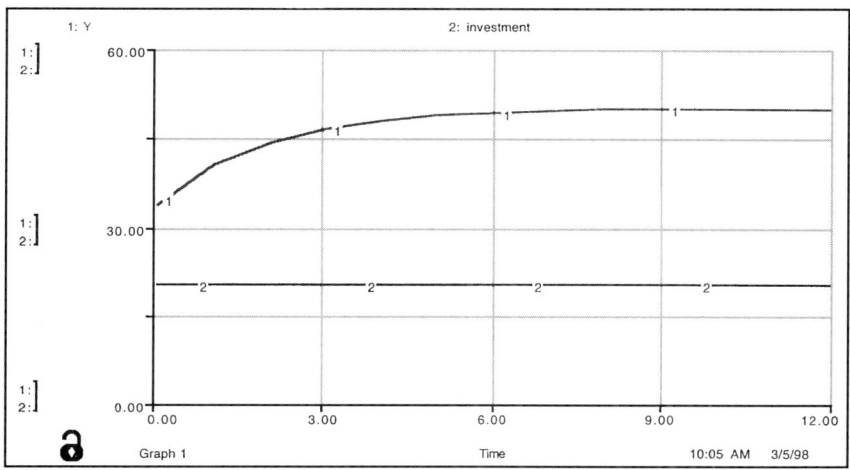

Figure 7.4.7 **The Keynesian Multiplier Effect**

However, in Chapter 5 we took a closer look at such populations, breaking them into subsets and deriving their overall growth rate from the fertility and mortality rates of their various age cohorts. In other words, we took account of the fact that a population can itself be regarded as a system, one whose dynamics are, at least in part, determined by the interaction among its components.

In this section we shall illustrate how Stella provides the option of regarding "rabbits" *either* as a homogeneous population whose growth rate is represented by (7.5.1) *or* as a system that is composed of interacting cohorts. This option is based on Stella's capacity for creating *submodels*, as described below.

In using Stella's icons to represent (7.5.1), we begin with a rectangular icon labeled "Rabbits," an inflow labeled "births" and an outflow labeled "deaths" (Figure 7.5.1). This time, however, in the dialog box for "Rabbits" we will specify that this rectangle is to constitute a submodel (an option made possible by first selecting a conveyor in place of a reservoir and then specifying that the conveyor is to become a submodel).

In response to such a designation, Stella provides us with an entirely new modeling space, one in which we are able to specify the interaction among various components of the submodel called Rabbits. In the case of Fibonacci's rabbits,

Rabbits

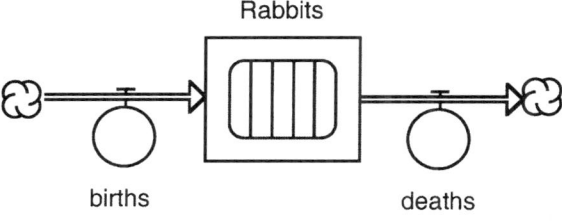

births   deaths

Figure 7.5.1 **A Population of Rabbits**

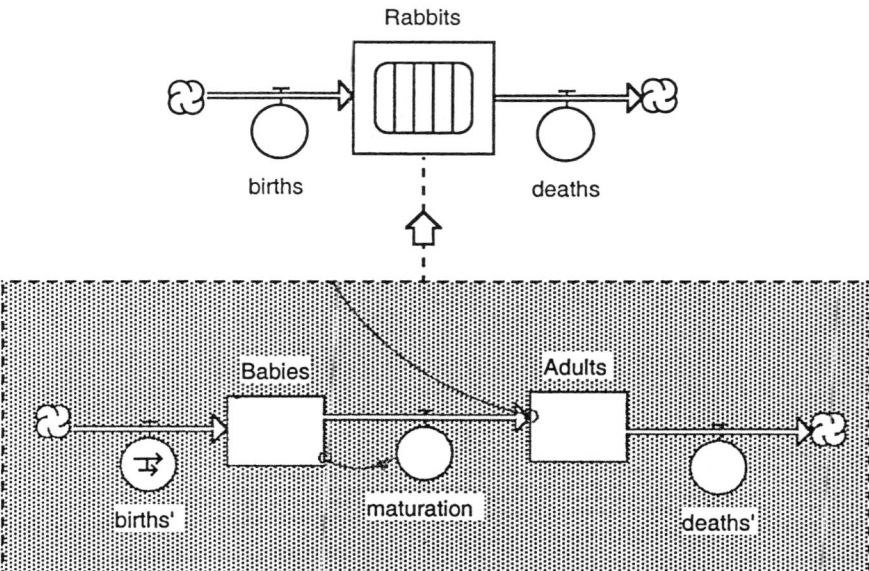

Figure 7.5.2 Fibonacci's Rabbits

these components consist of babies and of adults. Each of these cohorts is represented by its own reservoir, and there is a flow from "Babies" to "Adults" that is labeled "maturation." A connector indicates that the rate of the flow "maturation" depends on the number of babies (Figure 7.5.2).

The flow "births'" into "Babies" is derived from (and identical to) the flow "births" into "Rabbits."[1] Similarly, the flow "deaths'" emanating from "Adults" corresponds to the flow "deaths" emanating from "Rabbits" (Stella's reference manual provides detailed instructions for programming these derived flows).

Fibonacci's famous problem now calls for starting with one pair of babies and no adults, information that is readily programmed into dialog boxes for the corresponding reservoirs. To represent the fact that babies mature in 1 month and that adult pairs have 1 pair of babies each month we set DT = 1 and draw connectors from "Babies" to "maturation" and from "Adults" to "births." A 12-month census of the resulting population produces the sequence of Fibonacci numbers $1, 1, 2, 3, 5, 8, \ldots$ in which $x_{i+1} = x_i + x_{i-1}$. (See Figure 7.5.3.)

Problem 7.5.1 In Figure 7.5.2, replace the reservoir labeled Babies by a conveyor with transit time 1. Use this format to generate the Fibonacci numbers.

Problem 7.5.2 Create a Stella model for a rabbit population that starts with 1000 baby pairs and 1500 adult pairs. Assume 5% infant mortality and 10% adult

---

[1] The births' icon has a special symbol to indicate this dependence.

Time	Babies	Adults	Rabbits
0	1	0	1
1	0	1	1
2	1	1	2
3	1	2	3
4	2	3	5
5	3	5	8
6	5	8	13
7	8	13	21
8	13	21	34
9	21	34	55
10	34	55	89
11	55	89	144
Final	89	144	233

Figure 7.5.3 Stella Calculates the Fibonacci Numbers

mortality each month and that adult pairs average 0.15 baby pairs each month. Letting x_i denote the number of rabbit pairs at the beginning of the ith month, use the Stella format of Figure 7.5.4 to estimate the ratio x_{i+1}/x_i for large values of i.

Such submodels can lead to interesting refinements of models for interacting species. By way of example, consider a predator–prey system in which we distinguish between babies and adults in both rabbit and fox populations. By using conveyors to represent numbers of babies, we can allow for different maturation rates for the two species by specifying different transit times. Given this additional structure, it seems reasonable to assume that predation is based on random meetings between rabbits and adult foxes, rather than the entire fox population.

Problem 7.5.3 Modify the predator–prey system of Problem 7.3.1 to the case where rabbits mature in 1 month, foxes in 4 months. Suppose only adult foxes are engaged in predation and that harvesting affects only adult rabbits. Starting with an initial rabbit population of 50 babies and 60 adults and a fox population of 30 babies and 70 adults, model the interaction between these species for 3 years.

Thinking in terms of submodels can also help us develop a deeper understanding of environmental issues, such as the production of greenhouse gases generated by the burning of fossil fuels. At issue here is a human population whose birth rate exceeds its death rate by about 0.017 and whose global economy is highly dependent on what is sometimes termed a "throughput" of carbon-based fossil

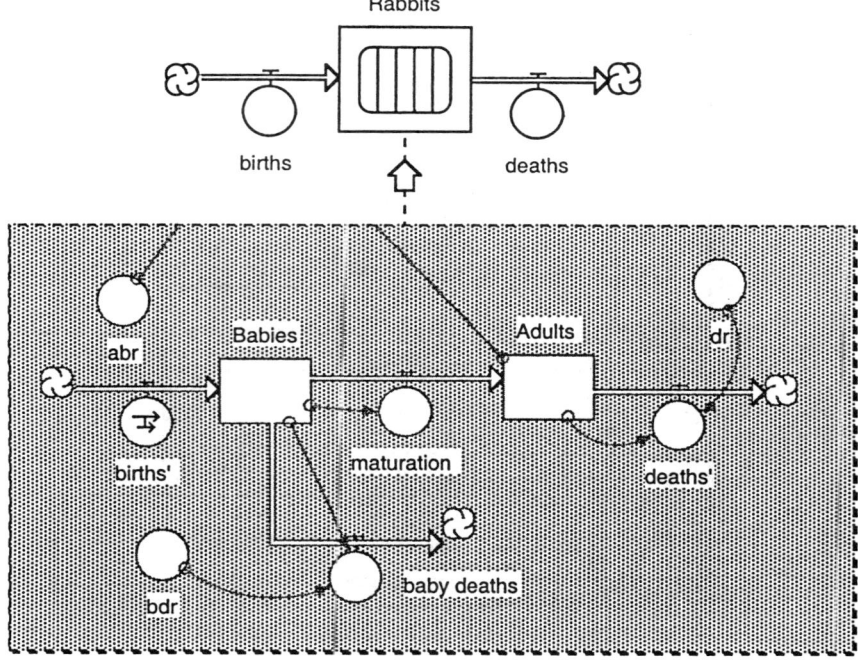

Figure 7.5.4 A generalization of Fibonacci's Problem

fuels. Stella's icons enable us to represent this situation by means of a diagram
such as that of Figure 7.5.5.

Problem 7.5.4 Global consumption of petroleum is about 25 billion barrels/
year. Of this amount, about 5 billion barrels are consumed in the United States.
Taking world population at 6 billion and United States population at 260 million,
estimate the daily per capita petroleum consumption both worldwide and in the
United States. [To express the answer in gallons/day, note that there are 42 gallons
in a barrel.]

To obtain further insights into the relationship between human population and
the global economy, it is possible to refine Figure 7.5.5. We have anticipated such
refinement by representing both Human Population and the Global Economy as
submodels. This enables us to inquire into both the structure of these individual
systems and the interactions between their components.

A starting point might be to model "Human Population" as a subsystem with
two adult components, "Heavy Users" (who tend to live in affluent industrialized countries) and "Light Users" (who tend to live in "third world" countries).
While these two adult groups tend to have different age structures, birth rates, and
death rates, both categories of adults emerge from a common pool of "Babies," as
represented in Figure 7.5.6 by a conveyor with transit time of 10 years.

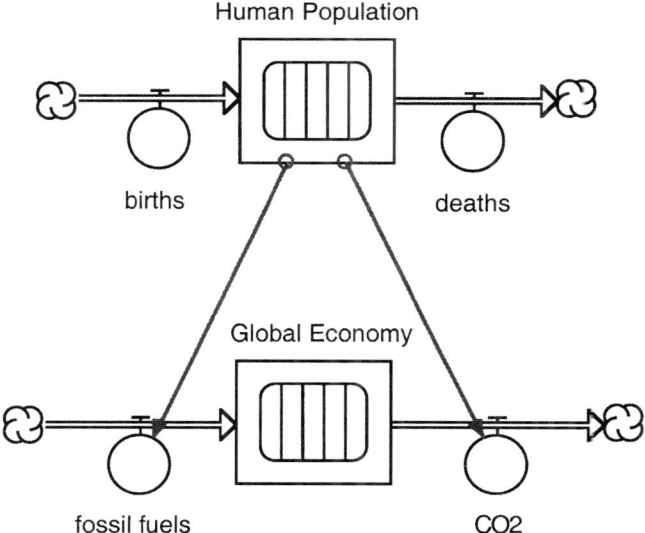

Figure 7.5.5 Greenhouse Gases and Human Population

The question of whether babies will become heavy or light users depends in part on the rate at which global industrialization proceeds. In the submodel below, a converter labeled "I index" determines the percentage of babies that progress into the "Light Users" reservoir via a flow with a special symbol suggesting "leakage." That is, babies becoming light users are regarded as an exception to those flowing into high-consuming populations that enjoy the benefits of economic development.

Thus the upper rectangular icon in Figure 7.5.5 has been programmed to embody the structure indicated in Figure 7.5.6.

Transforming "Global Economy" into a subsystem offers a different challenge. Here there are many competing views as to how the dynamics of the global economic development are best described. Having modeled the rudiments of one such theory in Section 7.4, we will take advantage of our prior efforts by simply "pasting" Figure 7.4.6 into the modeling space provided by the subsystem icon "Global Economy," as indicated in Figure 7.5.7.

A central question, now, is whether these two submodels can be connected in a meaningful way. Here we leave it to the reader to experiment with connectors between the subsystems of Figures 7.5.6. and 7.5.7. One possibility is to connect the reservoirs labeled "Heavy Users" and "Light Users" to the converter c representing a population's marginal propensity to consume. One might also draw a connector from "investment" to the converter "I index" that represents the percentage of babies that will become heavy users of resources.

While these connectors are only a start toward efforts to understand the link between population growth and the global economy, they do make an important

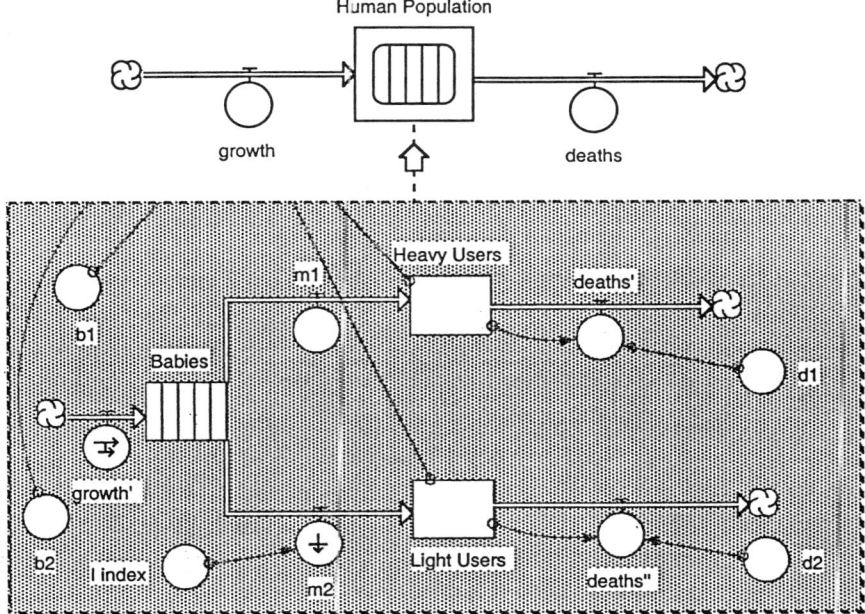

Figure 7.5.6 A Human Population Submodel

point. Classical economic theories tend to be detached from global environmental issues. Given that both economic development and environmental change are
central to our collective welfare, efforts to understand the connections between
them are deserving of serious attention.

Efforts to understand these connections will encounter a distinction between systems that are "open" vs. those that are "closed." It is this distinction that leads
to the topic of thermodynamics to be discussed in Section 7.6.

7.6. Closed Systems, Thermodynamics, and Entropy   An important
feature of Stella's icon-based representation is the appearance of "clouds" at the
ends of certain flows. When a flow originates and terminates in a reservoir, no
clouds appear. However, in the absence of an initiating and/or terminating reservoir, a cloud signals the fact that the source and/or destination of the flow is not
part of the system being modeled.

A flow that originates in a cloud is equivalent to one originating in a reservoir with
"infinite accumulation." That is, the flow will continue at the rate prescribed in
its dialog box without concern for the source. The alternative to such an implicit
assumption of infinite supply is to have the flow originate in a reservoir, in which
case a cloud does not appear.

Similar remarks apply at the flow's terminus. Clouds correspond to "sinks of infinite capacity," whereas reservoirs correspond to sinks in which the flow's accumulation is calculated within the system being modeled.

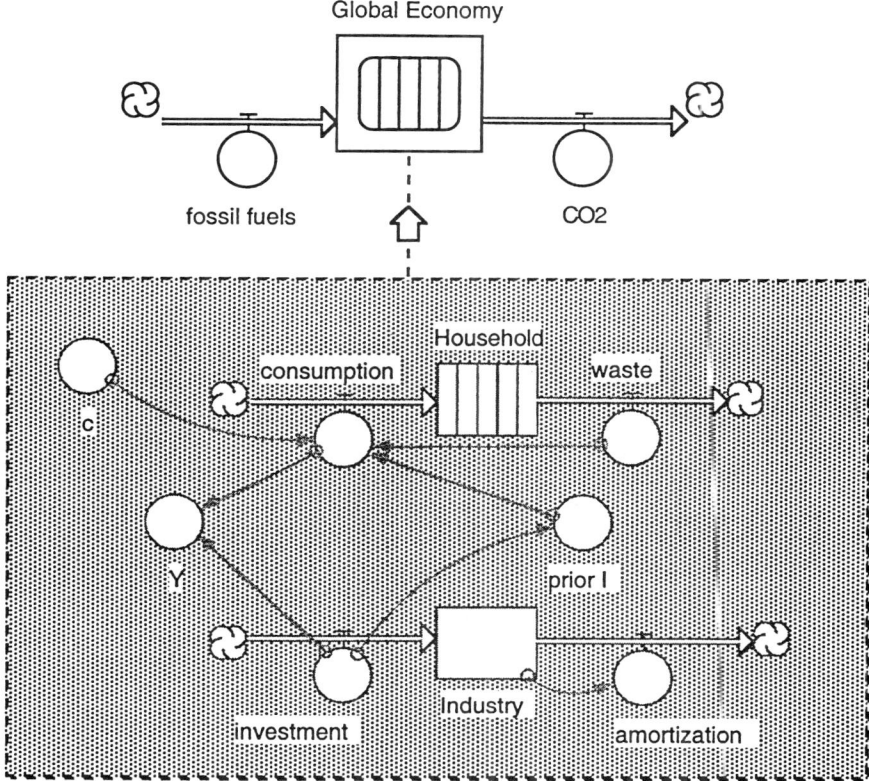

Figure 7.5.7 A Keynesian Economic Submodel

Based on these observations, there is a simple way of deciding whether a model represents a "closed system" or an "open system." Closed systems are those whose Stella representations contain no clouds, i.e., those in which all sources and sinks are included within the system. By contrast, the appearance of a cloud in a Stella model indicates that the system being modeled is in fact embedded in a larger system. Even though this larger system is finite, any limits that it may impose on the subsystem under study are, for the moment, being ignored.

One reason we rarely focus on the question of whether a particular system is open or closed may be the fact that most livelihoods are based on the study of *open* systems. When our car needs repair we seek a mechanic who understands it as a system unto itself, one that takes in fuel and emits exhaust (See Figure 7.6.1).

We do not expect the mechanic to be knowledgeable about how petroleum is produced or what happens to CO_2 after it is emitted into the atmosphere. Similar remarks apply to medicine, farming, manufacturing, and a host of other professions. Indeed, one can argue that human civilization is based on the collaborative efforts of individuals who contribute their understanding of a particular open system. Characteristic of such open systems is an inflow of "resources" that

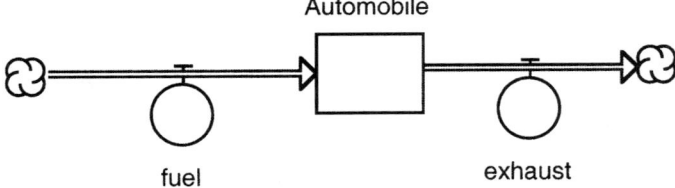

Figure 7.6.1 The Automobile as an Open System

originates in a cloud and an outflow of "goods and bads" that terminates in a cloud (Figure 7.6.2).

Why, then, should we be interested in *closed* systems. Our answer to this question will be based on the science of "thermodynamics," which has its roots in the study of steam engines. Here mechanically adept tinkerers used their intuition and experiential knowledge about the behavior of steam to build systems of pipes, valves, and cylinders that are able to "do work."[1] Given a boiler that produces steam and a drain into which to deposit the condensate, they were able to create systems that harness the thermal energy of steam in order to achieve desired mechanical ends, such as the turning of a wheel. However, questions regarding the generation of steam or the nature of the condensate were not central to early efforts to build steam engines.

While such tinkerers were able to perform remarkable feats, they did not possess an understanding of the scientific principles that underlie steam engines and impose theoretical limits on their ability to do work. Such understanding had to wait until steam engines were studied as *closed* systems, i.e., ones that include both the source of the steam and the sink for its condensate within the system under study. That is, the science of thermodynamics requires a "cloudless" model such as that of Figure 7.6.3.

Within the boundaries of this closed system (as with any fully contained process) two principles of great scientific importance hold sway:

 I. Energy is conserved.
 II. Entropy increases.

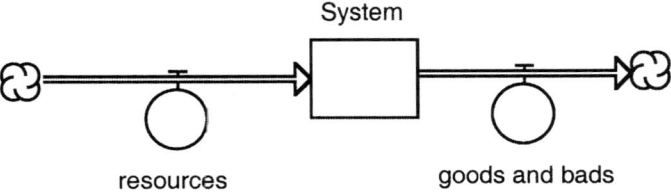

Figure 7.6.2 A Typical Open System

---

[1] Notably pumping water out of the coal mines that made possible the industrial revolution in nineteenth-century England.

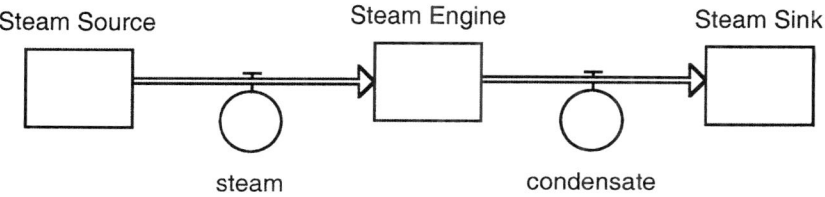

Figure 7.6.3 A Closed Thermodynamic System

Such a paraphrase of the first and second laws of thermodynamics raises two questions: "What is energy?" and "What is entropy?"

Neither of these fundamental questions has a simple answer. Although energy seems to be regarded as a familiar concept (e.g., we are admonished to "save energy"), physics does not provide a definition of what energy *is*.[2] This is sometimes obscured by focusing on what it does, e.g., "Energy is the capacity to do work," or by the laws that energy has been observed to satisfy, e.g., "In the functioning of a closed system, energy is conserved." However, such statements stop short of telling us what energy actually is.

Much the same is true of entropy, which is often defined as "a measure of disorder" within a particular system. The essence of the second law of thermodynamics is that in the functioning of any closed system, this measure of disorder increases. While a complete discussion of this concept is beyond the present scope, we will be able to provide some useful insights by means of two Stella models.

These models are based on Figure 7.6.3, with one exception. For simplicity we will begin by eliminating the Steam Engine, which is positioned between the Steam Source and Steam Sink. Instead, we will focus on the other two reservoirs and the flow of energy from Steam Source to Steam Sink. Once this process understood, it will be possible to reintroduce an intermediate steam engine and discuss its role.

The first model is called an Ehrenfest chain and provides insight into a probabilistic conception of both "disorder" and entropy. Corresponding to the Steam Generator, we consider a reservoir that is crowded with many water molecules. Being under extreme pressure, this reservoir corresponds to a source of hot steam. At the other end of the flow we have a completely empty reservoir, i.e., one with a total vacuum. Having no thermal activity, this reservoir corresponds to a Steam Condenser at $0°$ Kelvin (which is equivalent to about $-273°$ Celsius).

As stipulated in Newton's law of cooling, energy will flow from the hotter to the colder reservoir at a rate proportional to the temperature gradient. However, the Ehrenfest chain circumvents the concept of temperature by providing a probabilistic explanation for energy flows. Here it will be convenient to label the Steam Generator as A and the Steam Condenser as B. Starting with 1000 molecules in A and none in B, we assume that at specified intervals of time (e.g., each millisecond) a *randomly chosen* molecule makes the transition from the reservoir in

[2] See *The Feynman Lectures on Physics*, v. 1, p. 4–2 (Addison-Wesley, 1963).

which it happens to reside to the other reservoir. Since all the molecules start in
A, after one millisecond one molecule will surely have gone from A to B. In the
next millisecond it is highly likely (999/1000) that another molecule, chosen at
random, will flow from A to B, although there is a slight chance (1/1000) that the
molecule in B will return to A. More generally, if we identify the letters A and B
with the number of molecules in the corresponding reservoir, then the probability
of a unit flow from A to B is

(7.6.1a) $$\frac{A}{A + B},$$

whereas the probability of a unit flow from B to A is

(7.6.1b) $$1 - \frac{A}{A + B} = \frac{B}{A + B}.$$

As this process proceeds, there is a tendency for the number of molecules in A
and B to equalize, at which point the likelihood of a flow from A to B becomes
equal to the likelihood of a flow from B to A. This process of equalization corresponds to the temperature-based "sharing of wealth" considered in Chapter 6.

Figure 7.6.4 provides a basis for using Stella to model such a process and to generate a graphical description thereof. Stella's syntax for defining the probability-
based biflow flow from A to B is

(7.6.1c)   IF RANDOM(0, 1) $<$ A/(A + B) THEN 1 ELSE $-$ 1

A run of Figure 7.6.4 yields a graph like the one in Figure 7.6.5.

At the outset there is a clear tendency for molecules from A to migrate to B.
However, as the numbers of molecules in A and B begin to equalize, incidents
in which a molecule flows from B to A become more frequent. Eventually, the
process tends to become one of small variations about the "equilibrium value"
$A = B = 500$.

Problem 7.6.1 Create a Stella model of an Ehrenfest chain that corresponds
Figure 7.6.4. Run the model with initial values $A = 500$, $B = 0$ and with $A =
100$, $B = 0$.

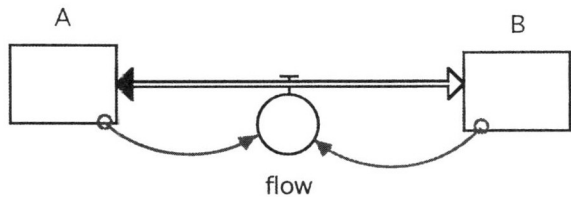

flow
Figure 7.6.4 Modeling an Ehrenfest Chain

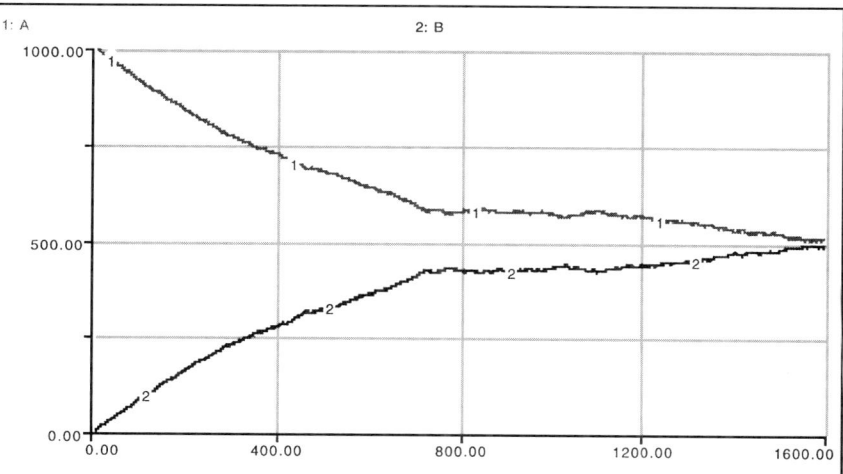

Figure 7.6.5 A Run of the Ehrenfest Chain

Problem 7.6.2 Create a spreadsheet model of an Ehrenfest chain that corresponds to Figure 7.6.4. [Hint: Excel's syntax for (7.6.1c) is "=IF(RAND() < A/(A + B), 1, −1)".]

While an initial distribution of $A = 1000$ and $B = 0$ conveys the underlying idea, the situations envisioned in such "statistical mechanics" involve much larger numbers of molecules. Avogadro's number sets the number of molecules in one mole of a perfect gas at 6.02×10^{23}. In dealing with such vast numbers of molecules we would expect the visible bumps in Figure 7.6.5 to be smoothed out.

The process described so far is suggestive of conservation of mass and perhaps of energy. But what does it have to do with disorder and with entropy? To answer these questions we consider a baby sitter who is asked to look after six children. These children, conveniently labeled {1, 2, 3, 4, 5, 6}, are able to move between two rooms labeled A and B. However, the wanderings of these children are controlled by the roll of a fair die. Once each minute the die is rolled, whereupon the child whose number corresponds to the outcome of the roll changes rooms. In other words, the baby sitter is being confronted by Figure 7.6.4 when the number of "molecules" is given by $N = 6$ (rather than $N = 1000$ or $N = 6.02 \times 10^{23}$).

Problem 7.6.3 Repeat Problem 7.6.1 with initial values $A = 6$, $B = 0$.

While the Ehrenfest chain for $N = 6$ no longer simulates an approach to equilibrium (i.e., to the case where the molecules become approximately equally distributed between the two reservoirs), it does point us toward a definition of complexity.

The least complex situation occurs when all the children are in one room, and so only two states are possible:

$$A = \{1, 2, 3, 4, 5, 6\} \text{ and } B \text{ is empty.}$$
$$A \text{ is empty and } B = \{1, 2, 3, 4, 5, 6\}.$$

After one roll of the die there will be 5 children in one room and one child in the other. Not knowing which child is alone or in which room, this leads to twelve possible states:

1.	$A = \{2, 3, 4, 5, 6\}$	and	$B = \{1\}$,
2.	$A = \{1, 3, 4, 5, 6\}$	and	$B = \{2\}$,
. . .			
6.	$A = \{1, 2, 3, 4, 5\}$	and	$B = \{6\}$,
7.	$A = \{1\}$	and	$B = \{2, 3, 4, 5, 6\}$,
. . .			
12.	$A = \{6\}$	and	$B = \{1, 2, 3, 4, 5\}$.

Before considering another roll of the die we would do well to find a more efficient way of counting the number of possible states. When there are K children in room A and $6 - K$ children in room B, every possible way of selecting the K children in room A corresponds to a distinct state. The number of such states equals "the number of ways of choosing the K children from among six." As shown in Problem 7.8.17, this number "six choose K" is given by the binomial symbol $_6C_K$ defined by

$$(7.6.2) \qquad\qquad _6C_K = \frac{6!}{K!(6-K)!}$$

and corresponds to the $(K+1)$st entry in the seventh row of the Pascal triangle:

$$1, 6, 15, 20, 15, 6, 1.$$

The symmetry inherent in $1, 6, 15, 20, 15, 6, 1$ (and in all other rows of Pascal's triangle) corresponds to the fact that there is an equal number of states with K children in A as with $6 - K$ children in A. It follows that the number of states with "K children in one room and $6 - K$ in the other" is just twice $_6C_K$.

Our measure of complexity for this situation will be $_6C_K$, i.e., half the number of possible states. This leads to

$$(7.6.3) \qquad\qquad \Omega(K, 6 - K) = {_6C_K} = \frac{6!}{K!(6-K)!},$$

where $\Omega(K, 6 - K)$ is the measure of complexity associated with having K children in one room and $6 - K$ in the other. An important consequence of this definition (see Problem 7.8.18) is that an equitable division of 3 children in each room corresponds to the highest level of complexity.

These considerations can be extended from $N = 6$ to $N = 1000$ and to $N = 6.02 \times 10^{23}$. The complexity associated with K molecules in one reservoir and $N - K$ in the other is given by $\Omega(K, N - K) = {}_NC_K$. As shown in Problem 7.8.18, the fact that the middle entry (or middle pair of entries) in any row of Pascal's triangle dominates all other entries in that row implies that an equal distribution of molecules gives rise to the highest measure of complexity.

Problem 7.6.4 Calculate ${}_4C_K$ for $K = 0, 1, 2, 3, 4$ and confirm that ${}_4C_2 \geq {}_4C_K$ for $0 \leq K \leq 4$.

Problem 7.6.5 Calculate ${}_7C_K$ for $K = 0, 1, 2, 3, 4, 5, 6, 7$. Confirm that ${}_7C_4 = {}_7C_3$ and that ${}_7C_4 \geq {}_7C_K$ for $0 \leq K \leq 7$.

Having arrived at a definition of complexity Ω, there remains the problem of defining the *entropy* associated with the various states generated by an Ehrenfest chain. This leads us to a milestone of modern science, namely Ludwig Boltzmann's definition of entropy as

(7.6.4) $$S = k \ln \Omega.$$

Here k is a constant of proportionality known as Boltzmann's constant[3] and S denotes the elusive concept of entropy.

Unfortunately, Stella does not have ${}_NC_K$ among its built-in functions, and this prevents us from including a converter for S in Figure 7.6.4. We can, however, revert to spreadsheets to model the same process and use Excel's syntax

$$=LN(COMBIN(N,K))$$

to calculate the increase in entropy that accompanies the Ehrenfest chain. See Figure 7.6.6.

Problem 7.6.6 Modify the spreadsheet of Problem 7.6.2 to calculate both the complexity and entropy of an Ehrenfest chain starting with N molecules in one reservoir and 0 in another.

Problem 7.6.6 corresponds to the second law of thermodynamics *only* when N is large, and this is something Boltzmann was keenly aware of. He noted that even with A very large and B relatively small, there is always a chance that a molecule, or a sequence of molecules, will go from B to A and thereby *reduce* entropy. Indeed, there is even the possibility (as tends to occur with six children) that *all* molecules will eventually find their way into the same reservoir!

But as to the likelihood of actually observing such behavior in the physical world, Boltzmann quipped, "You should live so long."

---

[3] The value of k depends on the units of energy underlying a particular development of thermodynamics. Ludwig Boltzmann (1844–1906) is buried in a Vienna cemetery where equation (7.6.4) is inscribed on his gravestone.

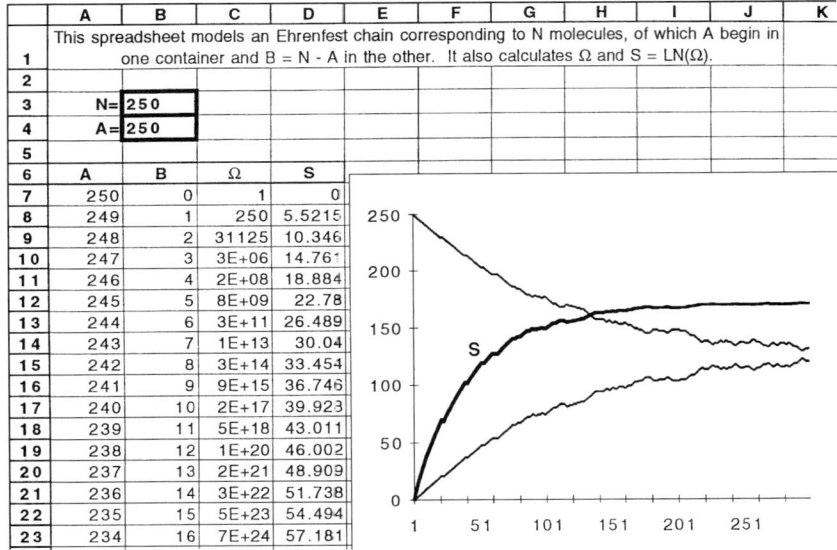

Figure 7.6.6 A Spreadsheet Representation of Entropy

In developing the Ehrenfest chain, we noted that it provides a basis for defining entropy without making use of the concept of *temperature*. If, however, we do have a measure of temperature at our disposal, then there is another way of demonstrating the increase of entropy that is associated with energy flow in a closed system. Here we consider heating a perfectly insulated room by means of a radiator. However, to make the room a "closed system," the radiator will be a tub of hot water (rather than an appliance that is connected to the outside by means of pipes). The process of heating the room can now be thought of as energy flow from tub to room within the closed system of Figure 7.6.7.

Such a Stella model will ask us to specify the initial energy in both tub and room. Note that setting both of these at 100 units does *not* ensure that the system will be in equilibrium. A large room with 100 units of thermal energy might be quite

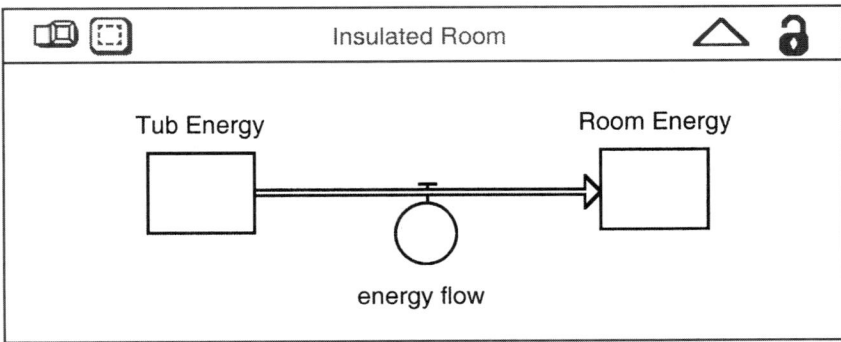

Figure 7.6.7 A Tub of Water Heating an Insulated Room

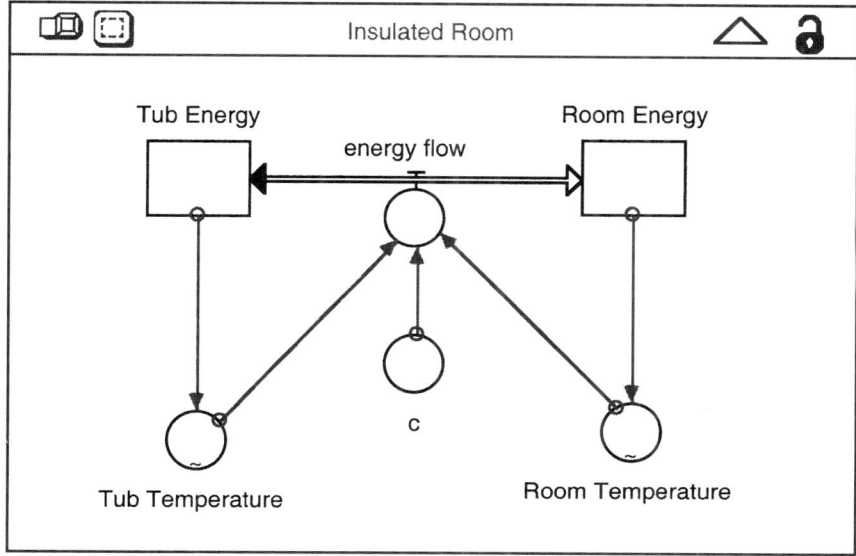

Figure 7.6.8 Temperature Determines Energy Flow

chilly, whereas a small container of water that embodies the same amount of energy may be quite hot. It is the *temperature* of both tub and room that determines the rate of energy flow from tub to room (or vice versa), a fact that can be modeled in the Stella format of Figure 7.6.8.

Here energy flow is specified to be proportional to the difference between Tub and Room temperature, where c is the constant of proportionality. The special mark inside "Tub Temperature" and "Room Temperature" indicates that we have used Stella's "make graph" option to specify the rule by which temperature depends on energy. In both, zero energy corresponds to $0°$ K and temperature increases with energy.[4] For the tub, an initial 100 units of energy corresponds to boiling water ($373°$ K), whereas the room's initial 100 units of energy corresponds to only $68°$ Fahrenheit ($293°$ K). See Figure 7.6.9.

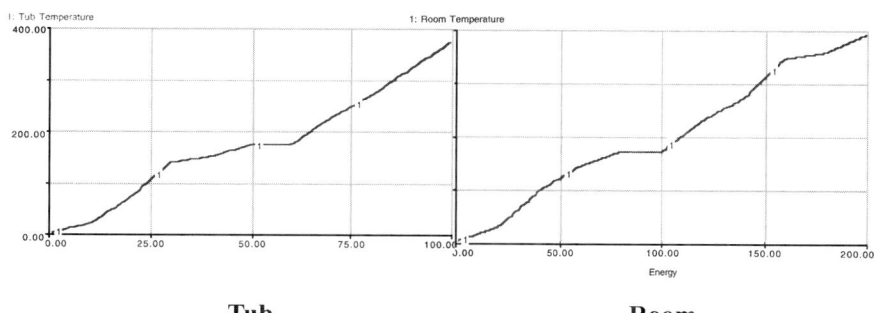

Tub   **Room**

Figure 7.6.9 Temperature is an Increasing Function of Energy

[4] There may, however, be flat places in the graph corresponding to a phase change.

So far, Figure 7.6.8 has provided a Stella representation of the first law of thermo-dynamics. Since all energy is flowing between two reservoirs in a closed (cloud-less) system, we have conservation of energy. This raises the question of why such a process should also result in an *increase* in entropy.

The answer lies in the size of the *two* entropy flows that correspond to the single energy flow in Figure 7.6.8. These two entropy flows are in the same direction as the energy flow, i.e., from hot to cold. This is because at $0°$ K there is no thermal activity, and this ultimate form of orderly behavior corresponds to $\Omega = 1$ and to $S = 0$. As temperature increases, so does thermal activity and the corresponding measures of disorder and entropy. As a result, a flow of heat from tub to room is accompanied by entropy flows in the same direction.

A fundamental law of thermodynamics asserts that energy flows are related to entropy flows by the differential equation

(7.6.5) $$\frac{dS}{dE} = \frac{1}{T},$$

where there are two choices for T. To calculate the decrease in entropy in the tub, we use the temperature of the tub. To calculate the increase of entropy in the room, we use the temperature of the room. It is the fact that the these tem-peratures are different that gives rise to *two* entropy flows.

The differential equation (7.6.5) can be approximated by the difference equation

(7.6.6) $$S_{i+1} - S_i = \frac{1}{T_i}(E_{i+1} - E_i),$$

where T_i depends on E_i in accordance with the rules implicit in the graphs of Figure 7.6.9.

It now remains to apply (7.6.6) to *both* the tub and the room. Even though ΔE_i for the tub equals $-\Delta E_i$ for the room, the tub and room are at different temperatures. As a result, they generate different values of $|\Delta S_i|$. Stella represents this as shown in Figure 7.6.10.

Except for units, the output generated by Stella in Figure 7.6.10, as represented in Figure 7.6.11 is remarkably similar to that obtained by combinatorial methods in Figure 7.6.6.

Problem 7.6.7 Create the Stella model of Figure 7.6.10 so that initial ener-gies of 100 units in both room and tub lead to a hotter tub than room. Confirm that Total Entropy (= Tub Entropy + Room Entropy) increases as a result of the subsequent energy flow.

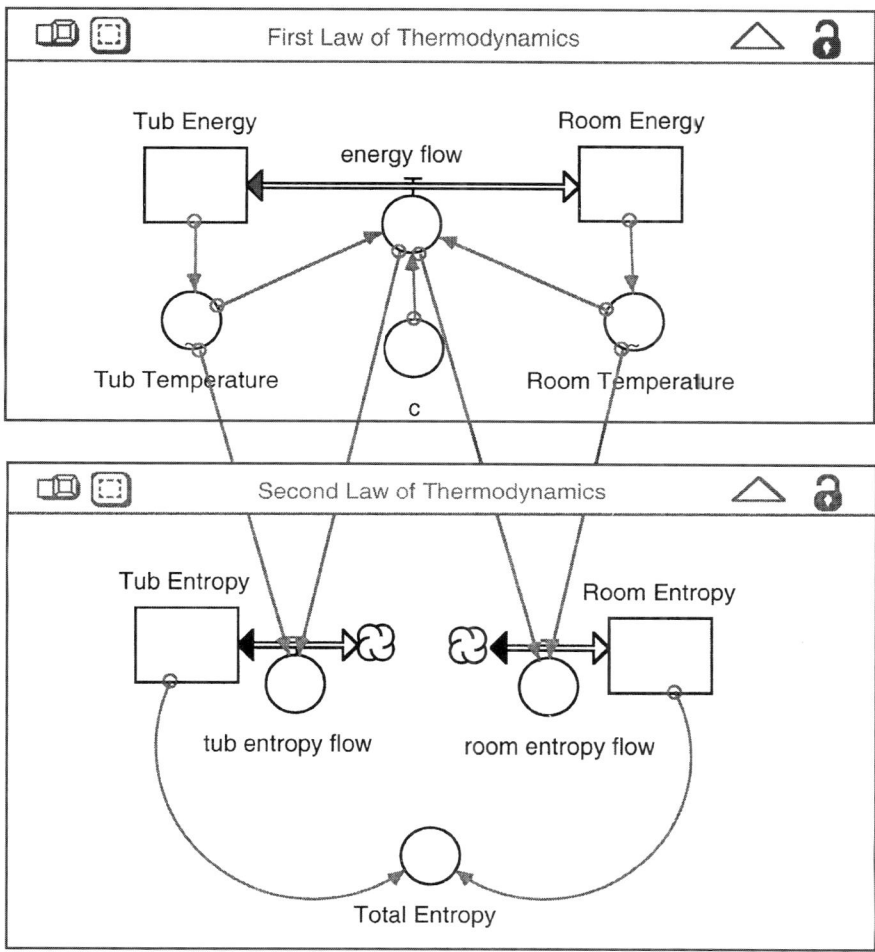

Figure 7.6.10 The First and Second Laws of Thermodynamics

Problem 7.6.8 To heat the room faster, we could invest the same 100 units of initial energy in a smaller tub of water. This would result in a higher initial Tub Temperature and a more rapid heat flow from Tub to Room. Modify Problem 7.6.7 in this way and confirm that "haste makes entropy."

Problem 7.6.9 Simulate bringing a tub of liquefied helium into a room by setting the initial Tub Energy well below 100 units (roughly at an initial temperature of 200° K). Confirm that such a cooling of the room also results in an increase in entropy.

Finally, we return to the steam engine that started this discussion but was then so unceremoniously eliminated from the subsequent analysis. Let us now bring such

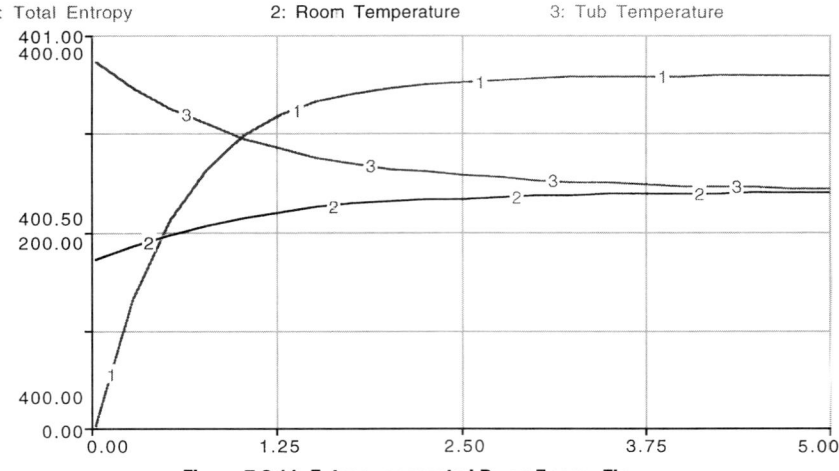

Figure 7.6.11 Entropy generated By an Energy Flow

a machine back into our insulated room and use the thermal energy in the tub to set it running. If this machine is harnessed to "do work," such as raising weights from the floor to a shelf, this can upset our conservation of *thermal* energy. However, taking into account the increase in *potential* energy associated with lifting weights to a shelf, we will find that the first law of thermodynamics remains in effect. That is, the decrease in thermal energy in the tub equals the increase in thermal energy in the room *plus* the potential energy embodied in the weights that have been lifted onto a shelf.

As for the second law, there is nothing fundamentally new to observe. The running of the steam engine will contribute to the cooling of the water in the tub and to the heating of the room. As the temperature of the tub approaches that of the room, the steam engine will slow and eventually cease to run. This corresponds to an approach to an equilibrium state in which the entropy of the system is maximized.

It is such a state of equilibrium toward which closed systems are inexorably drawn by the second law of thermodynamics.

7.7. The Global System   Our discussion of closed systems ended on a somewhat sour note. According to the second law of thermodynamics, the functioning of any closed system is accompanied by an increase in entropy, or "disorder." This means that as energy flows within a closed system, its capacity to do work is degraded. And despite the fact that energy is being conserved, such a system approaches an equilibrium state that is rather ominously referred to as "heat-death."

These observations raise some interesting questions about our global system, i.e., that consisting of the earth and its biosphere. What are the implications of the laws of thermodynamics for the earth and its inhabitants?

Here there is both good news and bad news. The good news is that while our global system is *materially* closed, it is open to energy flows. At a distance of 93 million miles from the sun, a one-meter × one-meter square that is oriented perpendicular to the sun's rays receives an energy flow of about 1,400 Joules/second. This is equivalent to 1.4 kilowatts, or the energy required to operate fourteen 100-watt light bulbs. In other words, one square meter of a 100% efficient solar panel that is located outside the earth's atmosphere would generate enough electricity to meet the energy requirements of an ordinary house.

Problem 7.7.1 Taking the earth's radius as 4,000 miles, estimate the number of kilowatt-hours of solar energy falling upon the earth each year. After determining the amount your local utility charges per kilowatt-hour delivered to your home, estimate the annual "dollar value" of this flow.

Thus the good news is that the earth is *not* a closed system in the thermodynamic sense.[1] Instead, it is the object of a massive energy flow, one that has over the eons supported the evolution of life forms whose remains have produced vast stores of carbon-based "fossil fuels." In other words, the earth has managed to store some of this energy flow, much like a steam engine harnessing the energy in a tub of hot water to lift weights to a shelf. And beyond having such an energy "savings account," the earth enjoys a large "annual income" of energy, as suggested by Problem 7.7.1 above.

So what (if any) is the bad news? From the perspective of human civilization, bad news may be hard to document. Our numbers have been growing exponentially, doubling every forty years for about the last 200 years. This period of growth has coincided with an industrial revolution made possible by withdrawals from the earth's "energy savings account." For many individuals, this has meant a life of ease and comfort compared to the labors and privations that were the norm in preindustrial times.

The bad news is mainly in the form of uncertainties regarding our ability to sustain the kind of growth that we have come to regard as "normal." To be sure, this is an area in which one should recall the adage, "Be careful about making predictions—especially about the future." However, a thermodynamic view of the global system points to two major areas of concern.

While the earth's annual solar "energy income" far exceeds the energy demands that our industrial civilization imposes, tapping into and/or storing this "renewable" energy source may not be easy. Any transition from fossil fuels to solar energy will entail substantial delays. And as such a transition proceeds, we will have to learn to forgo some of the "low entropy" forms of energy that fossil fuels so readily provide. Here a detached observer might suggest allocating a sizable amount of

---

[1] Some writers distinguish between "isolated sysytems" and "closed systems." Here a closed system allows energy flows but not material flows, while an isolated system allows neither. In this sense, we could refer to the earth as a closed system, but one that is *not* isolated.

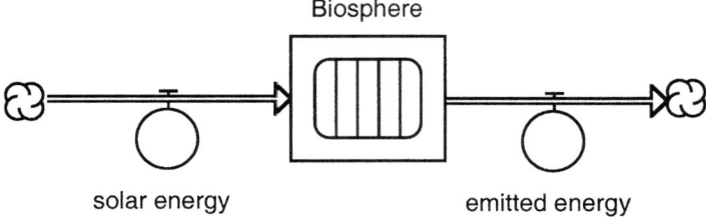

Figure 7.7.1 The Earth's Heat Exchange System

our stored energy reserves to ensuring a smooth transition to such "high entropy" forms of energy. This raises the question of whether the economic principles that govern our allocation of resources provide a basis for anticipating such future needs, or whether we face an "overshoot of carrying capacity" much like that suggested by the logistic equation with delays (see Figure 2.4.5).

Another area of concern deals with our very substantial annual energy income (see Problem 7.1.1). If the earth were permanently to retain thermal energy received from the sun, its temperature would rise dramatically. The evolution of existing life forms has been based on a system of heat exchange, one in which virtually all of the solar energy flow is eventually returned to space.

The average global temperature resulting from these energy flows represents an equilibrium value related to Figure 7.1.1, one about which small fluctuations occur. Some of these fluctuations have resulted in ice ages and others in periods of global warming. While such climate change may have resulted in the elimination of individual species, the earth's overall biological system has managed to adapt.

It is at this point that "greenhouse gases" enter the picture. Carbon dioxide (CO_2) can serve as a kind of "blanket" in the earth's atmosphere, affecting the percentage of the incoming energy income that is reflected. While vast amounts of CO_2 are produced by various natural processes on earth, plants in the earth's biological system also recycle CO_2 by a process of photosynthesis:

(7.7.1) CO_2 + water + sunlight → carbohydrates + oxygen.

It is photosynthesis that produces food and oxygen for the earth's animal species, and it is also the process by which the sun's energy has been converted into stored energy in the form of fossil fuel reserves.[2] The fact that photosynthesis uses solar energy and reduces the amount of CO_2 in the earth's atmosphere provides a basis for drawing a connector from "Biosphere" to "emitted energy" in Figure 7.7.1, yielding Figure 7.7.2.

While this heat exchange process is complicated (and not completely understood), there exists a growing body of concern about the extent to which the additional CO_2 generated by the burning of fossil fuels could alter the climatic

---

[2] It is on this basis that botanists sometimes refer to biology as "the study of plants and their parasites."

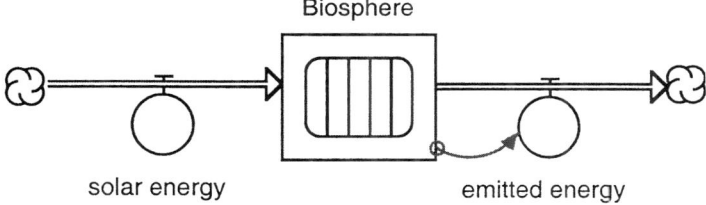

Figure 7.7.2 Atmospheric CO_2 Affects Emitted Flow

equilibrium established prior to the industrial revolution. Of the approximately six billion tons of carbon that reliance on fossil fuels injects into the atmosphere each year, only about half is recycled in accordance with (7.7.1). As indicated in Figure 7.7.3, this imbalance has resulted in significant increases in atmospheric CO_2 concentration.

The implications of Figure 7.7.3 are a topic of controversy, and we will not try to resolve such issues here. Critics of global climate change theories point out that such "anthropogenic" CO_2 production is less that 1% of that resulting from natural processes. They also suggest that any global warming would be accompanied by an increase in clouds. Since such clouds would shield the earth from sunlight, they hypothesize the existence of a connector from "earth energy" to the "solar energy" flow, one that could serve to stabilize such climate-change phenomena.

To the extent that the controversy sorrounding CO_2 production has created a new profession of "global climate change specialist," it is again based on the study of an open system (see Figure 7.6.2). But what about our effort to view the global system in a thermodynamic context? How can we relate Figure 7.7.4 to a closed system?

A starting point might be a Stella model for the global system that links submodels for "Global Economy" and "Photosynthesis." Here traditional economic theories are forced to take note of the "carbon cycle," i.e., the submodel that embodies recycling of CO_2 by the earth's diverse plant forms. The fact that many resources come from finite reservoirs is explicitly noted in this model, as is the finiteness of the atmospheric sink for CO_2 (see Figure 7.7.5).

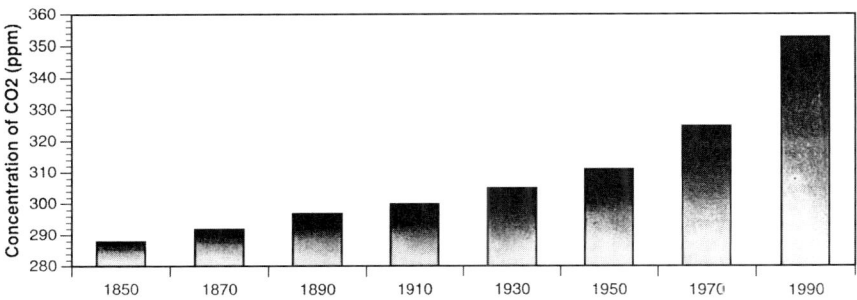

Figure 7.7.3 Atmospheric CO_2 Levels in Parts per Million

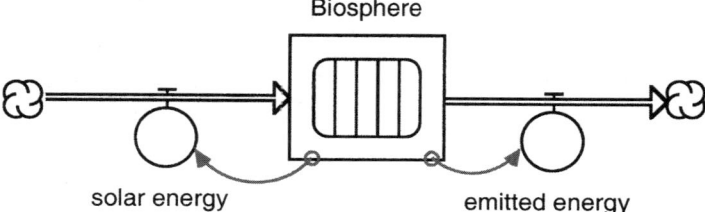

Figure 7.7.4 Atmospheric CO₂ Affects Solar Flow?

Within this system, we have included "temperature" as a converter labeled T that is the object of a connector from the reservoir "Atmospheric CO_2" and in turn connects to both "Global Economy" and to "Photosynthesis." Within the context of such a closed system, the simulation of various models for global climate change can be regarded as "scenarios." Underlying each such scenario is a rule embodied in the dialog box associated with T.

But is it really possible to model such complicated systems? And if we do, how well will they correspond to the real world? These are both important questions, ones that are likely to be asked in connection with any effort to confront global issues in a "closed system" context. However, as in the case of thermodynamics, doing good science requires a willingness to confront the world as it really is.

7.8. Problems, Exercises, and Projects   This chapter has dealt with a
simulation package that uses "reservoirs, flows, connectors, and converters" to

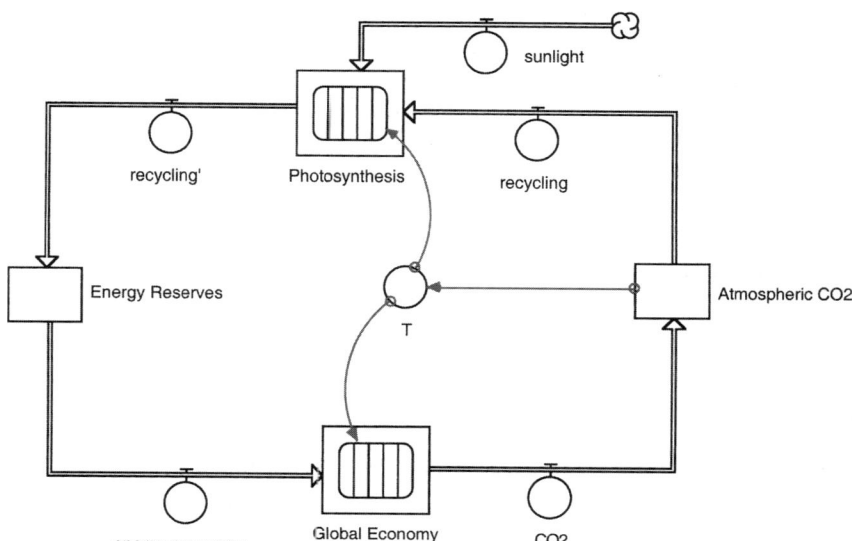

Figure 7.7.5 The Global Carbon Cycle

formulate iteration-based models for change. While other forms of computer technology (including spreadsheets) can also be used to implement the underlying iterative procedures, software such as Stella can reduce greatly the amount of effort required to create such models. Also, by enabling us to formulate models in terms of icons, Stella also provides a graphic representation of the system being modeled. The problems that follow take advantage of these features to elaborate on rules for change that have previously been studied by other means.

Flows in and out of Tanks

Problem 7.8.1 Given that the calculus-based solution of Problem 7.2.2 is $x(t) = t + t^3/3 - t^2$:

   (a) Use the calculus-based solution to find $x(8)$.

   (b) Use a spreadsheet in the format of Figure 7.2.2 to generate an approximating sequence $x_1, x_2, x_3, \ldots$, and

      (i) compare $x(8)$ with x_9 for $N = 1$;

      (ii) compare $x(8)$ with x_{33} for $N = 4$;

      (iii) compare $x(8)$ with x_{81} for $N = 10$.

   (c) Use Stella to calculate the accumulation in the tank at $t = 8$ when $DT = 1$; $DT = 0.25$; and $DT = 0.1$.

Problem 7.8.2 In a cylindrical tank, the height of water is proportional to the volume $x(t)$ accumulated at time t. Suppose that for a particular tank the height of the accumulated water satisfies $h = x/10$. At the same time as this tank is being filled by an inflow satisfying $f(t) = 1 + t^2$, it is also being drained by an outflow satisfying $g(t) = 2h(t)$. Assuming that $x(0) = 10$ and that the flow is monitored at times $t = 0, 1, 2, \ldots, 7$:

   (a) Create a spreadsheet model that approximates the accumulation of water in such a tank for $0 \leq t \leq 8$.

   (b) Create a Stella model that approximates the accumulation of water in such a tank for $0 \leq t \leq 8$ with $DT = 1$.

  *(c) Compare the answers from parts (a) and (b) with the calculus-based solution at $t = 8$.

Problem 7.8.3 A tank is being filled by a flow $f(t)$ that is measured in gallons per minute and that is increasing, in the sense that $f(t + k) > f(t)$ for every $k > 0$. Letting $f_i = f(i - 1)$ for $i = 1, 2, \ldots$:

   (a) Explain why the solution x_9 of (7.2.3) underestimates the actual accumulation $x(8)$ at time $t = 8$.

(b) Give a qualitative description (in words accompanied by a graph) of an increasing flow $f(t)$ that never exceeds 20 gallons/minute and for which $x(8) - x_9 > 19$ gallons.

(c) Suppose such a system is subject to a refinement of the form $\Delta t \mapsto \Delta t/4$ (i.e., DT = 1 $\mapsto$ DT = $\frac{1}{4}$). Give a qualitative description of an increasing flow $f(t)$ that never exceeds 20 gallons/minute but for which $x(8) - x_{33} > 4.5$ gallons.

(d) For the flow in part (c), is it possible that $x(8) - x_{33} > 5$ gallons? Explain.

The Logistic Equation Revisited

Problem 7.8.4 By including a converter for k, modify the Stella model of Problem 7.2.4 to represent $x_{i+1} - x_i = ax_i - bx_i^k$.

(a) Use this model to confirm that for $a = 0.1$, $b = 0.0005$, and $k = 2$ the iterates $x_1, x_2, x_3, \ldots$ approach 200.

(b) Predict the value approached for $k = \frac{3}{2}$ and use Stella to confirm your answer.

(c) Repeat part (b) for $k = 3$.

Problem 7.8.5 By including a converter for d, modify Problem 7.2.4 to represent the delay logistic equation $x_{i+1} - x_i = ax_i - bx_{i-d}^2$. With $a = 0.1$ and $b = 0.0005$, compare the behavior of solutions for $d = 4, 6$, and 8. [Hint: If the outflow corresponding to $-bx_{i-d}^2$ is named "damping," then the Stella syntax for programming a delay of d units of time is "DELAY(damping,d)."]

Problem 7.8.6 In Chapter 3 we saw how "antirefinements" of the form $\Delta t \mapsto M\Delta t$ are linked to the cyclic and chaotic behavior of solutions of the logistic equation. Create a Stella model that illustrates the "descent into chaos" for $x_{i+1} - x_i = 6x_i - 0.2x_i^2$. [Hint: Start with DT = 0.1. While recent versions of Stella allow values of DT greater than 1, older versions accept only DT $\leq$ 1.]

Another Mechanism for Damping

Problem 7.8.7 A population $x(t)$ can sustain an annual growth rate of 15/thousand by reliance on a nonrenewable resource $y(t)$. Specifically, by extracting petroleum at an annual rate equal to $f(t) = 6[x(t) - 2]$ barrels/year, this population can maintain an annual birth rate of 31/thousand and an annual death rate of 16/thousand. However, should this flow fall to $m[x(t) - 2]$, with $0 < m < 6$, the annual death rate would increase from 16/thousand to $[16 + 5(6 - m)]$/thousand. Given $x(0) = 6$ billion and $y(0) = 1,500$ billion barrels, model the dynamics of such a population:

(a) when the rate at which petroleum is extracted is the smaller of $6[x(t) - 2]$ and 5% of the remaining fuel supply.

(b) when the rate at which petroleum is extracted is the smaller of $6[x(t) - 2]$ and 10% of the remaining fuel supply.

(c) when petroleum is extracted at the rate $6[x(t) - 2]$ barrels/year until $y = 0$.

Using Stella to Model Predator–Prey Relationships

Problem 7.8.8 Elaborate on Problem 7.3.4 by including a converter to represent the rate at which rabbits are harvested. Rather than setting a constant harvest rate h, assume that h is dependent on the number of foxes.

(a) Devise three different rules $h_1(v)$, $h_2(v)$, and $h_3(v)$ that enable both species to survive at least 10 years.

(b) Determine which of the three rules produces the largest rabbit yield over the course of these 10 years.

Problem 7.8.9 Modify Problem 7.8.8 by considering rules that call for harvesting of rabbits only when the number of rabbits exceed the number of foxes. [Hint: This will require Stella's syntax for the "if . . . then" conditional.]

Epidemics

Problem 7.8.10 Use Stella to model the epidemic described in Problem 6.7.10, i.e., one in which survivors of the disease are immune and new infections are the result of random meetings between infected and susceptible individuals.

(a) Use Stella to repeat part (c) of Problem 6.7.10.

(b) Use Stella to repeat part (d) of Problem 6.7.10.

(c) Modify part (b) above to the case where in addition to the inflow of 300 susceptible individuals/week, the susceptible population has a birth rate of 2%.

Some Problems from Economics

Problem 7.8.11 Modify Figure 7.3.6 so that accumulation in the Warehouse can be used to determine price. Setting the initial accumulation in the Warehouse at $w(0) = 100$ units, find a price rule $x(w)$ and a value for DT for which the inventory remains in the range $90 < w(t) < 110$ for $0 \leq t \leq 100$.

Problem 7.8.12 Modify the Stella model of Problem 7.8.11 so that accumulation in the Warehouse determines price *change* rather than price itself. Setting the initial accumulation in the Warehouse at $w(0) = 100$ units, formulate a price change rule $f(w)$ for which $90 < w(t) < 110$ for $0 \leq t \leq 50$. [Hint: Be sure the flow corresponding to f is a *biflow*.]

Problem 7.8.13 Modify the Keynesian economy described in Figure 7.4.6 by introducing a converter labeled G for "government spending." Defining Y to be the sum $C + I + G$, use this model to address the role of government spending in moderating economic cycles, as posed in Problem 6.3.20.

More on Buffaloes

Problem 7.8.14 Use the "conveyor model of iteration" given in Figure 7.4.2 to create a Stella "cubic equation solver" for approximating solutions of the general cubic equation $ax^3 + bx^2 + cx + d = 0$.

Problem 7.8.15 Using conveyors with transit times of 1 year to represent the first two stages of life, construct a Stella model for a buffalo population consisting of babies, teens, and adults.

(a) Starting with one pair of babies, and assuming that each adult pair has one pair of babies each year, find the number of buffalo pairs at the end of 12 years.

(b) Letting x_i denote the number of buffalo pairs at time i, use the cubic equation solver of Problem 7.8.14 to approximate x_{i+1}/x_i for large values of i.

Problem 7.8.16 Modify the Stella model in Probem 7.8.15 to one that is based on 5% infant mortality, 2% teen mortality, and 10% adult mortality each year, and annual birth rates of 5% and 50% among teens and adults, respectively. Starting with 100 babies, 50 teens, and 200 adults, model the growth of this population for 20 years.

The Binomial Coefficients

In Section 7.6 the binomial coefficient $_NC_K$ was used to measure the "complexity" of a state in which there are K molecules in one reservoir and $N - K$ in another. This *binomial coefficient* $_NC_K$ was defined as the number of different ways of choosing a subset of K elements from a set having N elements. The next two problems show that

$$_NC_K = \frac{N!}{K!(N-K)!}$$

and examine the representation of these numbers in a triangular array known as Pascal's triangle.

Problem 7.8.17 Consider a set of N children, $A = \{C_1, C_2, \ldots, C_N\}$.

(a) Arranging K children from A in a row calls for choosing a "first child," a "second child," ..., and a "Kth child." Show that the number of ways of doing this is $N \times (N-1) \times (N-2) \times \cdots \times (N-K-1)$.

(b) Show that the number in part (a) is $\frac{N!}{(N-K)!}$

(c) Suppose that now, instead of counting arrangements of K children in a row, we seek to count *subsets* of K children chosen from among N children. Use the fact that each such subset gives rise to $K!$ arrangements in a row to show that there are $\frac{N!}{K!(N-K)!}$ such subsets. [Remark: It is on this basis that we sometimes refer to $_NC_K = \frac{N!}{K!(N-K)!}$ as "N choose K."]

(d) Using the convention $_NC_0 = 1$ for $N = 0, 1, 2, \ldots$, consider the triangular array in which the $(N+1)$st row is

$$_NC_0, \; _NC_1, \; _NC_2, \ldots \; _NC_{N-1}, \; _NC_N.$$

Confirm that the first six rows this array are as follows:

```
for N = 0:                        1
for N = 1:                      1   1
for N = 2:                    1   2   1
for N = 3:                  1   3   3   1
for N = 4:                1   4   6   4   1
for N = 5:              1   5   10   10   5   1
```

Calculate two more rows of this "Pascal's triangle" and verify that each "interior number" is the sum of the two nearest numbers in the preceding row. That is, verify that for $N \leq 7$ and $1 < K < N-1$,

$$_NC_K = \; _{N-1}C_{K-1} + \; _{N-1}C_K.$$

(e) Taking $_NC_K = \frac{N!}{K!(N-K)!}$ as the definition of binomial coefficients, show algebraically that the last relation in part (d) is true.

(f) Show that the rows of Pascal's triangle are symmetric, in the sense that $_NC_K = \; _NC_{N-K}$.

Problem 7.8.18 In connection with the measure of complexity (7.6.3), it was asserted that the middle entry (or middle pair of entries) in any row of Pascal's triangle dominates all other entries in that row. To establish this, show that:

(a) If $N \geq 2$ and $0 \leq K < (N-1)/2$, then $(K+1)!(N-K-1)! < K!(N-K)!$

(b) If $N \geq 2$ and $0 \leq K < (N-1)/2$, then $_NC_K < \; _NC_{K+1}$.

(c) If $N \geq 2$ and $K > (N-1)/2$, then $_NC_K > \; _NC_{K+1}$.

Index